TRANSCRIPTION FACTORS FOR ABIOTIC STRESS TOLERANCE IN PLANTS

TRANSCRIPTION FACTORS FOR ABIOTIC STRESS TOLERANCE IN PLANTS

Edited by

DR. SHABIR HUSSAIN WANI

Assistant Professor cum Scientist (Senior Scale),
Genetics and Plant Breeding,
Mountain Research Centre for Field Crops,
Khudwani, Sher-e-Kashmir University of Agricultural
Sciences and Technology, Kashmir, Srinagar,
Jammu and Kashmir, India

ELSEVIER

ACADEMIC PRESS

An imprint of Elsevier

Academic Press is an imprint of Elsevier
125 London Wall, London EC2Y 5AS, United Kingdom
525 B Street, Suite 1650, San Diego, CA 92101, United States
50 Hampshire Street, 5th Floor, Cambridge, MA 02139, United States
The Boulevard, Langford Lane, Kidlington, Oxford OX5 1GB, United Kingdom

Library of Congress Cataloging-in-Publication Data
A catalog record for this book is available from the Library of Congress

British Library Cataloguing-in-Publication Data
A catalogue record for this book is available from the British Library

ISBN 978-0-12-819334-1

For information on all Academic Press publications
visit our website at https://www.elsevier.com/books-and-journals

Publisher: Charlotte Cockle
Acquisitions Editor: Nancy Maragioglio
Editorial Project Manager: Emerald Li
Production Project Manager: Niranjan Bhaskaran
Cover Designer: Christian J. Bilbow

Typeset by SPi Global, India

Dedication

This book is dedicated to Dr. Manjit S. Kang

Dr. Manjit S. Kang, currently adjunct professor at Kansas State University-Manhattan, is an internationally recognized quantitative geneticist. He served as a professor of Quantitative Genetics at Louisiana State University, Baton Rouge for 21 years (1986–2007) and as a sugarcane geneticist at the Institute of Food and Agricultural Sciences (IFAS) at the University of Florida for 5 years (1981–1986). In April 2007, he was invited by the Government of Punjab to provide leadership to Punjab Agricultural University (PAU) at Ludhiana as vice chancellor (CEO) (April 2007–2011). He was research station manager and senior plant breeder (hybrid maize breeding) at Cargill, Inc., St. Peter, Minnesota (1977–1979).

Dr. Kang has published 166 peer-reviewed (refereed) journal articles in prestigious international journals, 43 book chapters/encyclopedia essays, and 135 other technical publications. He has served as technical editor of *Crop Science* and editor-in-chief of *Communications in Biometry and Crop Science*. He is currently the editor-in-chief of *Journal of Crop Improvement* and technical editor of *Crop Science*. He has edited or coedited and authored or coauthored 14 books, including *Genotype-by-Environment Interaction* (1996; CRC Press), *Quantitative Genetics, Genomics and Plant Breeding* (2002; CABI Publishing), *Crop Improvement: Challenges in the Twenty-First Century* (2002; Haworth Press), *Handbook of Formulas for Plant Geneticists and Breeder* (2003; Haworth Press), *GGE Biplot Analysis—A Graphical Tool for Breeders, Geneticists, and Agronomists* (2003; CRC Press), *Genetic and Production Innovations in Field Crop Technology* (2005; Haworth Press), *Agricultural and Environmental Sustainability: Considerations for the Future* (2007; Haworth Press), *Breeding Major Food Staples* (2007; Blackwell Publishing), *Water and Agricultural Sustainability Strategies* (2010; CRC Press), *Combating Climate Change: An Agricultural Perspective* (2013, CRC Press), *Punjab Agriculture: The Challenges Ahead* (2015; Aesthetics Publications), and second edition of *Quantitative Genetics, Genomics, and Plant Breeding* (CABI, UK) (2020).

Dr. Kang has lectured on quantitative genetics as applied to crop improvement in Hungary in 1992 (USDA/OICD sponsorship), in Malaysia in 1999 (Fulbright), at South Dakota State University in 2005 ("South Dakota Crop Improvement Association Lectureship"), at Yunnan Academy of Agricultural Sciences at Kunming in 2006, at International Rice Research Institute in the Philippines (1999), and at International Institute of Tropical Agriculture in Nigeria (2006, 2007). He has organized several successful international conferences/symposia, including *Genotype-by-environment Interaction in Plant Breeding* (1990), *Quantitative Genetics and Genomics* (2001), *Agricultural and Environmental Sustainability* (2005), *Water Management Strategies for Food Security and Environment Quality* (2007), and *Preparing Agriculture for Climate Change* (2011).

Dr. Kang has received several prestigious honors and awards from various organizations. He is a fellow of both American Society of Agronomy and Crop Science Society of America, and an honorary fellow of several professional societies of India, including Indian Society of Life Sciences and Crop Improvement Society of India. In 1999, he was selected as a US Fulbright Senior Scholar to Malaysia. Some of the other awards conferred on Dr. Kang are *Outstanding Agricultural Scientist* in 2007 (Association of Agricultural Scientists of Indian Origin), *Sigma Xi Distinguished Lecturer* in 2007–2009 (Sigma Xi [Scientific Research Society]), *Amity Academic Excellence Award* in 2010 (Amity University, Noida, India), *Bishan Singh Samundri Lectureship Award*, sponsored by S. Jaswant Singh Rai Memorial Trust in 2011 (Guru Nanak Dev University, Amritsar), *Excellence Award in Agriculture Development* in 2013 (Jain-Advisor Group), *Pride of Punjab* in 2018 (by Chief Minister of Punjab, India), and *Sri Guru Nanak Dev Ji Achievers Award* in November 2019, during the celebration of 550th Birthday of Sri Guru Nanak Dev Ji (Punjab Government, India).

Contents

Contributors

Muhammad Afzal
Plant Production Department, College of Food and Agricultural Sciences, King Saud University, Riyadh, Saudi Arabia

Usman Aslam
Department of Plant Breeding and Genetics, University of Agriculture Faisalabad, Faisalabad, Pakistan

Rana Muhammad Atif
Department of Plant Breeding and Genetics; Centre of Agricultural Biochemistry and Biotechnology, University of Agriculture Faisalabad, Faisalabad, Pakistan

Farrukh Azeem
Department of Bioinformatics & Biotechnology, Government College University, Faisalabad, Pakistan

Bijayalaxmi Badajena
Plant Molecular Biology and "OMICS" Laboratory, Regional Plant Resource Centre, Bhubaneswar, Odisha, India

Aditya Banerjee
Department of Biotechnology, St. Xavier's College (Autonomous), Kolkata, West Bengal, India

Khushboo Dasauni
Department of Biotechnology, Kumaun University Nainital, Bhimtal, India

Muhammad Qudrat Ullah Farooqi
Department of Soil Science, School of Agriculture and Environment, Faculty of Science, The University of Western Australia, Perth, WA, Australia

V.C. Dilukshi Fernando
Department of Biological Sciences, University of Manitoba, Winnipeg, MB, Canada

Vijay Gahlaut
Department of Plant Molecular Biology, University of Delhi, Delhi, India

Manu Pratap Gangola
Department of Plant Sciences, College of Agriculture and Bioresources, University of Saskatchewan, Saskatoon, SK, Canada

Mansour Ghorbanpour
Department of Medicinal Plants, Faculty of Agriculture and Natural Resources, Arak University, Arak, Iran

Selvakumar Gurunathan
DRDO-BU-Centre for Life Sciences, Bharathiar University Campus, Coimbatore, Tamil Nadu, India

B. Jagadhesan
Division of Plant Physiology, Indian Agricultural Research Institute, New Delhi, India

S.K. Jha
Division of Genetics, Indian Agricultural Research Institute, New Delhi, India

Ma Shi Jun
Department of Applied Plant Sciences, College of Agriculture and Life Sciences, Kangwon National University, Chuncheon, Republic of Korea

Chandra Kant
Department of Botany, D S College, Aligarh, India

Anuj Kumar
Advanced Centre for Computational and Applied Biotechnology, Uttarakhand Council for Biotechnology, Dehradun, Uttarakhand, India

Behnam Asgari Lajayer
Department of Soil Science, Faculty of Agriculture, University of Tabriz, Tabriz, Iran

M.G. Mallikarjuna
Division of Genetics, Indian Agricultural Research Institute, New Delhi, India

Sahil Mehta
International Centre for Genetic Engineering and Biotechnology, New Delhi, India

Pouya Motie-Noparvar
Department of Agricultural Biotechnology, Faculty of Agriculture, Tarbiat Modares University, Tehran, Iran

M. Nagaraju
Department of Genetics, Osmania University, Hyderabad, Telangana, India

Tapan K. Nailwal
Department of Biotechnology, Kumaun University Nainital, Bhimtal, India

Vimal Pandey
National Institute of Plant Genome Research, New Delhi, India

Amna Parveen
College of Pharmacy, Gachon University, Incheon, Republic of Korea

Anupam Patra
International Centre for Genetic Engineering and Biotechnology, New Delhi, India

Seema Pradhan
Institute of Life Sciences, Bhubaneswar, India

Bharathi Raja Ramadoss
Department of Plant Sciences, College of Agriculture and Bioresources, University of Saskatchewan, Saskatoon, SK, Canada

Aryadeep Roychoudhury
Department of Biotechnology, St. Xavier's College (Autonomous), Kolkata, West Bengal, India

Lekshmy Sathee
Division of Plant Physiology, Indian Agricultural Research Institute, New Delhi, India

Luqman Shahid
Department of Plant Breeding and Genetics, University of Agriculture Faisalabad, Faisalabad, Pakistan

Mayur Mukut Murlidhar Sharma
Department of Agriculture and Life Industry, Kangwon National University, Chuncheon, Republic of Korea

Pankaj Sharma
Department of Microbiology, CCS Haryana Agricultural University, Hisar, India

Komal Shoukat
Department of Plant Breeding and Genetics, University of Agriculture Faisalabad, Faisalabad, Pakistan

Baljinder Singh
National Institute of Plant Genome Research, New Delhi, India

Giridara-Kumar Surabhi
Plant Molecular Biology and "OMICS" Laboratory, Regional Plant Resource Centre, Bhubaneswar, Odisha, India

Manish Tiwari
National Institute of Plant Genome Research, New Delhi, India

Mohammad Behrouzi Varjovi
Department of Agronomy and Plant Breeding, Faculty of Agricultural and Natural Resources, University of Mohaghegh Ardabili, Ardabil, Iran

Medhavi Vashisth
Department of Molecular Biology, Biotechnology and Bioinformatics, CCS Haryana Agricultural University, Hisar, India

Muhammad Waqas
Department of Plant Breeding and Genetics, University of Agriculture Faisalabad, Faisalabad, Pakistan

Foreword

I am delighted to know that Dr. Shabir Hussain Wani has edited another volume titled "**Transcription Factors for Abiotic Stress Tolerance in Plants**" for the globally reputed publisher Academic Press, Elsevier. I know him since 2005 when he joined as a PhD student under my supervision in the Department of Plant Breeding, Genetics and Biotechnology, Punjab Agricultural University, Ludhiana, India. I found him very hard-working and an able researcher. His PhD research outcome was published in reputed high-impact research journals like Biologia Plantarum, Springer and New Biotechnology, Elsevier. Since then he did not look back and published many good research, review, and popular articles in journals of high repute and books under globally renowned publishers. In 2017, he successfully completed one year Raman postdoctoral fellowship program sponsored by the University Grants Commission, New Delhi, India at Department of Plant, Soil, and Microbial Sciences, Michigan State University, USA and worked on dissection of *Pythium* root rot resistance in soybean using molecular genetics approaches utilizing SNP markers. He has an excellent know-how to work in the area of plant biotechnology particularly molecular breeding approaches for the development of biotic and abiotic stress tolerance in plants. I value his zeal and commitment for science including research, teaching, and diffusion of scientific knowledge.

Global crop production is suffering severe losses due to widespread abiotic stresses mounting as a result of global climate change. Drought and salinity affect more than 10% of arable land, which results in more than 50% decline in the average yields of important crops worldwide. Among the various categories of genes which get stirred during abiotic stress are the regulatory proteins that further regulate the stress signal transduction and modify gene expression and thus possibly function in stress response. They comprise several transcription factors (TFs) highlighting the role of an array of transcriptional regulatory mechanisms in the stress signal transduction pathways, protein kinases (MAP kinase, CDP kinase, receptor protein kinase), and protein phosphatases and proteinases implicated in the regulation of stress signaling and gene expression. Genetic engineering technologies are one of the promising tools for improvement in plants with higher yields and resilience to biotic and abiotic stresses. Hence deciphering the molecular mechanisms and identifying the abiotic stress-associated genes that control plant response to abiotic stresses are vital requirements to develop plants with increased abiotic stress tolerance. ABA is the essential controller of abiotic stress tolerance in plants, and coordinates a complex gene regulatory network enabling plants to cope with decreased water availability. During the past 10 years, significant advances have been made in this area to discover the role of regulatory genes for abiotic stress tolerance in plants The discovery of novel genes and pathways, analyses of expression patterns, and the determination of function of these genes during abiotic stress adaptation in crop plants have opened new vistas for the development of genetically engineered plants species tolerant to abiotic stress. This volume edited by Dr. Wani will be a useful reference for future work in the area of abiotic stress tolerance in plants as it unravels the transcriptional regulatory networks in plants.

Dr. Wani has done an exceptional attempt by editing this volume which includes high-quality chapters from the international as well as national level experts in various research fields. The chapters included in this volume are well written by prospective scientists and researchers who are experts in their fields. I applaud Dr. Wani for editing this volume which will be a valuable reference material for scientists, postgraduate and graduate students, and researchers working on abiotic stress tolerance in plants using state-of-the-art biotechnological tools.

Satbir S. Gosal
Punjab Agricultural University, Ludhiana, India

Preface

Global human population is mounting at a worrisome rate and assumed to go beyond 9.7 billion by 2050; however, the crop production and productivity are declining due to the emergent abiotic stresses as a consequence of global climate change. Drought and salinity influence about 10% of arable land, which results in more than 50% reduction in the average yields of imperative crops worldwide. More than one abiotic stress such as extreme temperature and osmotic stress can occur at once hence rising the intricacy of these environmental stresses. Among the range of genes which get stirred during abiotic stress are the regulatory proteins that control the stress signal transduction and amend gene expression and perhaps function in stress response. They consist of several transcription factors (TFs) describing the role of a variety of transcriptional regulatory mechanisms in the stress signal transduction pathways, protein kinases (MAP kinase, CDP kinase, receptor protein kinase), and protein phosphatases and proteinases responsible for the regulation of stress signaling and gene expression. Contemporary genetic engineering techniques are one of the promising tools for the improvement in plants with proficient yields and resilience to various kinds abiotic stresses like extreme temperatures, soil salinity, drought, and heavy metal stress. Therefore, cracking molecular mechanisms and discovering abiotic stress-associated genes that govern plant response to abiotic stresses are critical necessities to develop plants with increased abiotic stress tolerance. Since last few decades numerous novel developments have taken place in this area to ascertain the role of regulatory genes for abiotic stress tolerance in plants These findings of novel genes and pathways, analyses of expression patterns, and the determination of function of these genes during abiotic stress adaptation in crop plants have given new outlooks on the development of genetically engineered plant species tolerant to abiotic stress. Through the book "*Transcription Factors for Abiotic Stress Tolerance in Plants*" an effort has been made to include chapters from potential authors describing the role of transcription factors in deciphering transcriptional regulatory networks in plants and development of abiotic stress tolerance in important plant species.

This book provides organized and inclusive reference material for researchers, teachers, and graduate students involved in deciphering

transcriptional regulatory networks involved in abiotic stress tolerance in plants. The chapters are written by international and national level reputed researchers and academicians in the field of plant biotechnology and plant stress biology. First of all, I am highly grateful to Almighty Allah, who gave me the spirit and intellect to complete this book project. I express sincere thanks and appreciation to my revered authors, without their persistent efforts this book project would not have been possible. I am also thankful to Academic Press, Elsevier for providing such break to complete this book project. Last but not the least I thank all my family members for their support during the entire book project completion.

Shabir Hussain Wani

CHAPTER 1

Role of DNA-binding with one finger (Dof) transcription factors for abiotic stress tolerance in plants

Muhammad Waqas[a,*], Luqman Shahid[a,*], Komal Shoukat[a], Usman Aslam[a], Farrukh Azeem[b], and Rana Muhammad Atif[a,c]
[a]Department of Plant Breeding and Genetics, University of Agriculture Faisalabad, Faisalabad, Pakistan
[b]Department of Bioinformatics & Biotechnology, Government College University, Faisalabad, Pakistan
[c]Centre of Agricultural Biochemistry and Biotechnology, University of Agriculture Faisalabad, Faisalabad, Pakistan

1. Introduction

Plants grown under diverse environmental conditions (being sessile) face various abiotic stresses including drought, heat, cold, and salinity. These stresses are the major limiting factors of plant growth, yield, and quality (Singh et al., 2002; Corrales et al., 2014). To endure these stresses, plants undergo several metabolic and physiological changes via regulation of specific stress-responsive genes. Various genes especially that encode transcription factors (TFs) have been reported to take part in concerted regulation of stress responses in plants. TFs are regulatory proteins which control the transcriptional activity through binding with specific DNA sequences in promoter regions of genes during stress management and various developmental phases in plants (Karin, 1990; Latchman, 1997; Yamaguchi-Shinozaki and Shinozaki, 2006). These TFs also ensure the activation and regulation of several stress-responsive genes during stressed environmental conditions (Nakashima and Yamaguchi-Shinozaki, 2009). These TFs normally contain highly conserved DNA-binding domains (DBDs) that regulate the spatiotemporal expression of target genes involved in stress resistance mechanisms in plants.

Several stress-responsive TFs have been found through studies regarding the regulation of gene expression at the transcriptional level, for instance,

[*] Joint first author.

Transcription Factors for Abiotic Stress Tolerance in Plants
https://doi.org/10.1016/B978-0-12-819334-1.00001-0

bZIP, DREB, DOF, NAC, and WRKY etc. Among them, Dof (DNA-binding with one finger) represents plant-specific TF family (Yanagisawa, 2004) that play a substantial role in resistance mechanisms against biotic and abiotic stress tolerance in plants (Yanagisawa, 2004; Noguero et al., 2013). Dof TFs are widely distributed in the entire plant kingdom. The Dof proteins occur throughout the plant kingdom; from lower plants such as unicellular green alga (*Chlamydomonas reinhardtii* containing single *DOF* gene and *Physcomitrella patens carrying 19 DOF* genes) to the higher plants including both angiosperm and gymnosperm (Yanagisawa, 2002). Dof TFs also show the widespread distribution in monocots and dicots. The first Dof protein (ZmDof1) was discovered and characterized in maize using maize leaf cDNA library screening (Yanagisawa, 2002; Yanagisawa and Izui, 1993). Dof TFs have a variety of functions in plant growth and development including seed maturation and germination, phytohormones and light-mediated regulation as well as plant responses to biotic and abiotic stresses.

The ubiquitous functions of genes for biotic stresses have been well reported in plants, however, the functional modulation of *Dof* encoding genes for abiotic stress response regulation is reported only in few plant species. Now the recent advancements in the plant genomic sequencing tools have greatly enabled the genome-wide identification of *Dof* genes in plant species with complex genomes that will expand our understanding of their potential roles in plant stress regulation.

2. Dof domain structure and its targeted site

The Dof proteins belong to zinc finger family which mostly comprises highly conserved 200–400 amino-acid residues. DOF TFs possess two major domain structures, one highly conserved structure at N-terminal region called Dof DNA-binding domain and a less conserved transcriptional regulation domain at C-terminal region (Gupta et al., 2015; Kushwaha et al., 2011). Dof DNA-binding domain consists of 50–52 amino-acid residues with four conserved cysteines (CX_2-CX_{21}-CX_2-C) and able to form single zinc (Zn^{2+}) finger (Fig. 1), hence called DNA-binding with one finger (Yanagisawa, 2004; Noguero et al., 2013). The Dof DNA-binding domain recognizes a *cis*-regulatory element with a common core sequence 5′AAAG3′ at the promoter region of target genes with the exception of the pumpkin Dof protein (AOBP) that recognizes 5′AGTA3′ (Umemura et al., 2004). The Dof domain is a bifunctional domain that can mediate both DNA-protein as well as protein-protein interactions (Yanagisawa, 1997). The first

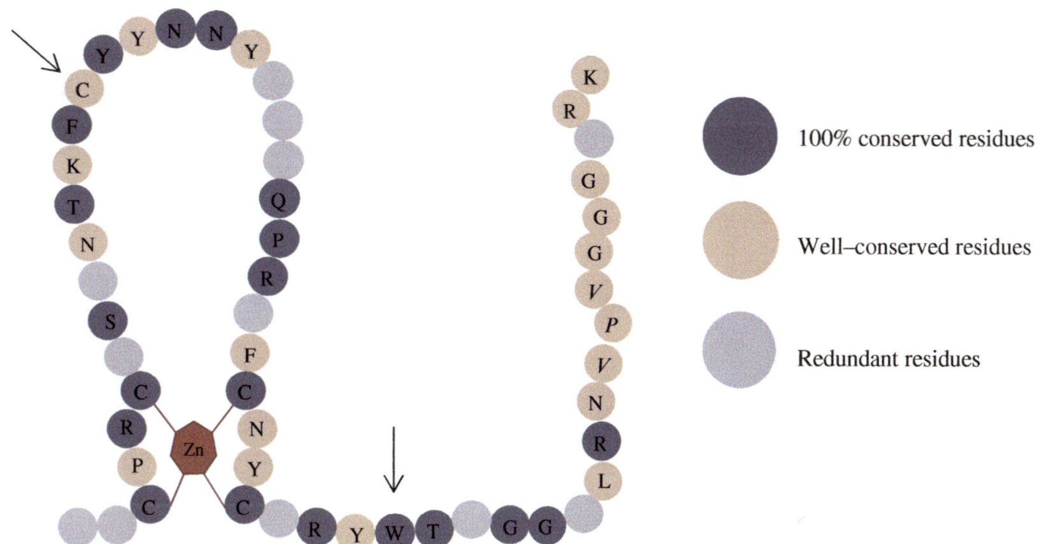

Fig. 1 Graphical representation of conserved DOF domain structure of 52 amino-acids. Fully conserved amino-acids are indicated in *blue-gray*, well-conserved residues are shown in *pink* and redundant residues are shown in *blue*. Two *black arrows* represent the nonzinc finger cysteine and the tryptophan residues at the C-terminal of the DOF domain.

protein-protein interaction of a Dof protein was reported in *Arabidopsis thaliana* in which AtOBP1 interacted with bZIP to mediate stress response. Later on, prolamin-box-binding factor (PBF) was found in maize which is a member of Dof TF family that can switch on the expression of cereal storage protein genes through binding to 7-bp prolamin-box (P-box or TGTAAAG) in the prolamine gene promoter (Yanagisawa, 1997; Vicente-Carbajosa et al., 1997). Similarly, many Dof TFs have been reported to regulate seed development by controlling the expression of seed storage genes such as ZmPBF by interacting with Opaque2 (O2) in maize, OsDof3 by interacting with GAMYB in rice, and Dof 3.2 by interacting with TCP14 in *A. thaliana*. Another important feature of Dof TFs is a highly conserved nuclear localization signal (NLS) that direct these proteins to the nucleus. This 17 amino-acids long NLS is typically bipartite in nature that is partly linked with the Dof DNA-binding domain at its downstream end and this property has allowed the identification of *Dof* genes in several plant species.

3. Role of Dof TFs in abiotic stresses tolerance

The Dof TFs are well-recognized protein family known for their diverse functioning in plants. They play a crucial role in various plant physiological processes such as carbon and nitrogen metabolism, phytochrome signaling, hormonal signaling, and cell cycle (Ward et al., 2005; Skirycz et al., 2008; Venkatesh and Park, 2015; Xu and Dai, 2016). Dof TFs are also involved in various biological mechanisms, such as photosynthesis, seed storage protein synthesis, and development and germination of seed (Cai et al., 2013).

These TFs detect specific *cis*-regulatory elements of DNA sequences and control the spatiotemporal expression of targeted genes (Ma et al., 2015). Very first Dof TFs were functionally characterized in Arabidopsis (Papi et al., 2000; Gualberti et al., 2002) followed by various crops including rice (Washio, 2003), tomato (Cai et al., 2013), soybean (Guo and Qiu, 2013), and rapeseed (Xu and Dai, 2016). The functional studies of Dof TFs in plants were mostly focused on their role in growth, development, and biotic stress responses. But during recent years, some researchers reported the role of Dof TFs for abiotic stress tolerance as well. These TFs are indirectly involved in abiotic stress tolerance as activators and repressors of several abiotic stress-responsive genes. Up till now several Dof TFs have been functionally characterized in regulating biotic (Kang et al., 2016) and abiotic stress (Zhang et al., 2018; Liu et al., 2019) tolerance in plants (Table 1). In poplar trees, the pine transcriptional regulator *Dof5* affects carbon and nitrogen

Table 1 Functional characterization of Dof genes involved in abiotic stress tolerance in different plant species.

Sr.#	Gene name	Function	Scientific name	Common name	Reference
1.	ThDof1.4/ThZFP1	Regulate salt or osmotic stress	*Tamarix hispida*	Tamarisk	Zang et al. (2017)
2.	PtrDof (14, 16, 25, 27, 28, 30, 37, 39, and)	Resistance to abiotic stress	*Populus trichocarpa*	poplar	Wang et al. (2017)
3.	GhDof1	Improved salt and cold tolerance and seed oil contents	*Gossypium hrisutum*	cotton	Su et al. (2017)
4.	TDDF1	Flowering accelerator and protector against various stresses	*Solanum lycopersicum*	Tomato	Ewas et al. (2017)
5.	CDF3	Regulation of flowering time and abiotic stress responses	*Arabidopsis thaliana*	Arabidopsis	Corrales et al. (2017)
6.	AhDof or AhERF	Increase water deficit and salt-stress tolerance	*Arabidopsis thaliana*	Arabidopsis	Massange-Sánchez et al. (2016)
7.	SlDof22	Involved in ascorbate accumulation and salinity stress	*Solanum lycopersicum*	Tomato	Cai et al. (2016)
8.	SlCDF1 and SlCDF3	Salt and drought tolerance	*Solanum lycopersicum*	Tomato	Corrales et al. (2014)
9.	SCAP1; AtDof5.7	Stomatal maturation and functioning	*Arabidopsis thaliana*	Arabidopsis	Negi et al. (2013)

assimilation (Rueda-López et al., 2017). In transgenic *Arabidopsis*, the expression of *Dof1*TF of maize is responsible for the upregulation of several genes, involved in the production of carbon-skeleton and improved N-assimilation (Yanagisawa et al., 2004; Kurai et al., 2011). In *Arabidopsis*, under high nitrogen supply, the expression of *OsDof25* gene from rice modulated the C and N metabolism (Santos et al., 2012). The *ZmDof1* and *ZmDof2* genes from maize regulate expression of genes associated with the metabolism of carbohydrate including (C4PEPC); C_4 key photosynthetic gene, i.e., phosphoenolpyruvate carboxylase (Yanagisawa, 2000). Dof TFs are also involved in the regulation of lipid metabolisms. For example, 27 putative Dof proteins were recognized in the soybean, two genes (*GmDof4* and *GmDof11*) of which have been reported to identified the biosynthesis of fatty acids to increase lipid content in the transgenic *Arabidopsis* and *Chlorella ellipsoidea* (Wang et al., 2007; Zhang et al., 2014). Various Dof TFs are also involved in the synthesis and regulation of storage proteins, however, some proteins are only distinct during the development of endosperm in cereals at the seed-filling stage (Noguero et al., 2013). In rice, the Dof transcriptional *RPBF* highly expressed during the maturation stage of endosperm and interacts with *RISBZ1* to control the expression of genes involved in regulation of seed storage proteins (Yamamoto et al., 2006).

4. Salt, cold, and drought stress

Sallinity and drought are major limiting factors in crop production and their intensity has been increasing due to global climate change. Dof TF are one of the key factors that physiologically and bionomically assist plants to combat these stresses. Several Dof TFs have been reported to regulate these stresses. A Dof TF (GhDof1) from *Gossypium hirsutum* (upland cotton) is highly expressed in leaves, roots, and shoots, and showed to upregulate salt and cold stress-responsive genes such as *GhMYB*, *GhSOD*, and *GhP5CS* (Su et al., 2017). The increase in the regulation of these *Dof* encoding genes was induced by both, salt and cold treatments. The overexpressing transgenic lines showed more salinity and low-temperature tolerance as compared to the wild type. Under salt and cold stress, more proline accumulation was also detected in transgenic lines and it is hypothesized that *GhDof1* can also improve the ROS (reactive oxygen species) scavenging ability of plants. Similarly, the Cyclic Dof factor TF (*CDF3*) enhances the photosynthetic activity resulting in the accumulation of higher biomass in tomato under salt stress (Renau-Morata et al., 2017). The overexpression of *AtCDF3* and

SlCDF3 was involved in the upregulation of the genes which are involved in cell growth, primary metabolism through nitrogen assimilation, and photosynthetic activity under salt stress. The improvement in the tomato fruit quality due to the overexpression of *CDF3*, provide a clue about the role of CDFs for the enhancement of the tomato production under salt stress. A *Brassica napus* Dof gene (*BnCDF1*), homologs of *Arabidopsis CDF1*, considered as a nuclear factor, transcriptionally regulate the freezing tolerance and mutually control the photoperiodicity of flowering. The expression analysis of BnCDF1 through qRT-PCR revealed its response to the low temperature suggesting its involvement in upregulation of cold-responsive TFs (Xu and Dai, 2016). Transcriptome profiling demonstrates that several abiotic stress tolerance, oxidative, and osmoprotectant stress-responsive transcription factors (CBF, ZAT12s and DREB2A) are regulated by *CDF3*. The metabolomic profiling of overexpressing CDF3 also shows an increase in the accumulation of proline, sucrose, and glucose concentrations in plant tissues after exposure to high temperature and osmotic stress, i.e., drought stress (Hussain et al., 2011).

The functional studies of *CDF3* in *Arabidopsis* have demonstrated their role in the upregulation of numerous drought, cold, and osmotic stress-responsive genes (Corrales et al., 2017). In another study, the transgenic approach was used to evaluate the functions of *Amaranthus hypochondriacus* Dof TFs under drought and salt treatment. Two TFs (AhDOF and AhERF) overexpressed in *Arabidopsis* showed tolerance against drought and salt stress. Microarray and gene ontology (GO) analysis revealed that *AhDof* was involved in ABA-dependent and independent salt stress tolerance (Massange-Sánchez et al., 2016). The expression of some members of the Dof TFs family in *Triticum aestivum* is differentially regulated under drought, salt, heat, and light stress (Shaw et al., 2009). The expression of all the TaDof TFs encoding genes was decreased except two, i.e., *TaDof14* and *TaDof15* under drought stress. These TFs were considered as putative genes for drought tolerance in wheat. While *TaDof1* was recognized as salt-responsive TF. The complex regulatory pathways and mode of action of Dof protein family for abiotic stress are still at initial stages. Recently, an evolutionary and expression study of Dof TFs in *Malus domestica* (Apple) indicated that some MaDof TFs encoding genes also respond to different abiotic stress stimuli (Zhang et al., 2018). The transcriptional activity and sensitivity of *MaDof* in response to polyethylene glycol, salt, low temperature, and ABA treatment, suggested that these TFs were involved in drought, salinity, cold, and other abiotic stresses. The expression profile of *Populus trichocarpa* (poplar, an important timber tree) *Dof* genes shows that seven *Dof* genes

(*PtrDof14*, *-16*, *-25*, *-27*, *-28*, *-37*, and *PtrDof39*) were expressed under ABA and osmotic stress (Wang et al., 2017). The constitutive expression in leaves and roots under a stressed environment shows their crucial role in abiotic stress tolerance. Interestingly, *DcDof-B2-2* exhibited the upregulated expression under drought stress (Huang et al., 2016). The expression profile also revealed the transcriptional variation under the same environmental conditions. Some Dof TFs have been involved in the regulation of ascorbic acid, which is an important antioxidant, growth regulator, and enzyme cofactor of both animal and plant cells (Alhagdow et al., 2007). The Dof TF of rice (SlDof22), interacts with the three major salt-response pathways such as SOS1, SOS2, and SOS3 through ascorbate accumulation and control the salt responses in tomato (Cai et al., 2016). The expression and repression of these Dof TFs encoding genes affect the ascorbate biosynthesis, which increases or decreases the salt tolerance through binding with the promoter of the *SlSOS1*. The lower-expression of *SlDof22* results in the downregulation of *SlSOS1* ultimately reduces the plant tolerance to salt stress. The protein of *Juglans regia Dof3* was seemed to be binding with the *cis*-regulatory element, i.e., DOFCOR-EZM, present in the promoter region of JrGRAS2 TF encoding gene. The abiotic stress response study of *Daucus carota* (carrot) shows that Dof TFs have a significant positive role in abiotic stress tolerance. The stress response validated by RT-qPCR revealed that the expression of *DcDofC1-1*, *DcDof-B2-2*, and *DcDof-C2.1-5* was upregulated under cold stress. The Dof TF encoding genes of *Capsicum annum* such as *CaDof10*, *-13*, *-15*, *-16*, *-24*, and *CaDof25* were upregulated under salt stress. The expression analysis of *StDof* gene family in *Solanum tuberosum* (potato) reported the role of putative *Dof* genes for a wide range of abiotic stresses such as drought, salt, and ABA treatment (Venkatesh and Park, 2015). The detailed functional and expression analysis of these *Dof* genes can be used to unravel the hidden potential of StDof Tfs against abiotic stresses.

5. Temperature and heat stress

Temperature intensity is a crucial parameter and sensitive to seed germination, transition between vegetative and reproductive phases, flower induction and maturity in plants. Even slight changes in the temperatures can cause sever damage to plant physiology at important stages. Dof TFs play important roles in modulating the heat stress responses in plants. Three Dof encoding genes (*TaDof5*, *TaDof17* and *TaDof19*) were downregulated in leaves of wheat under prolonged exposure to heat stress. Likewise, in *J. regia* (walnut) the expression of *JrGRAS2* (a heat shock protein expression controlling gene) was upregulated by *JrDof3* and induce heat tolerance (Yang et al., 2018). The Dof genes of *Docus carota* (*DcDof-C1-1*, *DcDof-A-1*, and *DcDof-B2-2*) were involved in the responsiveness to heat treatment. The genome-wide identification and expression profile study of Dof TFs in *Capsicum annuum* (peppers) revealed that some of the Dof TFs encoding genes can be used as a biomarker for salt and heat stress (Wu et al., 2016). The transcriptional activity of some genes from *C. annuum* (*CaDof10*, *-13*, *-16*, *-25*, and *CaDof28*) is responsive to heat stress tolerance. But the functional characterization and activation mechanisms are not confirmed yet for heat stress.

6. Role of Dof TFs in response to photoperiod

Arabidopsis Dof TFs are observed to be involved in suppressing the expression of CONSTANS (CO), an important transcriptional regulator that promotes flowering based on photoperiodic responses (Fornara et al., 2009). The transgenic and overexpression studies of *CDF3* in Arabidopsis produced a similar expression pattern and phenotype (Corrales et al., 2017). The subfamily of Dof transcription factors, i.e., CDFs contain 52 amino-acid residues and play a very important role during photoperiod in different plant species. Due to overexpression of *CDF1,2,3*, and *5* during conditions of long days, the phenomenon of flowering delayed by repression of CO transcription in Arabidopsis. Early flowering which is photoperiod insensitive is potentially caused by loss of allele features in one of these four genes (Fornara et al., 2009). The PbDof9.2 responds to photoperiod and circadian rhythms and activates promoters (PbTFL1a and PbTFL1b) due to which regulation of flowering time suppressed (Liu et al., 2020).

In rice, *OsDof4* gene played an important role in the regulation of flowering time but its overexpression led to earlier flowering under long-days whereas late flowering under short-days (Wu et al., 2017). The expression analysis revealed that many Dof TFs were seemed to be upregulated in response to the light cycle and involved in the photosynthetic process.

7. Protein-protein interactions of Dof TFs with other proteins

The Dof TFs work in genetic pathways with diverse range of interacting partners. For example the ATDOF5.8 (a Dof protein of *Arabidopsis*) was reported as an upregulator for the *ANAC06* and both are involved in the abiotic stress tolerance (He et al., 2015). Salt-induced regulation and binding of the

ATDOF5.8 protein to the Dof motif of the promoter of *ANAC069* initiate the regulation of a series of stress-responsive genes against abiotic stresses. The interaction of ATDOF5.8 and ANAC069 confirmed by Y1H (yeast one-hybrid) were further validated by transient expression assay in tobacco. Interestingly, the protein and its targeted gene interactional cascade produce a similar expression pattern under salt, drought, and ABA treatments. In Barley, the GAMYB protein is activated in aleurone cells because of gibberellin supply associated with TF BPBF, which activates endosperm-specific genes during the development of seed (Diaz et al., 2002). The Dof TF OBP3 from Arabidopsis showed interaction with light and hormone signaling (Ward et al., 2005). The phytochromes (phyA and phyB) positively regulate the expression of *Arabidopsis COG1* gene-encoded Dof-domain TF, but in contrast, *COG1* gene negatively regulates the light-signaling pathway (Park et al., 2003).

Dof TF recognizes a sequence (T/GTAAAG) of prolamin box present in the promoter region of the target genes (Isabel-LaMoneda et al., 2003). Experiments for deletion and point mutation showed that the expression of specific genes of endosperm is regulated by PB (Diaz et al., 2005; Mena et al., 1998). Many factors involving Dof and bZIP family members function together during the development of seed for storage proteins expression regulation (Mena et al., 2002). BiFC biomolecular fluorescent technology confirmed that SAD, a DOF protein of barley interacts with HvGAMYB protein in plant nuclei (Diaz et al., 2005). Studies demonstrated that Dof proteins bind with *cis*-regulatory elements in the promoters of target genes and mediate the interaction between different proteins by its C-terminal and Dof domain.

8. Future perspectives

Although several researchers had identified and elucidated the abiotic stress tolerance in plants as regulated by Dof TFs, studies are still needed to explore the additional interaction factors and regulatory pathways for abiotic stresses through Dof TFs. The versatile abiotic stress-related role of the Dof family is not evident even in the model plants. A few recent studies have demonstrated that Dof TFs play a positive role in different environmental stress tolerance in plants. Therefore, extensive and comprehensive functional studies are needed to explore as well as characterize the dynamic networks, regulatory system, and associated pathways of Dof TFs and respective genes with special reference to abiotic stresses.

References

Alhagdow, M., Mounet, F., Gilbert, L., Nunes-Nesi, A., Garcia, V., Just, D., et al., 2007. Silencing of the mitochondrial ascorbate synthesizing enzyme L-galactono-1,4-lactone dehydrogenase affects plant and fruit development in tomato. Plant Physiol. 145 (4), 1408–1422.

Cai, X., Zhang, Y., Zhang, C., Zhang, T., Hu, T., Ye, J., et al., 2013. Genome-wide analysis of plant-specific Dof transcription factor family in tomato. J. Integr. Plant Biol. 55 (6), 552–566.

Cai, X., Zhang, C., Shu, W., Ye, Z., Li, H., Zhang, Y., 2016. The transcription factor SlDof22 involved in ascorbate accumulation and salinity stress in tomato. Biochem. Biophys. Res. Commun. 474 (4), 736–741.

Corrales, A.-R., Nebauer, S.G., Carrillo, L., Fernández-Nohales, P., Marqués, J., Renau-Morata, B., et al., 2014. Characterization of tomato Cycling Dof Factors reveals conserved and new functions in the control of flowering time and abiotic stress responses. J. Exp. Bot. 65 (4), 995–1012.

Corrales, A.R., Carrillo, L., Lasierra, P., Nebauer, S.G., Dominguez-Figueroa, J., Renau-Morata, B., et al., 2017. Multifaceted role of cycling DOF factor 3 (CDF3) in the regulation of flowering time and abiotic stress responses in Arabidopsis. Plant Cell Environ. 40 (5), 748–764.

Diaz, I., Vicente-Carbajosa, J., Abraham, Z., Martínez, M., Isabel-La Moneda, I., Carbonero, P., 2002. The GAMYB protein from barley interacts with the DOF transcription factor BPBF and activates endosperm-specific genes during seed development. Plant J. 29 (4), 453–464.

Diaz, I., Martinez, M., Isabel-LaMoneda, I., Rubio-Somoza, I., Carbonero, P., 2005. The DOF protein, SAD, interacts with GAMYB in plant nuclei and activates transcription of endosperm-specific genes during barley seed development. Plant J. 42 (5), 652–662.

Ewas, M., Khames, E., Ziaf, K., Shahzad, R., Nishawy, E., Ali, F., et al., 2017. The tomato DOF daily fluctuations 1, TDDF1 acts as flowering accelerator and protector against various stresses. Sci. Rep. 7 (1), 10299.

Fornara, F., Panigrahi, K.C., Gissot, L., Sauerbrunn, N., Rühl, M., Jarillo, J.A., et al., 2009. Arabidopsis DOF transcription factors act redundantly to reduce CONSTANS expression and are essential for a photoperiodic flowering response. Dev. Cell 17 (1), 75–86.

Gualberti, G., Papi, M., Bellucci, L., Ricci, I., Bouchez, D., Camilleri, C., et al., 2002. Mutations in the Dof zinc finger genes DAG2 and DAG1 influence with opposite effects the germination of Arabidopsis seeds. Plant Cell 14 (6), 1253–1263.

Guo, Y., Qiu, L.-J., 2013. Genome-wide analysis of the Dof transcription factor gene family reveals soybean-specific duplicable and functional characteristics. PLoS One. 8(9), e76809.

Gupta, S., Malviya, N., Kushwaha, H., Nasim, J., Bisht, N.C., Singh, V., et al., 2015. Insights into structural and functional diversity of Dof (DNA binding with one finger) transcription factor. Planta 241 (3), 549–562.

He, L., Su, C., Wang, Y., Wei, Z., 2015. ATDOF5.8 protein is the upstream regulator of ANAC069 and is responsive to abiotic stress. Biochimie 110, 17–24.

Huang, W., Huang, Y., Li, M.-Y., Wang, F., Xu, Z.-S., Xiong, A.-S., 2016. Dof transcription factors in carrot: genome-wide analysis and their response to abiotic stress. Biotechnol. Lett. 38 (1), 145–155.

Hussain, S.S., Kayani, M.A., Amjad, M., 2011. Transcription factors as tools to engineer enhanced drought stress tolerance in plants. Biotechnol. Prog. 27 (2), 297–306.

Isabel-LaMoneda, I., Diaz, I., Martinez, M., Mena, M., Carbonero, P., 2003. SAD: a new DOF protein from barley that activates transcription of a cathepsin B-like thiol protease gene in the aleurone of germinating seeds. Plant J. 33 (2), 329–340.

Kang, W.-H., Kim, S., Lee, H.-A., Choi, D., Yeom, S.-I., 2016. Genome-wide analysis of Dof transcription factors reveals functional characteristics during development and response to biotic stresses in pepper. Sci. Rep. 6, 33332.
Karin, M., 1990. Too many transcription factors: positive and negative interactions. New Biol. 2 (2), 126–131.
Kurai, T., Wakayama, M., Abiko, T., Yanagisawa, S., Aoki, N., Ohsugi, R., 2011. Introduction of the ZmDof1 gene into rice enhances carbon and nitrogen assimilation under low-nitrogen conditions. Plant Biotechnol. J. 9 (8), 826–837.
Kushwaha, H., Gupta, S., Singh, V.K., Rastogi, S., Yadav, D., 2011. Genome wide identification of Dof transcription factor gene family in sorghum and its comparative phylogenetic analysis with rice and Arabidopsis. Mol. Biol. Rep. 38 (8), 5037–5053.
Latchman, D.S., 1997. Transcription factors: an overview. Int. J. Biochem. Cell Biol. 29 (12), 1305–1312.
Liu, J., Cheng, Z., Xie, L., Li, X., Gao, J., 2019. Multifaceted role of PheDof12-1 in the regulation of flowering time and abiotic stress responses in moso bamboo (*Phyllostachys edulis*). Int. J. Mol. Sci. 20 (2), 424.
Liu, X., Liu, Z., Hao, Z., Chen, G., Qi, K., Zhang, H., et al., 2020. Characterization of Dof family in *Pyrus bretschneideri* and role of PbDof9. 2 in flowering time regulation. Genomics 112 (1), 712–720.
Ma, J., Li, M.-Y., Wang, F., Tang, J., Xiong, A.-S., 2015. Genome-wide analysis of Dof family transcription factors and their responses to abiotic stresses in Chinese cabbage. BMC Genomics 16 (1), 33.
Massange-Sánchez, J.A., Palmeros-Suárez, P.A., Espitia-Rangel, E., Rodríguez-Arévalo, I., Sánchez-Segura, L., Martínez-Gallardo, N.A., et al., 2016. Overexpression of grain amaranth (*Amaranthus hypochondriacus*) AhERF or AhDOF transcription factors in *Arabidopsis thaliana* increases water deficit-and salt-stress tolerance, respectively, via contrasting stress-amelioration mechanisms. PLoS One. 11(10), e0164280.
Mena, M., Vicente-Carbajosa, J., Schmidt, R.J., Carbonero, P., 1998. An endosperm-specific DOF protein from barley, highly conserved in wheat, binds to and activates transcription from the prolamin-box of a native B-hordein promoter in barley endosperm. Plant J. 16 (1), 53–62.
Mena, M., Cejudo, F.J., Isabel-Lamoneda, I., Carbonero, P., 2002. A role for the DOF transcription factor BPBF in the regulation of gibberellin-responsive genes in barley aleurone. Plant Physiol. 130 (1), 111–119.
Nakashima, K., Yamaguchi-Shinozaki, K., 2009. Promoters and transcription factors in abiotic stress-responsive gene expression. In: Abiotic Stress Adaptation in Plants. Springer, pp. 199–216.
Negi, J., Moriwaki, K., Konishi, M., Yokoyama, R., Nakano, T., Kusumi, K., et al., 2013. A Dof transcription factor, SCAP1, is essential for the development of functional stomata in Arabidopsis. Curr. Biol. 23 (6), 479–484.
Noguero, M., Atif, R.M., Ochatt, S., Thompson, R.D., 2013. The role of the DNA-binding One Zinc Finger (DOF) transcription factor family in plants. Plant Sci. 209, 32–45.
Papi, M., Sabatini, S., Bouchez, D., Camilleri, C., Costantino, P., Vittorioso, P., 2000. Identification and disruption of an Arabidopsis zinc finger gene controlling seed germination. Genes Dev. 14 (1), 28–33.
Park, D.H., Lim, P.O., Kim, J.S., Cho, D.S., Hong, S.H., Nam, H.G., 2003. The Arabidopsis COG1 gene encodes a Dof domain transcription factor and negatively regulates phytochrome signaling. Plant J. 34 (2), 161–171.
Renau-Morata, B., Molina, R.V., Carrillo, L., Cebolla-Cornejo, J., Sánchez-Perales, M., Pollmann, S., et al., 2017. Ectopic expression of CDF3 genes in tomato enhances biomass production and yield under salinity stress conditions. Front. Plant Sci. 8, 660.
Rueda-López, M., Pascual, M.B., Pallero, M., Henao, L.M., Lasa, B., Jauregui, I., et al., 2017. Overexpression of a pine Dof transcription factor in hybrid poplars: a comparative

study in trees growing under controlled and natural conditions. PLoS One. 12(4), e0174748.

Santos, L.A., de Souza, S.R., Fernandes, M.S., 2012. OsDof25 expression alters carbon and nitrogen metabolism in Arabidopsis under high N-supply. Plant Biotechnol. Rep. 6 (4), 327–337.

Singh, K.B., Foley, R.C., Oñate-Sánchez, L., 2002. Transcription factors in plant defense and stress responses. Curr. Opin. Plant Biol. 5 (5), 430–436.

Skirycz, A., Radziejwoski, A., Busch, W., Hannah, M.A., Czeszejko, J., Kwaśniewski, M., et al., 2008. The DOF transcription factor OBP1 is involved in cell cycle regulation in *Arabidopsis thaliana*. Plant J. 56 (5), 779–792.

Su, Y., Liang, W., Liu, Z., Wang, Y., Zhao, Y., Ijaz, B., et al., 2017. Overexpression of GhDof1 improved salt and cold tolerance and seed oil content in *Gossypium hirsutum*. J. Plant Physiol. 218, 222–234.

Umemura, Y., Ishiduka, T., Yamamoto, R., Esaka, M., 2004. The Dof domain, a zinc finger DNA-binding domain conserved only in higher plants, truly functions as a Cys2/Cys2 Zn finger domain. Plant J. 37 (5), 741–749.

Venkatesh, J., Park, S.W., 2015. Genome-wide analysis and expression profiling of DNA-binding with one zinc finger (Dof) transcription factor family in potato. Plant Physiol. Biochem. 94, 73–85.

Vicente-Carbajosa, J., Moose, S.P., Parsons, R.L., Schmidt, R.J., 1997. A maize zinc-finger protein binds the prolamin box in zein gene promoters and interacts with the basic leucine zipper transcriptional activator Opaque2. Proc. Natl. Acad. Sci. U. S. A. 94 (14), 7685–7690.

Wang, H.W., Zhang, B., Hao, Y.J., Huang, J., Tian, A.G., Liao, Y., et al., 2007. The soybean Dof-type transcription factor genes, GmDof4 and GmDof11, enhance lipid content in the seeds of transgenic Arabidopsis plants. Plant J. 52 (4), 716–729.

Wang, H., Zhao, S., Gao, Y., Yang, J., 2017. Characterization of Dof transcription factors and their responses to osmotic stress in poplar (*Populus trichocarpa*). PLoS One. 12(1), e0170210.

Ward, J.M., Cufr, C.A., Denzel, M.A., Neff, M.M., 2005. The Dof transcription factor OBP3 modulates phytochrome and cryptochrome signaling in Arabidopsis. Plant Cell 17 (2), 475–485.

Washio, K., 2003. Functional dissections between GAMYB and Dof transcription factors suggest a role for protein-protein associations in the gibberellin-mediated expression of the RAmy1A gene in the rice aleurone. Plant Physiol. 133 (2), 850–863.

Wu, Z., Cheng, J., Cui, J., Xu, X., Liang, G., Luo, X., et al., 2016. Genome-wide identification and expression profile of dof transcription factor gene family in pepper (*Capsicum annuum* L.). Front. Plant Sci. 7, 574.

Wu, Q., Liu, X., Yin, D., Yuan, H., Xie, Q., Zhao, X., et al., 2017. Constitutive expression of OsDof4, encoding a C_2-C_2 zinc finger transcription factor, confesses its distinct flowering effects under long-and short-day photoperiods in rice (*Oryza sativa* L.). BMC Plant Biol. 17 (1), 166.

Xu, J., Dai, H., 2016. *Brassica napus* Cycling Dof Factor1 (BnCDF1) is involved in flowering time and freezing tolerance. Plant Growth Regul. 80 (3), 315–322.

Yamaguchi-Shinozaki, K., Shinozaki, K., 2006. Transcriptional regulatory networks in cellular responses and tolerance to dehydration and cold stresses. Annu. Rev. Plant Biol. 57, 781–803.

Yamamoto, M.P., Onodera, Y., Touno, S.M., Takaiwa, F., 2006. Synergism between RPBF Dof and RISBZ1 bZIP activators in the regulation of rice seed expression genes. Plant Physiol. 141 (4), 1694–1707.

Yanagisawa, S., 1997. Dof DNA-binding domains of plant transcription factors contribute to multiple protein-protein interactions. Eur. J. Biochem. 250 (2), 403–410.

Yanagisawa, S., 2000. Dof1 and Dof2 transcription factors are associated with expression of multiple genes involved in carbon metabolism in maize. Plant J. 21 (3), 281–288.

Yanagisawa, S., 2002. The Dof family of plant transcription factors. Trends Plant Sci. 7 (12), 555–560.
Yanagisawa, S., 2004. Dof domain proteins: plant-specific transcription factors associated with diverse phenomena unique to plants. Plant Cell Physiol. 45 (4), 386–391.
Yanagisawa, S., Izui, K., 1993. Molecular cloning of two DNA-binding proteins of maize that are structurally different but interact with the same sequence motif. J. Biol. Chem. 268 (21), 16028–16036.
Yanagisawa, S., Akiyama, A., Kisaka, H., Uchimiya, H., Miwa, T., 2004. Metabolic engineering with Dof1 transcription factor in plants: improved nitrogen assimilation and growth under low-nitrogen conditions. Proc. Natl. Acad. Sci. U. S. A. 101 (20), 7833–7838.
Yang, G., Gao, X., Ma, K., Li, D., Jia, C., Zhai, M., et al., 2018. The walnut transcription factor JrGRAS2 contributes to high temperature stress tolerance involving in Dof transcriptional regulation and HSP protein expression. BMC Plant Biol. 18 (1), 367.
Zang, D., Wang, L., Zhang, Y., Zhao, H., Wang, Y., 2017. ThDof1.4 and ThZFP1 constitute a transcriptional regulatory cascade involved in salt or osmotic stress in *Tamarix hispida*. Plant Mol. Biol. 94 (4–5), 495–507.
Zhang, J., Hao, Q., Bai, L., Xu, J., Yin, W., Song, L., et al., 2014. Overexpression of the soybean transcription factor GmDof4 significantly enhances the lipid content of *Chlorella ellipsoidea*. Biotechnol. Biofuels 7 (1), 128.
Zhang, Z., Yuan, L., Liu, X., Chen, X., Wang, X., 2018. Evolution analysis of Dof transcription factor family and their expression in response to multiple abiotic stresses in *Malus domestica*. Gene 639, 137–148.

CHAPTER 2

WRKY transcription factors for biotic and abiotic stress tolerance in plants

Manu Pratap Gangola and Bharathi Raja Ramadoss
Department of Plant Sciences, College of Agriculture and Bioresources, University of Saskatchewan, Saskatoon, SK, Canada

1. Background

Plants are continuously exposed to environmental fluctuations including abiotic and biotic stresses that cause metabolic, morphological, physiological, and molecular changes in plants affecting growth, development, and ultimately production (Agarwal et al., 2011; Finatto et al., 2018). Abiotic stresses includes drought, heat, cold, high soil salinity, heavy metal toxicity, oxidative stress, and other adverse environmental conditions limiting plant performance whereas biotic stress can be triggered by living organisms, viz., fungi, bacteria, viruses, nematodes, insects, arachnids, and weeds (Suzuki et al., 2014). A compound effect of different types of biotic and abiotic stresses adversely impact agricultural production and is a major cause of yield ceiling in various crops (Suzuki et al., 2014), which is one of major constraints in feeding a world population of approximately 7 billion, expected to be over 9 billion by 2050 (Béné et al., 2015). Therefore, it is critical to improve plant tolerance against abiotic and biotic stresses causing significant reduction in agricultural production worldwide (Pandey et al., 2017). In response to these stresses, plants have evolved various defense systems and mechanisms to cope with abiotic and biotic challenges (Schwachtje et al., 2019). Consequently, stress response in plants initiates with the perception of stress clues from abiotic or biotic factors at cell wall and/or membrane which is perceived by secondary messengers (like calcium ions, reactive oxygen species, and phytohormones) and transferred as intracellular signal to downstream signal transduction pathways like kinases or phosphatases (Erpen et al., 2018; Novaković et al., 2018). The signal transduction pathways tailor the expression of transcription factors which in return modulate

Transcription Factors for Abiotic Stress Tolerance in Plants
https://doi.org/10.1016/B978-0-12-819334-1.00002-2

the expression of stress responsive genes in plant cells (Erpen et al., 2018; Lan Thi Hoang et al., 2017). Thus, stress response in plants is a highly complex mechanism where transcription factors (TFs) are one of major players. TFs are important protein complexes for a cell or plant to regulate expression of genes based on environmental or developmental cues, therefore have been attributed to diverse, important, and unique regulatory role during growth, development, hormone signaling, and primary/secondary metabolism (Agarwal et al., 2011; Sukumari Nath et al., 2019). TFs based on their DNA-binding domains have been categorized into various groups, viz., DREB/ERF, MYB, NAC, and WRKY (Agarwal et al., 2011). Targeting TFs to improve abiotic and biotic stress tolerance in plants has been an extensively cultivated thought.

WRKY TFs, named after their conserved amino acid sequence of DNA-binding domain (Song et al., 2018), represent one of the largest families of transcriptional regulators in plants. WRKYs are referred to as plant-specific TFs and are also present in flagellated protozoan (*Giardia lamblia*) and soil-living amoeba like slime mold (*Dictyostelium discoideum*) suggesting their long evolutionary history (Mingyu et al., 2012; Phukan et al., 2016). The first WRKY cDNA-encoding DNA-binding protein SPF1 (Sweet Potato Factor 1) was isolated from 5′ upstream region of three genes synthesizing sporamin and α-amylase in tuberous roots of sweet potato (*Ipomoea batatas* L.) (Ishiguro and Nakamura, 1994). Thereafter, WRKY TFs has been widely studied in various crop species especially for their role during abiotic and biotic stress tolerance that has been summarized in this chapter.

2. WRKY structure and distribution among crop species

WRKY transcription factors contain 60 amino acids long 4-stranded β-sheet DNA-binding domain with a highly conserved heptapeptide (WRKYGQK) signature motif on N-terminus and zinc finger-like motif on C-terminus (Bakshi and Oelmüller, 2014; Jiang et al., 2017). WRKYs based on number of DNA-binding domains and type of zinc finger-like motif are categorized into three main groups: (i) group I contain two DNA-binding domains and zinc finger-like motif pattern (C-X4-5-C-X22-23-H-X1-H), (ii) group II possess only one DNA-binding domain but has zinc finger-like motif pattern like group I WRKYs, and (iii) group III also comprises single DNA-binding domain but has different zinc finger-like motif pattern (C-X7-C-X23-H-X1-C) (Agarwal et al., 2011; Finatto et al., 2018). Group II is further dissected into following five subgroups: IIa, IIb, IIc, IId, and IIe based on their phylogenetic relationship or primary amino acids sequence (Chen et al., 2017). Some reports also suggest a group IV WRKY TFs containing WRKY domain but no zinc finger-like motif; however, they might be either pseudogenes or some in silico error (Jimmy and Babu, 2019). WRKY TFs are localized all over the genome and highly diverse in different crop species including adzuki bean (*Vigna angularis* Willd.), cassava (*Manihot esculenta* Crantz), chickpea (*Cicer arietinum* L.), common bean (*Phaseolus vulgaris* L.), cotton (*Gossypium hirsutum* L.), grapevine (*Vitis vinifera* L.), mouse-ear cress (*Arabidopsis thaliana* L.), mung bean (*Vigna radiata* L.), pigeon pea (*Cajanus cajan* L. Millsp.), quinoa (*Chenopodium quinoa* Wild.), rapeseed (*Brassica napus* L.), rice (*Oryza sativa* L. ssp. *indica*/*O. sativa* L. ssp. *japonica*), sesame (*Sesamum indicum* L.), soybean (*Glycine max* L. Merr.), tea (*Camellia sinensis* L.), tomato (*Solanum lycopersicum* L.), and wheat (*Triticum aestivum* L.) as summarized in Table 1.

3. WRKY TFs play diverse roles in plant cell

WRKY TFs are import member of multifaceted signaling network required from germination to senescence or seed production in any plant's life cycle. However, their role especially during abiotic and biotic stresses has been studied extensively to improve plant tolerance against them (Finatto et al., 2018; Chen et al., 2017), therefore various transgenic plants with tailored WRKY expression or activity have been developed to understand the underlying molecular, biochemical, or signaling mechanisms.

4. WRKY TFs substantially improve abiotic and biotic stress tolerance in plants

WRKY TFs have been attributed to improve tolerance against heat, drought, salt, freezing, UV radiation, cadmium stress, and disease-causing organisms in different crop species including *Arabidopsis thaliana* L., *Arachis hypogaea* L., *Capsicum annum* L., *Cucumis sativus* L., *Dendranthema grandiflorum* (Ramat.) Kitam., *O. sativa* L., *Nicotiana benthamiana* Domin, *Nicotiana tabacum* L., *Solanum pimpinellifolium* (L.) Mill., and *Triticum aestivum* L. as summarized in Table 2.

4.1 WRKY TFs during abiotic stress tolerance

WRKY TFs have been attributed to regulate root architecture or morphology, downstream stress-responsive gene expression or their corresponding

Table 1 Distribution of WRKY transcription factors in plants.

Crop	Total WRKY[a] (Group I + II + III)	Reference
Adzuki bean (*Vigna angularis* Willd.)	77 (15 + 53 + 9)	Song et al. (2018)
Cassava (*Manihot esculenta* Crantz)	85 (17 + 56 + 12)	Wei et al. (2016)
Chickpea (*Cicer arietinum* L.)	58 (12 + 39 + 7) 70 (14 + 48 + 8)	Song et al. (2018) Waqas et al. (2019)
Common bean (*Phaseolus vulgaris* L.)	88 (14 + 62 + 12)	Song et al. (2018)
Cotton (*Gossypium hirsutum* L.)	102 (20 + 72 + 10)	Dou et al. (2014)
Grapevine (*Vitis vinifera* L.)	59 (12 + 39 + 6)	Wang et al. (2014)
Mouse-ear cress (*Arabidopsis thaliana* L.)	66 (27 + 26 + 13)	Wang et al. (2011)
Mung bean (*Vigna radiata* L.)	76 (16 + 48 + 12)	Song et al. (2018)
Pigeon pea (*Cajanus cajan* L. Millsp.)	92 (16 + 63 + 13)	Song et al. (2018)
Quinoa (*Chenopodium quinoa* Wild.)	92 (16 + 62 + 14)	Yue et al. (2019)
Rapeseed (*Brassica napus* L.)	287 (121 + 158 + 51)	He et al. (2016a)
Rice (*Oryza sativa* L. ssp. *indica*/ *Oryza sativa* L. ssp. *japonica*)	102 (13 + 42 + 34)/ 98 (13 + 47 + 32)	Ross et al. (2007)
Sesame (*Sesamum indicum* L.)	71 (12 + 48 + 7)	Li et al. (2017)
Soybean (*Glycine max* L. Merr.)	178 (30 + 124 + 24)	Song et al. (2018)
Tea (*Camellia sinensis* L.)	56 (12 + 40 + 4)	Wang et al. (2019a)
Tomato (*Solanum lycopersicum* L.)	81 (15 + 52 + 11)	Huang et al. (2012)
Wheat (*Triticum aestivum* L.)	171 (31 + 95 + 45)	Ning et al. (2017)

[a]If the total does not match with the addition of the three groups, the remaining WRKY genes did not group with any of them.

enzyme activity, ion or redox or reactive oxygen species homoeostasis, chlorophyll accumulation in leaves, and biosynthesis of various signaling molecules, secondary metabolites, or antioxidants, thus are one of the key factors modulating abiotic stress tolerance in plants. Moreover, their cross talk with other signaling pathways enables them to expand their reach to control various mechanisms in plant system. Niu et al. (2012) identified two group I *WRKY* genes, *TaWRKY2* and *TaWRKY19*, in wheat whose overexpression improved tolerance against salt, drought, and abscisic acid (ABA) treatment stress in arabidopsis plants by either increasing soluble sugars biosynthesis that reduced membrane injury or modulating expressions of

Table 2 Summary of transgenic studies establishing importance of *WRKY* transcription factors in regulating abiotic and biotic stresses in plants.

Crop	Transgene	Source	Regulates tolerance/resistance against	Reference
Arabidopsis thaliana L.	*TaWRKY2* (Group I) and *TaWRKY19* (Group I)	*Triticum aestivum* L.	Salt, drought, and freezing stresses	Niu et al. (2012)
	TaWRKY93 (Group II)	*Triticum aestivum* L.	Salt, drought, low temperature, and osmotic stress	Qin et al. (2015)
	GhWRKY34 (Group III)	*Gossypium hirsutum* L.	Salt stress	Zhou et al. (2015)
	TaWRKY1 (Group III) and *TaWRKY33* (Group II)	*Triticum aestivum* L.	Heat and drought stress	He et al. (2016b)
	PcWRKY33 (Group I)	*Polygonum cuspidatum* Sieb. et Zucc.	Salt stress	Bao et al. (2018)
	WRKY40 (Group II)	*Cicer arietinum* L.	*Pseudomonas syringae*	Chakraborty et al. (2018)
	ZmWRKY106 (Group II)	*Zea mays* L.	Heat and drought stress	Wang et al. (2018)
	WRKY12 (Group II)	*Arabidopsis thaliana* L.	Cadmium stress	Han et al. (2019)
	GhWRKY33 (Group III)	*Gossypium hirsutum* L.	Drought stress	Wang et al. (2019b)
	VvWRKY30 (Group III)	*Vitis vinifera* L.	Salt stress	Zhu et al. (2019)
Arachis hypogaea L.	*MuWRKY3* (Group I)	*Macrotyloma uniflorum* Lam. Verdc.	Drought stress	Kiranmai et al. (2018)

Continued

Table 2 Summary of transgenic studies establishing importance of *WRKY* transcription factors in regulating abiotic and biotic stresses in plants—cont'd

Crop	Transgene	Source	Regulates tolerance/resistance against	Reference
Capsicum annum L.	*CaWRKY40* (Group II)	*Capsicum annum* L.	Heat stress and *Ralstonia solanacearum* infection	Dang et al. (2013)
	CaWRKY58 (Group I)	*Capsicum annum* L.	*Ralstonia solanacearum* infection	Wang et al. (2013)
Cucumis sativus L.	*CsWRKY46* (Group II)	*Cucumis sativus* L.	Cold stress	Zhang et al. (2016)
	CsWRKY50 (Group II)	*Cucumis sativus* L.	*Pseudoperonospora cubensis*	Luan et al. (2019)
Dendranthema grandiflorum (Ramat.) Kitam.	*DgWRKY5* (Group I)	*Dendranthema grandiflorum* (Ramat.) Kitam.	Salt stress	Liang et al. (2017)
Oryza sativa L.	*OsWRKY89* (Group III)	*Oryza sativa* L.	UV-B radiation stress, blast disease, and whitebacked planthopper	Wang et al. (2007)
	OsWRKY31 (Group III)	*Oryza sativa* L.	Blast disease	Zhang et al. (2008)
	OsWRKY22 (Group III)	*Oryza sativa* L.	Blast disease	Abbruscato et al. (2012)
	OsWRKY67 (Group I)	*Oryza sativa* L.	Blast and bacterial blight disease	Liu et al. (2018)
Nicotiana benthamiana Domin	*GhWRKY39-1* (Group II)	*Gossypium hirsutum* L.	Salt/Oxidative stress and *Ralstonia solanacearum/Rhizoctonia solani* infection	Shi et al. (2014)
	GhWRKY41 (Group III)	*Gossypium hirsutum* L.	Salt and drought stress	Chu et al. (2015)
	GhWRKY25 (Group I)	*Gossypium hirsutum* L.	Salt/drought stress and *Botrytis cinerea* infection	Liu et al. (2016)

Nicotiana tabacum L.	*GhWRKY15* (Group II)	*Gossypium hirsutum* L.	Infection of cucumber, mosaic virus, *Colletotrichum gossypii*, and *Phytophthora parasitica*	Yu et al. (2012)
	CaWRKY40 (Group II)	*Capsicum annum* L.	Heat stress and *Ralstonia solanacearum* infection	Dang et al. (2013)
	DnWRKY11 (Group II)	*Dendrobium nobile* Lindl.	Salt and drought stress	Xu et al. (2014)
	GhWRKY17 (Group II)	*Gossypium hirsutum* L.	Salt and drought stress	Yan et al. (2014)
	MbWRKY1 (Group II)	*Malus baccata* (L.) Borkh	Drought stress	Han et al. (2018)
Solanum pimpinellifolium (L.) Mill.	*SpWRKY6* (Group I)	*Solanum pimpinellifolium* (L.) Mill.	Late blight by *Phytophthora infestans*	Hong et al. (2018)
Triticum aestivum L.	*TaWRKY2* (Group I)	*Triticum aestivum* L.	Drought stress	Gao et al. (2018)

downstream stress-responsive genes. However, cold treatment of transgenic arabidopsis plants induced expression of only *TaWRKY19*. Overexpression of group II WRKY wheat gene *TaWRKY93* in arabidopsis improved salt, drought, low temperature, and osmotic stress tolerance by maintaining membrane stability, sustaining osmotic balance, supporting root development, and tailoring expression of downstream stress-responsive genes (Qin et al., 2015). Overexpression of *C. sativus* group II *WRKY* gene *CsWRKY46* also enhanced cold tolerance in transgenic *C. sativa* plants via ABA-dependent pathway (Zhang et al., 2016).

Transgenic arabidopsis plants overexpressing cotton group III WRKY gene GhWRKY34 showed sustained Na^+/K^+ homeostasis, higher seed germination, green cotyledons, enhanced root development, higher chlorophyll content, and increased activity of other stress-responsive signaling molecules compared to the wild type, thus showing higher tolerance against salt stress (Zhou et al., 2015). Salt stress has been negatively (Bao et al., 2018) and positively (Zhu et al., 2019) regulated by a *Polygonum cuspidatum* group I WRKY TF *PcWRKY33* and *Vitis vinifera* group III WRKY TF *VvWRKY30* in transgenic arabidopsis plants, respectively, by modulating root growth, seed germination, chlorophyll concentration, Na^+/K^+ or reactive oxygen species homeostasis, carbohydrate metabolism, proline accumulation, and activity of antioxidants or stress-responsive genes (Zhou et al., 2015; Zhu et al., 2019). Similarly, *WRKY* genes including Chrysanthemum group I (Liang et al., 2017), cotton group II/II/III *GhWRKY39-1* (Shi et al., 2014)/*GhWRKY17* (Yan et al., 2014)/*GhWRKY41* (Chu et al., 2015), *Dendrobium nobile* group II *DnWRKY11* (Xu et al., 2014) positively regulated salt tolerance by reducing reactive oxygen species concentration, improving plant morphology, and increasing antioxidant activities.

Transgenic arabidopsis plants overexpressing group II WRKY gene *ZmWRKY106* from maize showed significantly increased heat and drought tolerance by modulating ABA-signaling pathway and increased antioxidants [superoxide dismutase (SOD), peroxide dismutase (POD), and catalase (CAT)] activities thus regulating expression of stress-responsive genes (Wang et al., 2018). *GhWRKY33* in *G. hirsutum* has been characterized as a negative regulator of drought tolerance as a result of its association with ABA-signaling pathway (Wang et al., 2019b). Likewise, overexpression of *Capsicum annum* group II *WRKY* gene *CaWRKY40* improved heat tolerance in transgenic tobacco plants (Dang et al., 2013), whereas transgenic groundnut, tobacco, and wheat plants overexpressing *WRKY* genes from *Macrotyloma uniflorum* (*MuWRKY3*, group I) (Kiranmai et al., 2018), *Malus baccata* (*MbWRKY1*, group II) (Han et al., 2018), and *Triticum aestivum* (*TaWRKY2*, group I) (Gao et al., 2018) showed enhanced drought tolerance.

A negative regulator of cadmium stress in arabidopsis *WRKY12* has been identified by Han et al. (2019). *WRKY12* knocked out by transgenic arabidopsis plants has been associated with induced glutathione-dependent phytochelatin synthesis pathway, a key player in cadmium detoxification in plants. Constitutive expression of *WRKY89* in rice improved its tolerance against UV-B radiation by promoting wax deposition, lignin accumulation, and redox homeostasis (Wang et al., 2007).

Some WRKY TFs shows differential expressions when exposed to different types of abiotic stresses. He et al. (2016b) characterized two wheat WRKY genes *TaWRKY1* (group III) and *TaWRKY33* (group II) in transgenic arabidopsis plants. *TaWRKY1* was upregulated by heat and downregulated by low temperature whereas *TaWRKY33* had high responses against high/low temperature and treatment of ABA/jasmonic acid methylester (MeJA). Overexpression of cotton group I *WRKY* gene *GhWRKY25* in transgenic *N. benthamiana* Domin established that *GhWRKY25* was a positive and negative regulator of salt and drought stress tolerance (Liu et al., 2016).

4.2 WRKY TFs during biotic stress tolerance

WRKY TFs are essential components of both pattern- and effector-triggered immunity in plants, therefore can regulate plant tolerance against different biotic factors (Dang et al., 2013; Chakraborty et al., 2018; Luan et al., 2019). WRKY via promoting accumulation of physical barriers, histone modification, transcriptional regulation of pathogenesis or other defense-related genes, ROS signaling interaction, and cross talk with salicylic (SA) or jasmonic (JA) acid or other phytohormones regulate disease initiation, progression, and severity positively or negatively (Bakshi and Oelmüller, 2014). Consequently, various *WRKY* genes have been shown to regulate disease resistance mechanisms in different crops. Overexpression of *WRKY* genes *OsWRKY89* (group III) (Wang et al., 2007), *OsWRKY31* (group III) (Zhang et al., 2008), *OsWRKY22* (group III) (Abbruscato et al., 2012), and *OsWRKY67* (group I) (Liu et al., 2018) have been shown to improve resistance against blast disease caused by a fungus *Magnaporthe grisea* in transgenic rice plants by altering the expression of defense-related genes including SA-dependent pathway and auxin-response genes. Additionally, overexpression of *OsWRKY67* and *OsWRKY89* has also enhanced resistance against another important disease of rice, bacterial blight caused by *Xanthomonas oryzae* (Liu et al., 2018) and white-backed plant hopper (Wang et al., 2007), respectively. *Capsicum annum* WRKY gene *CaWRKY58* (group I) (Wang et al., 2013)/*CaWRKY40* (group II) (Dang et al., 2013) and *G. hirsutum* group II *WRKY* gene *GhWRKY39-1* (Shi et al., 2014) have been reported to regulate bacterial blight disease caused by *Ralstonia solanacearum*. *CaWRKY40* and *GhWRKY39-1* positively where as *CaWRKY58* negatively regulated the disease resistance in transgenic *Capsicum annum/N. tabacum*, *N. benthamiana*, and *Capsicum annum* plants, respectively. *GhWRKY39-1* has also enhanced resistance against fungal pathogen *Rhizoctonia solani* in transgenic *N. benthamiana* plants (Shi et al., 2014). *G. hirsutum* group II *WRKY* gene *GhWRKY15* has been overexpressed in transgenic *N. tabacum* plants that showed enhanced resistance against viral (cucumber mosaic virus) and fungal (*Colletotrichum gossypii* and *Phytophthora parasitica*) infections (Yu et al., 2012). Cotton group I *WRKY* gene *GhWRKY25* adversely influenced SA, JA, and ethylene pathways, thus has been established as a negative regulator of fungal pathogen *Botrytis cinereal* (Liu et al., 2016). Similarly, resistance against *Phytophthora infestans* (late blight), *Pseudomonas syringae* and *Pseudoperonospora cubensis* have been improved by the overexpression of *SpWRKY6* (group I *WRKY* from *S. pimpinellifolium*) (Hong et al., 2018), *WRKY40* (group II *WRKY* from *C. arietinum*) (Chakraborty et al., 2018), and *CsWRKY50* (group II WRKY from *C. sativus*) (Luan et al., 2019) in transgenic *S. pimpinellifolium*, *Arabidopsis thaliana*, and *C. sativus* plants, respectively.

5. Conclusion

WRKY transcription factors are essential components of plant defense system. They are key players to enhance abiotic and biotic stress tolerance or resistance in different crop species, therefore are being targeted to improve plant performance and production to feed the growing population worldwide. However, various underlying mechanisms of their effect on downstream genes or pathways still needs to be explored in future studies.

References

Abbruscato, P., Nepusz, T., Mizzi, L., Del Corvo, M., Morandini, P., Fumasoni, I., Michel, C., Paccanaro, A., Guiderdoni, E., Schaffrath, U., Morel, J.B., 2012. *OsWRKY22*, a monocot *WRKY* gene, plays a role in the resistance response to blast. Mol. Plant Pathol. 13 (8), 828–841.

Agarwal, P., Reddy, M.P., Chikara, J., 2011. *WRKY*: its structure, evolutionary relationship, DNA-binding selectivity, role in stress tolerance and development of plants. Mol. Biol. Rep. 38 (6), 3883–3896.

Bakshi, M., Oelmüller, R., 2014. *WRKY* transcription factors: jack of many trades in plants. Plant Signal. Behav. 9(2):e27700.

Bao, W., Wang, X., Chen, M., Chai, T., Wang, H., 2018. A *WRKY* transcription factor, *PcWRKY33*, from *Polygonum cuspidatum* reduces salt tolerance in transgenic *Arabidopsis thaliana*. Plant Cell Rep. 37 (7), 1033–1048.

Béné, C., Barange, M., Subasinghe, R., Pinstrup-Andersen, P., Merino, G., Hemre, G.I., Williams, M., 2015. Feeding 9 billion by 2050—putting fish back on the menu. Food Sec. 7 (2), 261–274.

Chakraborty, J., Ghosh, P., Sen, S., Das, S., 2018. Epigenetic and transcriptional control of chickpea *WRKY40* promoter activity under *Fusarium* stress and its heterologous expression in *Arabidopsis* leads to enhanced resistance against bacterial pathogen. Plant Sci. 276, 250–267.

Chen, F., Hu, Y., Vannozzi, A., Wu, K., Cai, H., Qin, Y., Mullis, A., Lin, Z., Zhang, L., 2017. The *WRKY* transcription factor family in model plants and crops. Crit. Rev. Plant Sci. 36 (5–6), 311–335.

Chu, X., Wang, C., Chen, X., Lu, W., Li, H., Wang, X., Hao, L., Guo, X., 2015. The cotton *WRKY* gene *GhWRKY41* positively regulates salt and drought stress tolerance in transgenic *Nicotiana benthamiana*. PLoS ONE 10(11):e0143022.

Dang, F.F., Wang, Y.N., Yu, L.U., Eulgem, T., Lai, Y.A., Liu, Z.Q., Wang, X.U., Qiu, A.L., Zhang, T.X., Lin, J., Chen, Y.S., 2013. *CaWRKY40*, a *WRKY* protein of pepper, plays an important role in the regulation of tolerance to heat stress and resistance to *Ralstonia solanacearum* infection. Plant Cell Environ. 36 (4), 757–774.

Dou, L., Zhang, X., Pang, C., Song, M., Wei, H., Fan, S., Yu, S., 2014. Genome-wide analysis of the *WRKY* gene family in cotton. Mol. Gen. Genomics. 289 (6), 1103–1121.

Erpen, L., Devi, H.S., Grosser, J.W., Dutt, M., 2018. Potential use of the *DREB/ERF, MYB, NAC* and *WRKY* transcription factors to improve abiotic and biotic stress in transgenic plants. Plant Cell Tissue Organ Cult. 132 (1), 1–25.

Finatto, T., Viana, V.E., Woyann, L.G., Busanello, C., Maia, L.C., Oliveira, A.C., 2018. Can *WRKY* transcription factors help plants to overcome environmental challenges? Genet. Mol. Biol. 41 (3), 533–544.

Gao, H., Wang, Y., Xu, P., Zhang, Z., 2018. Overexpression of a *WRKY* transcription factor *TaWRKY2* enhances drought stress tolerance in transgenic wheat. Front. Plant Sci. 9, 997.

Han, D., Ding, H., Chai, L., Liu, W., Zhang, Z., Hou, Y., Yang, G., 2018. Isolation and characterization of *MbWRKY1*, a *WRKY* transcription factor gene from *Malus baccata* (L.) Borkh involved in drought tolerance. Can. J. Plant Sci. 98 (5), 1023–1034.

Han, Y., Fan, T., Zhu, X., Wu, X., Ouyang, J., Jiang, L., Cao, S., 2019. *WRKY12* represses *GSH1* expression to negatively regulate cadmium tolerance in *Arabidopsis*. Plant Mol. Biol. 99 (1–2), 149–159.

He, Y., Mao, S., Gao, Y., Zhu, L., Wu, D., Cui, Y., Li, J., Qian, W., 2016a. Genome-wide identification and expression analysis of *WRKY* transcription factors under multiple stresses in *Brassica napus*. PLoS ONE 11(6):e0157558.

He, G.H., Xu, J.Y., Wang, Y.X., Liu, J.M., Li, P.S., Chen, M., Ma, Y.Z., Xu, Z.S., 2016b. Drought-responsive *WRKY* transcription factor genes *TaWRKY1* and *TaWRKY33* from wheat confer drought and/or heat resistance in *Arabidopsis*. BMC Plant Biol. 16 (1), 116.

Hong, Y., Cui, J., Liu, Z., Luan, Y., 2018. *SpWRKY6* acts as a positive regulator during tomato resistance to *Phytophthora infestans* infection. Biochem. Biophys. Res. Commun. 506 (4), 787–792.

Huang, S., Gao, Y., Liu, J., Peng, X., Niu, X., Fei, Z., Cao, S., Liu, Y., 2012. Genome-wide analysis of *WRKY* transcription factors in *Solanum lycopersicum*. Mol. Gen. Genomics. 287 (6), 495–513.
Ishiguro, S., Nakamura, K., 1994. Characterization of a cDNA encoding a novel DNA-binding protein, *SPF1*, that recognizes SP8 sequences in the 5′ upstream regions of genes coding for sporamin and β-amylase from sweet potato. Mol. Gen. Genet. 244 (6), 563–571.
Jiang, J., Ma, S., Ye, N., Jiang, M., Cao, J., Zhang, J., 2017. *WRKY* transcription factors in plant responses to stresses. J. Integr. Plant Biol. 59 (2), 86–101.
Jimmy, J.L., Babu, S., 2019. Variations in the structure and evolution of rice *WRKY* genes in indica and japonica genotypes and their co-expression network in mediating disease resistance. Evol. Bioinforma. 15, 1–12.
Kiranmai, K., Lokanadha Rao, G., Pandurangaiah, M., Nareshkumar, A., Amaranatha Reddy, V., Lokesh, U., Venkatesh, B., Anthony Johnson, A.M., Sudhakar, C., 2018. A novel *WRKY* transcription factor, *MuWRKY3* (*Macrotyloma uniflorum* lam. Verdc.) enhances drought stress tolerance in transgenic groundnut (*Arachis hypogaea* L.) plants. Front. Plant Sci. 9, 346.
Lan Thi Hoang, X., Du Nhi, N.H., Binh Anh Thu, N., Phuong Thao, N., Phan Tran, L.S., 2017. Transcription factors and their roles in signal transduction in plants under abiotic stresses. Curr. Genomics 18 (6), 483–497.
Li, D., Liu, P., Yu, J., Wang, L., Dossa, K., Zhang, Y., Zhou, R., Wei, X., Zhang, X., 2017. Genome-wide analysis of *WRKY* gene family in the sesame genome and identification of the *WRKY* genes involved in responses to abiotic stresses. BMC Plant Biol. 17 (1), 152.
Liang, Q.Y., Wu, Y.H., Wang, K., Bai, Z.Y., Liu, Q.L., Pan, Y.Z., Zhang, L., Jiang, B.B., 2017. Chrysanthemum *WRKY* gene *DgWRKY5* enhances tolerance to salt stress in transgenic chrysanthemum. Sci. Rep. 7 (1), 4799.
Liu, X., Song, Y., Xing, F., Wang, N., Wen, F., Zhu, C., 2016. *GhWRKY25*, a group I *WRKY* gene from cotton, confers differential tolerance to abiotic and biotic stresses in transgenic *Nicotiana benthamiana*. Protoplasma 253 (5), 1265–1281.
Liu, Q., Li, X., Yan, S., Yu, T., Yang, J., Dong, J., Zhang, S., Zhao, J., Yang, T., Mao, X., Zhu, X., 2018. *OsWRKY67* positively regulates blast and bacteria blight resistance by direct activation of *PR* genes in rice. BMC Plant Biol. 18 (1), 257.
Luan, Q., Chen, C., Liu, M., Li, Q., Wang, L., Ren, Z., 2019. *CsWRKY50* mediates defense responses to *Pseudoperonospora cubensis* infection in *Cucumis sativus*. Plant Sci. 279, 59–69.
Mingyu, Z., Zhengbin, Z., Shouyi, C., Jinsong, Z., Hongbo, S., 2012. *WRKY* transcription factor superfamily: structure, origin and functions. Afr. J. Biotechnol. 11 (32), 8051–8059.
Ning, P., Liu, C., Kang, J., Lv, J., 2017. Genome-wide analysis of *WRKY* transcription factors in wheat (*Triticum aestivum* L.) and differential expression under water deficit condition. Peer J. 5:e3232.
Niu, C.F., Wei, W.E., Zhou, Q.Y., Tian, A.G., Hao, Y.J., Zhang, W.K., Ma, B., Lin, Q., Zhang, Z.B., Zhang, J.S., Chen, S.Y., 2012. Wheat *WRKY* genes *TaWRKY2* and *TaWRKY19* regulate abiotic stress tolerance in transgenic *Arabidopsis* plants. Plant Cell Environ. 35 (6), 1156–1170.
Novaković, L., Guo, T., Bacic, A., Sampathkumar, A., Johnson, K., 2018. Hitting the wall—sensing and signaling pathways involved in plant cell wall remodeling in response to abiotic stress. Plan. Theory 7 (4), 89.
Pandey, P., Irulappan, V., Bagavathiannan, M.V., Senthil-Kumar, M., 2017. Impact of combined abiotic and biotic stresses on plant growth and avenues for crop improvement by exploiting physio-morphological traits. Front. Plant Sci. 8, 537.
Phukan, U.J., Jeena, G.S., Shukla, R.K., 2016. *WRKY* transcription factors: molecular regulation and stress responses in plants. Front. Plant Sci. 7, 760.

Qin, Y., Tian, Y., Liu, X., 2015. A wheat salinity-induced *WRKY* transcription factor *TaWRKY93* confers multiple abiotic stress tolerance in *Arabidopsis thaliana*. Biochem. Biophys. Res. Commun. 464 (2), 428–433.

Ross, C.A., Liu, Y., Shen, Q.J., 2007. The *WRKY* gene family in rice (*Oryza sativa*). J. Integr. Plant Biol. 49 (6), 827–842.

Schwachtje, J., Whitcomb, S.J., Firmino, A.A., Zuther, E., Hincha, D.K., Kopka, J., 2019. Induced, imprinted and primed metabolic responses of plants to changing environments: do plants store and process information by metabolic imprinting and metabolic priming? Front. Plant Sci. 10, 106.

Shi, W., Hao, L., Li, J., Liu, D., Guo, X., Li, H., 2014. The *Gossypium hirsutum WRKY* gene *GhWRKY39-1* promotes pathogen infection defense responses and mediates salt stress tolerance in transgenic *Nicotiana benthamiana*. Plant Cell Rep. 33 (3), 483–498.

Song, H., Sun, W., Yang, G., Sun, J., 2018. *WRKY* transcription factors in legumes. BMC Plant Biol. 18 (1), 243.

Sukumari Nath, V., Kumar Mishra, A., Kumar, A., Matoušek, J., Jakše, J., 2019. Revisiting the role of transcription factors in coordinating the defense response against citrus bark cracking viroid infection in commercial hop (*Humulus lupulus* L.). Viruses 11 (5), 419.

Suzuki, N., Rivero, R.M., Shulaev, V., Blumwald, E., Mittler, R., 2014. Abiotic and biotic stress combinations. New Phytol. 203 (1), 32–43.

Wang, H., Hao, J., Chen, X., Hao, Z., Wang, X., Lou, Y., Peng, Y., Guo, Z., 2007. Overexpression of rice *WRKY89* enhances ultraviolet B tolerance and disease resistance in rice plants. Plant Mol. Biol. 65 (6), 799–815.

Wang, Q., Wang, M., Zhang, X., Hao, B., Kaushik, S.K., Pan, Y., 2011. *WRKY* gene family evolution in *Arabidopsis thaliana*. Genetica 139 (8), 973.

Wang, Y., Dang, F., Liu, Z., Wang, X., Eulgem, T., Lai, Y., Yu, L., She, J., Shi, Y., Lin, J., Chen, C., 2013. *CaWRKY58*, encoding a group I WRKY transcription factor of *Capsicum annuum*, negatively regulates resistance to *Ralstonia solanacearum* infection. Mol. Plant Pathol. 14 (2), 131–144.

Wang, L., Zhu, W., Fang, L., Sun, X., Su, L., Liang, Z., Wang, N., Londo, J.P., Li, S., Xin, H., 2014. Genome-wide identification of *WRKY* family genes and their response to cold stress in *Vitis vinifera*. BMC Plant Biol. 14 (1), 103.

Wang, C.T., Ru, J.N., Liu, Y.W., Li, M., Zhao, D., Yang, J.F., Fu, J.D., Xu, Z.S., 2018. Maize *WRKY* transcription factor *ZmWRKY106* confers drought and heat tolerance in transgenic plants. Int. J. Mol. Sci. 19 (10), 3046.

Wang, P., Yue, C., Chen, D., Zheng, Y., Zhang, Q., Yang, J., Ye, N., 2019a. Genome-wide identification of *WRKY* family genes and their response to abiotic stresses in tea plant (*Camellia sinensis*). Genes Genomics 41 (1), 17–33.

Wang, N.N., Xu, S.W., Sun, Y.L., Liu, D., Zhou, L., Li, Y., Li, X.B., 2019b. The cotton *WRKY* transcription factor (*GhWRKY33*) reduces transgenic *Arabidopsis* resistance to drought stress. Sci. Rep. 9 (1), 724.

Waqas, M., Azhar, M.T., Rana, I.A., Azeem, F., Ali, M.A., Nawaz, M.A., Chung, G., Atif, R.M., 2019. Genome-wide identification and expression analyses of *WRKY* transcription factor family members from chickpea (*Cicer arietinum* L.) reveal their role in abiotic stress-responses. Genes Genomics 41 (4), 467–481.

Wei, Y., Shi, H., Xia, Z., Tie, W., Ding, Z., Yan, Y., Wang, W., Hu, W., Li, K., 2016. Genome-wide identification and expression analysis of the *WRKY* gene family in cassava. Front. Plant Sci. 7, 25.

Xu, X.B., Pan, Y.Y., Wang, C.L., Ying, Q.C., Song, H.M., Wang, H.Z., 2014. Overexpression of *DnWRKY11* enhanced salt and drought stress tolerance of transgenic tobacco. Biologia 69 (8), 994–1000.

Yan, H., Jia, H., Chen, X., Hao, L., An, H., Guo, X., 2014. The cotton *WRKY* transcription factor *GhWRKY17* functions in drought and salt stress in transgenic *Nicotiana benthamiana* through ABA signaling and the modulation of reactive oxygen species production. Plant Cell Physiol. 55 (12), 2060–2076.

Yu, F., Huaxia, Y., Lu, W., Wu, C., Cao, X., Guo, X., 2012. *GhWRKY15*, a member of the *WRKY* transcription factor family identified from cotton (*Gossypium hirsutum* L.), is involved in disease resistance and plant development. BMC Plant Biol. 12 (1), 144.

Yue, H., Chang, X., Zhi, Y., Wang, L., Xing, G., Song, W., Nie, X., 2019. Evolution and identification of the *WRKY* gene family in quinoa (*Chenopodium quinoa*). Genes 10 (2), 131.

Zhang, J., Peng, Y., Guo, Z., 2008. Constitutive expression of pathogen-inducible *OsWRKY31* enhances disease resistance and affects root growth and auxin response in transgenic rice plants. Cell Res. 18 (4), 508.

Zhang, Y., Yu, H., Yang, X., Li, Q., Ling, J., Wang, H., Gu, X., Huang, S., Jiang, W., 2016. *CsWRKY46*, a *WRKY* transcription factor from cucumber, confers cold resistance in transgenic-plant by regulating a set of cold-stress responsive genes in an ABA-dependent manner. Plant Physiol. Biochem. 108, 478–487.

Zhou, L., Wang, N.N., Gong, S.Y., Lu, R., Li, Y., Li, X.B., 2015. Overexpression of a cotton (*Gossypium hirsutum*) *WRKY* gene, *GhWRKY34*, in *Arabidopsis* enhances salt-tolerance of the transgenic plants. Plant Physiol. Biochem. 96, 311–320.

Zhu, D., Hou, L., Xiao, P., Guo, Y., Deyholos, M.K., Liu, X., 2019. *VvWRKY30*, a grape *WRKY* transcription factor, plays a positive regulatory role under salinity stress. Plant Sci. 280, 132–142.

CHAPTER 3

Zinc finger proteins: Novel sources of genes for abiotic stress tolerance in plants

Khushboo Dasauni and Tapan K. Nailwal
Department of Biotechnology, Kumaun University Nainital, Bhimtal, India

1. Introduction

Abiotic stress factors, viz., drought, flooding/water logging, extremes of temperatures (cold, chilling, frost, and heat), salinity, mineral deficiency, and toxicity cause changes in plant metabolism, growth and development, and productivity of crops, which leads to huge economic loss to farmers. It has been projected that abiotic stresses may reduce crop production by as much as 70% and many crops perform only at 30% of their genetic potential under abiotic stress conditions. Greater genetic gains are possible by combining classical genetic improvement with advanced molecular biology techniques. The understanding of mechanisms triggered by plants to meet conditions of stress is of fundamental importance for the explanation of these processes. Currently, genetically modified crops help to reduce the effects of these stresses, increasing genetic gains in order to supply the agricultural market and the demand for better quality food across the world. To obtain safe genetically modified organisms for planting and consumption, a thorough grasp of the routes and genes that act in response to these stresses is necessary. This study was conducted in order to collect important information about essential TF gene families for transcriptional control under abiotic stress responses. Nowadays, zinc finger proteins (ZFP), one of the biotechnological tools, have been shown to be useful to overcome this stress related problems in plants (Singh et al., 2015, 2019; Haak et al., 2017).

Zinc finger proteins (ZFP) are among the most abundant proteins in eukaryotic genomes; they have variable secondary structures and enormous functional diversity. All of them have at least one common motif zinc finger which was first identified in *Xenopus laevis* (frog) TF3 (Miller et al., 1985), more commonly known as IIIA (TFIIIA).

Transcription Factors for Abiotic Stress Tolerance in Plants
https://doi.org/10.1016/B978-0-12-819334-1.00003-4

Zinc finger protein was first recognized 30 years ago in *Xenopus* transcription factor IIIA, which activates the transcription of 5S RNA gene in African clawed toad *Xenopus*. Zinc interacts with different domains of alpha helix, beta sheets, loops, and is important to maintain different structure of motif. Whole moiety has amino acids which protrude out and interact with DNA and RNA. Different types of amino acids present in motif have similar functionality even in the absence of metal ion. Folding is important from the functional prospective of proteins, but the idea is that the metal ion should be positively charged, because DNA and RNA backbone is negatively charged which improves the interaction. Zinc finger motif starts to bind with mRNA structure at different places. Zinc finger is present in transcription factors; it is simply a DNA-, RNA-, and protein-interacting domain. Normally, inside the cell, it is important for regulating transcription, translation, cellular process but in synthetic way we can design and modify the amino acid present in active site of structure. The primary structure of TFIIIA contains nine repeat sequences of about 30 amino acids and each repeat contains two Cys and His and three hydrophobic residues in conserved positions. One repeat sequences consist of Tyr/Phe-X-Cys-X_{2-4}-Cys-X_3-Phe-X_5-Leu-X_2-His-$X_{3,4}$-His-X_5, where X is a variable amino acid. These units form a tetrahedral coordination with zinc atom generating peptide domain; these peptide domains have typical compact folding and DNA binding loop formed by the coordination of these ligands with zinc has protruding finger like shape, as shown in Fig. 1 (Lyu and Cao, 2018; Rubin et al., 2000).

Zinc finger has tetrahedral structure including C2H2-like fingers, treble clef fingers, and zinc. It plays many important regulatory roles in plants; the percentage of ZFP transcription factor in total transcription factors is

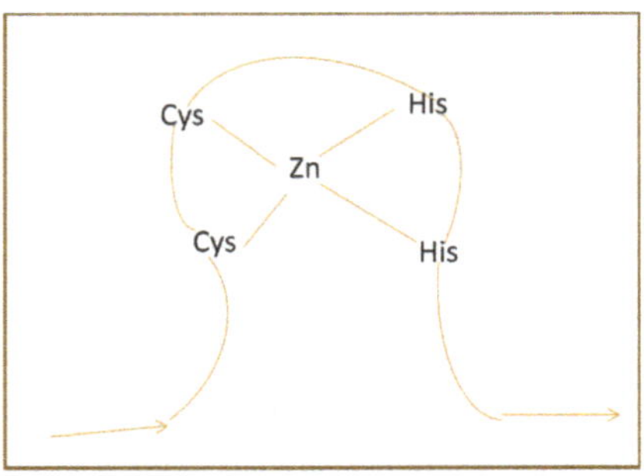

Fig. 1 Schematic representation of the tetrahedral structure of zinc finger motif (Latchman, 1997). *(Modified from David S. Latchman, Science Direct Journal 2008.)*

similar among plant species. Structurally ZFP can be applied in Biotechnology as a tool for inducing site-specific mutagenesis and base substitution for plant genetic improvement (Sun et al., 2019; Wang et al., 2019; Szczepek et al., 2007; Wright et al., 2005).

Zinc finger nuclease (ZFN) is a synthetic protein composed of a DNA-binding domain and a DNA-cleaving domain. The DNA-binding domain consists of two finger modules that bind together to form a zinc finger protein, each of which has a specificity of ≥24 bp (Dreier et al., 2001). ZFN specifies a cleavage of DNA duplexes on chromosomes and promotes site-specific mutagenesis and base substitution, which can modify the expression of specific gene, such as gene knockdown (Wang et al., 2019; Szczepek et al., 2007; Wright et al., 2005); ZFN was successfully used in *Arabidopsis*, maize, and tobacco (Sun et al., 2019; Wright et al., 2005; Shukla et al., 2009).

2. Biological functions of ZFPs

Biological functions of ZFPs in plants are related to plant structure, development, stress, and phytohormone response as shown in Fig. 2A. In plant structure, it controls leaf and tiller angle, while in phytohormone response it regulates abscisic acid and brassinosteroids. It has been found that most ZFPs have a particular role in biological function, however, some ZFPs, such as AZF2, OsDSG1, OsBIRF1, and PtaRHE1, have two or three important biological functions.

Most of the genes in C2H2 and CCCH types have plant developmental functions, while the genes in C3HC4 type are primarily associated with abiotic and biotic stress tolerance. As for the relationship between ZFP and plant hormones, the *ZFP* genes were mainly related to abscisic acid. The percentage of ZFP transcription factor in total transcription factors is similar in all plant species and functions of ZFP have been retained in plant evolution as shown in Fig. 2B (Sun et al., 2019; Wang et al., 2019; Mazumdar et al., 2017; Klug, 2010; Yun et al., 2002; Molnár et al., 2002; Huang et al., 2002; Chrispeels et al., 2000; Kubo et al., 2000; Luo et al., 1999; Van DerKrol et al., 1999; Sakai et al., 1995).

3. Zinc finger protein: A tool in biotechnology

Zinc finger proteins play a critical role in many cellular functions, including transcriptional regulation, RNA binding, regulation of apoptosis, and

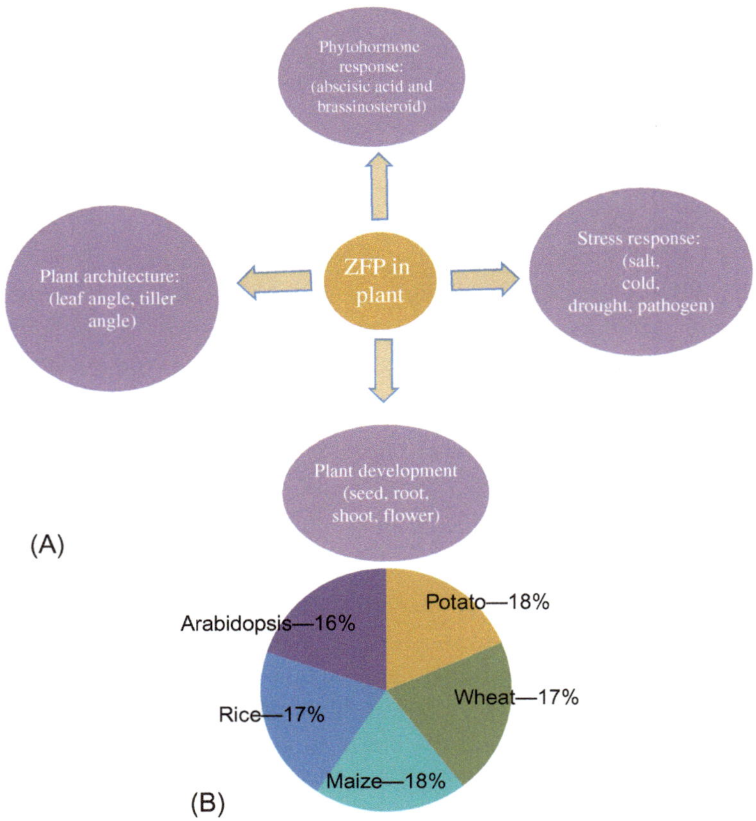

Fig. 2 (A) Biological functions of ZFP in plants. (B) The percentage of ZFP transcription factor in all transcription factors of some plants.

protein-protein interactions. They are classified into different types such as C2H2, C2C2, C2HC, C2C2C2C2, and C2HCC2C2 based on the number and order of the Cys and His residues that bind with Zinc ion in the secondary structure of the finger. Among the different zinc finger types, C2H2-type zinc finger proteins are one of the best studied and most abundant in TF eukaryotes, which have a sequence motif of cysteine and/or histidine that coordinates zinc ions to form specific peptide structures. Zinc finger protein is nothing but a DNA-RNA-protein-loop-turns interaction. Redesigning of zinc protein domain can then be applied on DNA and RNA as shown in Fig. 3A. Expression of target gene can be controlled by modifying zinc finger controlling protein through the designated section of gene (Novak, 2019; Szczepek et al., 2007).

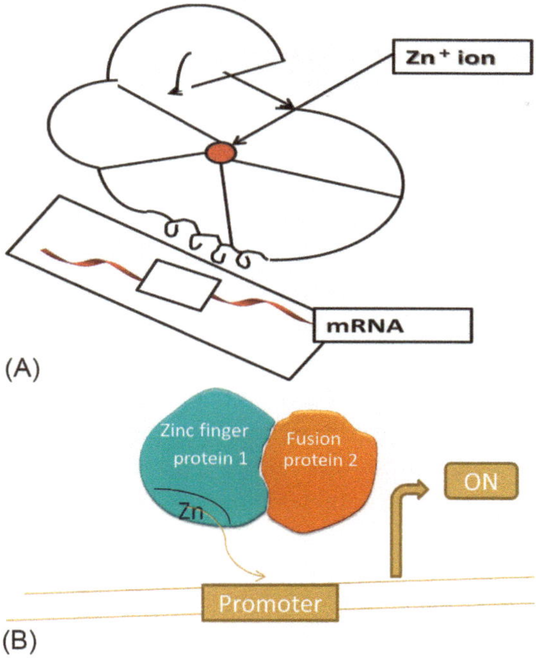

Fig. 3 (A) Schematic representation of zinc finger protein where Zn^{+} interact with protein, loops, turns, DNA, and mRNA. (B) Zinc finger domain binds to fusion protein and regulates transcription.

When zinc finger-containing proteins bind to fusion protein, it provides a particular region of promoter gene to turn ON gene transcription, but it is not possible in vitro, hence redesigning of zinc finger domain is required which gets attached to fusion protein which further controls the process of transcription in such a way that it will design and bind to the promoter. Thus, we can control expression of gene from outside (Fig. 3B) and this opens a new window of metabolomics. Protein-protein interaction, protein-RNA interaction, and protein-DNA interaction are possible due to zinc finger domain making it a useful tool for generating stress tolerance in plants.

4. Role of zinc finger protein in the functioning of transcription factors

Zinc is crucial for the functioning of transcription factors and it was observed that loss of zinc in TF3A leads to a significant change in protein conformation as well as loss of DNA binding ability. This led to the conclusion that the role of zinc in TFIIIA was to hold protein in proper conformation so that specific amino acid side chains make specific contacts on the DNA binding site. Zinc motif brings alpha helix and beta sheet closer together forming a hydrophobic core, which is important for the stability of the beta-alpha domain of zinc finger due to the cross-linking by coordination with zinc. The structure of very small protein domain is generally stabilized by the formation of disulfide bonds or by binding to metal ions most commonly zinc. The binding of metal increases the thermal and conformational stability of small domain, though not directly involved in their functions. Disulfide bridges can be easily cleaved by the reduction in the reducing atmosphere of the cell and the nucleus, hence nature has chosen zinc rather than disulfide bridges for the stabilization of structure of zinc fingers. Zinc is an ideal cation for structural role in protein because of many reasons like high natural abundance and its nontoxic nature. Zn is high in Irvig-Williams's series, and thus forms stable complexes especially with sulfur and nitrogen donors. Zn ions have filled d-orbital and hence do not have any redox activity associated with it. Zn^{2+} can adopt tetrahedral coordination where zinc ion is present in the middle and it holds the structure together contributing to the stability of the domain, like Fe, Mg, and Co^{2+} (Laity et al., 2001).

Structure as shown in Fig. 4 was confirmed by spectroscopy. Single zinc finger peptide has been prepared and purified. Titration experiment revealed that zinc ion bind more strongly than other ions. UV spectroscopy on the cobalt complex has been performed and has showed tetrahedral coordination for cobalt CD spectra of the peptide. Zinc or cobalt absence show nonformation of well-defined structure and addition of either one of the metals lead to a change in CD spectra.

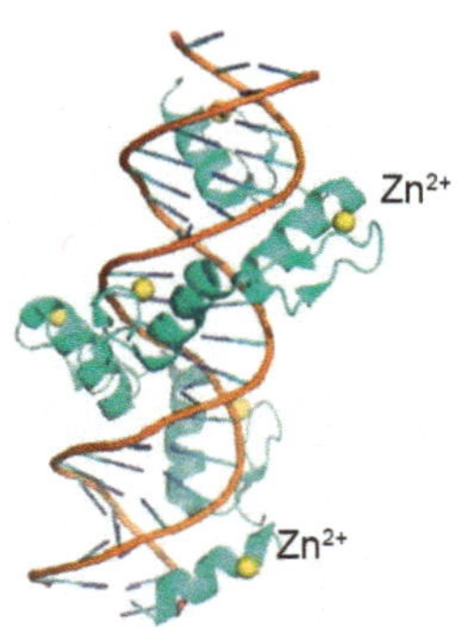

Fig. 4 DNA bound structure of a zinc finger protein (ZFP).

5. ZFP: A protein for making abiotic stress-tolerant varieties

Zinc fingers are structurally diverse and are present in proteins that perform a wide range of functions in various cellular processes, viz., growth and development, stress, and phytohormone response. ZFP families like C2H2, CCCH, C3HC4, and C4 are involved in flower, shoot, seed, pistil, and leaf development (Yuan et al., 2018; Yun et al., 2002; Chrispeels et al., 2000; Kubo et al., 2000; Sakai et al., 1995; Luo et al., 1999). In *Arabidopsis thaliana*, 600 ZFP genes were identified. C2H2-type belongs to zinc finger transcription factors (ZFTFs), which have a sequence motif of cysteine and/or histidine coordinating with zinc atoms to form a specific peptide structure, and has role in many metabolic pathways as well as in stress response and defense activation in plants. It has repression activity via ethylene-responsive element-binding factor (ERF)-associated amphiphilic repression (EAR) repressor domain that is important for genes that regulate both biotic and abiotic stresses (Reyes and Chua, 2007). EAR motif-mediated transcriptional repression is emerging as one of the main mechanisms of plant gene regulation. The EAR motif was the first active repression motif reported in plants. It was first identified as a small motif in a subset of class II ERFs and TFIIIA-type zinc finger proteins a decade ago (Table 1).

Several plant mutant studies revealed the importance of this family of transcription factors against various abiotic stresses in rice. OsISAP1 gene, isolated from a zinc finger domain, was found to have a high transcription rate after being stressed by cold, salt, dehydration, and heavy metals. In *Arabidopsis*, the expression of Zat12 indicates that it produces transcripts

Table 1 Proteins and enzymes involved in plant responses to different types of stress.

Proteins and enzymes	Stress
ERF family	Cold, drought, pathogen infection, wounding
bZIPs family(e.g., ABF1, ABF2)	Drought, temperature, salt
WRKY	Pathogen, wounding, salinity, temperature, drought
NAC	Drought, salinity, cold

Modified from Perez-Clemente, R.M. Vives, V., Zandalinas, S.I., Lopez, M.F., Munoz, V., Gomez, A., 2013. Biotechnological approaches to study plant responses to stress. BioMed Res. Int., 1–10, Article ID 654120.

during oxidation, infiltration, saline, and heat stress. When the same gene is constitutively expressed by genetic transformation, a large number of gene transcripts that respond to oxidation and mild stress are affected (Dai et al., 2012).

Arabidopsis rhl41 gene when placed in front of a constitutive promoter exhibited efficiency for tolerating high-intensity light and increased anthocyanin and chlorophyll contents, and is critical for plant adaptation under dramatic changes in light intensity (Agarwal et al., 2011). The AZF1, AZF2, AZF3, and STZ proteins with repressor domains are effective in inhibiting the expression of other transcription factors (TFs). AZF2 and STZ are strongly induced when plants are subjected to dehydration, salinity, cold, and ABA stress (Ding et al., 2009; Nakano et al., 2006).

6. Types of ZFP families

ZFP are structurally differentiated based on the combinations and order of Cys and His for coordination with divalent Zn ions (Dai and Liew, 1998; Berg and Shi, 1996), and C2H2 represented by TFIIIA (Miller et al., 1985) C3HC4, called RING finger (Barlow et al., 1994; Joshi et al., 2016), C4 (Omichinski et al., 1993), CCCH (DuBois et al., 1990), C4HC3 (Gabig et al., 1994), and C2HC5 (Freyd et al., 1990). There are two types of C2H2 families, C and Q. Type Q with conserved motif (QALGGH) is plant specific, while type C is also present in other organisms (Agarwal et al., 2007). The finger conforms to the proposed pattern of TFIIIA type finger \$-X-X-X2, 4, 5-C-X3-\$-X5-\$-X2-H-X3, 4-H, wherein, X is any amino acid and \$ is a hydrophobic residue (Klug and Schwabe, 1995). Based on the different amino acid spacing numbers between C and H in zinc finger motif, the CCCH family is divided into 18 groups (Wang et al., 2008). There is a conserved motif (C-X6-14-C-X4-6-C-X3-H) for ZFP in the CCCH family (Berg and Shi, 1996; Wang et al., 2008). RING motifs can form a binding site for two zinc atoms, including a consensus sequence (C-X2-C-X9-39-C-X1-3-H-X2-3-C/H-X2-C-X4-48-C-X2-C).

Based on the fifth coordination site (C/H), the RING family can be divided into seven types with different conserved motifs: RING-H2, RING-HC, RING-v, RING-D, RING-S/T, RING-G, and RING-C2 (Lim et al., 2010; Stone et al., 2005; Kosarev et al., 2002). Most proteins with RING domain belong to E3 Ubiquitin ligases (Alam et al., 2019; Deshaies and Joazeiro, 2009). In addition, for ZFP in the C4 family, the structure of binding zinc atoms is complicated. This family includes seven series of ZFP types, GATA, FYVE, Tim10/DDP, and most ZFPs contain one type of zinc finger, which can be divided into different types based on these significant structural features.

6.1 C1 family

The C1 family of plant ZFP constitutes 64 members and is one of the largest and transmutative zinc finger families. C1 family members contain either one isolated or two to five dispersed C2H2-type zinc fingers, and are classified according to the number of zinc fingers: 1i donates one finger, 2i two fingers, etc. The C1 subclasses comprise C1–1i (33 members), C1–2i (20 members), C1–3i (8 members), C1–4i (2 members), and C1–5i (1 member) (Englbrecht et al., 2004). Among the subclasses, members of C1–1i and C1–2i are the most investigated plant C2H2-type zinc finger proteins, and are transcriptional activators or repressors that bind DNA.

7. Applications of zinc finger proteins for the improvement of plants

Similarly, DREB2-type genes were also shown to improve abiotic tolerance in plants (Table 2). In maize, *ZmDREB2A* under the control of constitutive or stress-inducible promoter resulted in enhanced drought tolerance (Qin et al., 2007). Overexpression of a constitutively active form of *AtDREB2A*, like wheat *TaAIDFa* from DREB2-type with a deletion at a Ser/Thr-rich motif, activated transcription of downstream genes and improved drought and osmotic stress tolerance in transgenic *Arabidopsis* plants (Xu et al., 2008; Sakuma et al., 2006), although *AtDREB2A* and *OsDREB2A* did not trigger downstream genes under normal growth conditions (Dubouzet et al., 2003). However, DREB2-type genes have been studied in only a limited number of plant species (Agarwal et al., 2006). Therefore, studies on additional DREB2-type genes are necessary to get a more comprehensive picture of DREB regulations (Han et al., 2019; Ijaz et al., 2019; Malik, 2019; Imran et al., 2018).

8. Future scenario of zinc finger proteins for abiotic stress-tolerant varieties

The challenges of abiotic stress on plant growth and development are evident among the emerging ecological impacts of climate change (Bellard et al., 2012), and the constraints on crop production exacerbated with the increasing human population competing for environmental resources (Wallace et al., 2003). Climate change is predicted to affect agricultural production the most,

Table 2 List of transgenic research for abiotic stress tolerance through engineering genes for zinc finger transcription factors.

Gene engineered for ZFTFs	Transgenic host	Source organism	Trait improved	References
MtWRKY76	*Medicago truncatula*	*Medicago truncatula*	Drought tolerance Salinity tolerance	Liu et al. (1998)
OSMYB91	*Oryza sativa*	Rice	Salinity tolerance	Zhu et al. (2016)
HRD	*T. alexandrinum*	*Arabidopsis*	Drought and salt tolerance	Abogadallah et al. (2011)
DREB1C	*Medicago truncatula* and China Rose	*M. truncatula*	Freezing tolerance	Chen et al. (2010)
GhDREB	Wheat	*Gossypium hirsutum*	Drought, salt, and freezing tolerance	Gao et al. (2009)
DREB1A	Chrysanthemum	*Arabidopsis*	Heat stress tolerance	Hong et al. (2009)
HRD	Rice	*Arabidopsis*	Drought tolerance	Karaba et al. (2007)
DREB1A	Chrysanthemum (*Dendranthema vestitum*)	*Arabidopsis*	Drought, salt, and low-temperature tolerance	Hong et al. (2006)
DREB1A	Rice	*Arabidopsis*	Abiotic stress tolerance	Oh et al. (2005)
DREB1A	Tobacco	*Arabidopsis*	Drought and low-temperature tolerance	Kasuga et al. (2004)
DREB1A	Wheat	*Arabidopsis*	Drought tolerance	Pellegrineschi et al. (2004)
DREB1A	Tomato	*Arabidopsis*	Chilling tolerance	Hsieh et al. (2004)

primarily at low latitudes and populated developing countries, facing adverse effects of increasing carbon dioxide and high temperature. These constraints of global food supply and an unpredictable climate change encourage research and development of climate smart crops, resilient to climate change (Wheeler and VonBraun, 2013).

The field of plant abiotic stress encompasses all studies on abiotic factors or stressors from the environment that can impose stress on a variety of species (Sulmon et al., 2015). These stressors include extreme levels of light (high and low), radiation (UV-B and UV-A), temperature (high and low (chilling, freezing)), water (drought, flooding, and submergence), chemical factors (heavy metals and pH), salinity due to excessive Na^+, deficient or excess of essential nutrients, gaseous pollutants (ozone, sulfur dioxide), mechanical factors, and other less frequently occurring stressors. Since combinations of these stresses such as heat and drought frequently occur under field conditions, and can cause unique effects that cannot be predicted from individual stressors (Suzuki et al., 2014), a multiplicity of physiological interactions can be expected, needing individual novel solutions.

Plants which are rooted in the field have to adapt to the changing conditions brought about by the multitude of environmental factors, with extreme levels of eliciting abiotic stresses. A major challenge in abiotic stress biology is to decipher how plants perceive different stressors, how early signals are transduced within the plant, what is the diversity of response pathways elicited by them, and how are they genetically determined (Yoshida et al., 2014). Beyond model plants and reference genotypes, the challenge is to identify how signaling pathways have evolved within a species to program a suite of responses differing in signals and regulatory networks, and constitute genotypes that are adapted to specific stressful environments. Further studies are required to compare few genotypes, such as tolerance and sensitivity within a species, for analyzing differential responses to a defined stress. Since, these responses can be due to differences in sets of genes, an understanding of the diversity in signaling pathways can be attained by making a system level study of the differences between genotypes. Such comparative studies offer a challenge for the integration of diverse functional genomics datasets of gene expression, metabolomics, and stress physiological responses to make comparisons in the network of responses across genotypes.

New challenges will come from genome-wide analyses of stress-tolerant genotypes from multiple plant species that will probably reveal novel tolerance and selective mechanisms in natural populations. Supporting technologies from next-generation sequencing to GWAS are available for many plant species, and much research is concentrated in this area for stress tolerance. This is, therefore, an area for future discoveries which will reveal the evolution of diverse mechanisms for stress tolerance that could be valuable for the design of crop improvement strategies to deal with the challenges of climate change.

A fruitful strategy for the identification of stress tolerance genes has been the reverse genetics analyses of candidate genes identified through gene expression studies and other bioinformatics methods. The biological role of such candidate genes has been most often tested by the analysis of overexpression, knockout/knockdown genotypes in model, and crop plants (Todaka et al., 2015). Studies involving overexpression of transcription factors and other regulatory genes have been conducted in transgenic crops, with the objective of improving their stress tolerance and productivity (Mickelbart et al., 2015), and enabling applications across plant species. Potential redundancy of stress tolerance genes remains a challenge, since studies involving overexpression might not represent the natural function of genes in the plant. Nevertheless, all studies testing the potential phenotype of genes for alterations in stress response provide useful information on gene function as well as its applications. Gene expression analysis of plants in response to abiotic stresses revealed that a large fraction of the genome can be perturbed, reflecting the plasticity in stress response and protection. The complexity of plant response to abiotic stress factors, in interaction with its genetic constitution, provides a multitude of morphophysiological, biochemical, gene expression, and other molecular responses that can best be described by networks of response pathways leading to expression of tolerance and adaptation to the environment.

9. Conclusion

ZFP plays important biological role in plants. It can be classified into many families based on the conserved "zinc finger" motif and/or other notable structures. At the same time, ZFP is part of a large class of transcription factors. The percentage of ZFP transcription factor in total transcription factors are similar in all plant species, revealing that ZFP may have been retained during plant evolution. ZFP plays an important role in plant, such as, growth and development, and stress and phytohormone responses. Meanwhile, each family has a major biological function, indicating the presence of family-specific functions in plants. However, ZFP genes with similar expression pattern are not clustered together based on similarity in amino acid sequence, and the expression of some genes had tissue-specific biological functions. Analysis of ZFP gene expression indicates that the regulation of ZFP control in various tissues is significantly correlated with each other, and significantly correlated with responses to different abiotic stresses. Meanwhile, the regulation of ZFP control in certain specific tissue is related to the regulation of stress-specific responses. Future research should focus on (i) in-depth biological analysis of different ZFP families on the basis of their characteristics, (ii) functional relationships between growth and development, and between stress and phytohormone responses, (iii) ZFP expressing organisms with the use of biology, and (iv) studies on the development of ZFP-based DNA/RNA recognition sites and biological functions.

References

Abogadallah, G.M., Nada, R.M., Malinowski, R., Quick, P., 2011. Overexpression of HARDY, an AP2/ERF gene from Arabidopsis, improves drought and salt tolerance by reducing transpiration and sodium uptake in transgenic *Trifolium alexandrinum* L. Planta 233 (6), 1265–1276.

Agarwal, P.K., Agarwal, P., Reddy, M.K., Sopory, S.K., 2006. Role of DREB transcription factors in abiotic and biotic stress tolerance in plants. Plant Cell Rep. 25 (12), 1263–1274.

Agarwal, P., Arora, R., Ray, S., Singh, A.K., Singh, V.P., Takatsuji, H., Kapoor, S., Tyagi, A.K., 2007. Genome-wide identification of C2H2 zinc-finger gene family in rice and their phylogeny and expression analysis. Plant Mol. Biol. 65 (4), 467–485.

Agarwal, P., Reddy, M.P., Chikara, J., 2011. WRKY: its structure, evolutionary relationship, DNA-binding selectivity, role in stress tolerance and development of plants. Mol. Biol. Rep. 38 (6), 3883–3896.

Alam, I., Batool, K., Cui, D.L., Yang, Y.Q., Lu, Y.H., 2019. Comprehensive genomic survey, structural classification and expression analysis of C2H2 zinc finger protein gene family in *Brassica rapa* L. PLoS One 14(5) e0216071.

Barlow, P.N., Luisi, B., Milner, A., Elliott, M., Everett, R., 1994. Structure of the C3HC4 domain by 1H-nuclear magnetic resonance spectroscopy. A new structural class of zinc-finger. J. Mol. Biol. 237 (2), 201–211.

Bellard, C., Bertelsmeier, C., Leadley, P., Thuiller, W., Courchamp, F., 2012. Impacts of climate change on the future of biodiversity. Ecol. Lett. 15 (4), 365–377.

Berg, J.M., Shi, Y., 1996. The galvanization of biology: a growing appreciation for the roles of zinc. Science 271 (5252), 1081–1085.

Chen, J.R., Lu, J.J., Liu, R., Xiong, X.Y., Wang, T.X., Chen, S.Y., Guo, L.B., Wang, H.F., 2010. DREB1C from *Medicago truncatula* enhances freezing tolerance in transgenic *M. truncatula* and China rose. Plant Growth Regul. 60 (3), 199–211.

Chrispeels, H.E., Oettinger, H., Janvier, N., Tague, B.W., 2000. AtZFP1, encoding Arabidopsis thaliana C2H2 zinc-finger protein 1, is expressed downstream of photomorphogenic activation. Plant Mol. Biol. 42 (2), 279–290.

Dai, K.S., Liew, C.C., 1998. Characterization of a novel gene encoding zinc finger domains identified from expressed sequence tags (ESTs) of a human heart cDNA database. J. Mol. Cell. Cardiol. 30 (11), 2365–2375.

Dai, X., Wang, Y., Yang, A., Zhang, W.H., 2012. OsMYB2P-1, an R2R3 MYB transcription factor, is involved in the regulation of phosphate-starvation responses and root architecture in rice. Plant Physiol. 159 (1), 169–183.

Deshaies, R.J., Joazeiro, C.A., 2009. RING domain E3 ubiquitin ligases. Annu. Rev. Biochem. 78, 399–434.

Ding, Z., Li, S., Liu, X., Qin, H., Wang, D., 2009. Transgenic expression of MYB15 confers enhanced sensitivity to abscisic acid and improved drought tolerance in Arabidopsis thaliana. J. Genet. Genomics 36 (1), 17–29.

Dreier, B., Beerli, R.R., Segal, D.J., Flippin, J.D., Barbas, C.F., 2001. Development of zinc finger domains for recognition of the 5′-ANN-3′ family of DNA sequences and their use in the construction of artificial transcription factors. J. Biol. Chem. 276 (31), 29466–29478.
DuBois, R.N., McLane, M.W., Ryder, K., Lau, L.F., Nathans, D., 1990. A growth factor-inducible nuclear protein with a novel Cys/His repetitive sequence. J. Biol. Chem. 265 (31), 19185–19191.
Dubouzet, J.G., Sakuma, Y., Ito, Y., Kasuga, M., Dubouzet, E.G., Miura, S., Seki, M., Shinozaki, K., Yamaguchi, K., Shinozaki, K., 2003. OsDREB genes in rice, *Oryza sativa* L., encode transcription activators that function in drought-, high-salt-and cold-responsive gene expression. Plant J. 33 (4), 751–763.
Englbrecht, C., Schoof, H., Bohm, S., 2004. Conservation, diversification and expansion of C2H2 zinc finger proteins in the *Arabidopsis thaliana* genome. BMC Genomics 5 (1), 39.
Freyd, G., Kim, S.K., Horvitz, H.R., 1990. Novel cysteine-rich motif and homeodomain in the product of the *Caenorhabditis elegans* cell lineage gene lin-II. Nature 344 (6269), 876.
Gabig, T.G., Mantel, P.L., Rosli, R., Crean, C.D., 1994. Requiem: a novel zinc finger gene essential for apoptosis in myeloid cells. J. Biol. Chem. 269 (47), 29515–29519.
Gao, S.Q., Chen, M., Xia, L.Q., Xiu, H.J., Xu, Z.S., Li, L.C., Zhao, C.P., Cheng, X.G., Ma, Y.Z., 2009. A cotton DRE-binding transcription factor gene, GhDREB, confers enhanced tolerance to drought, high salt, and freezing stresses in transgenic wheat. Plant Cell Rep. 28 (2), 301–311.
Haak, D.C., Grene, T., Hua, R., Ivanov, Z., Perrella, R., Li, S., 2017. Multilevel regulation of abiotic stress responses in plants. Front. Plant Sci. 8, 1564.
Han, G., Wang, M., Yuan, F., Guo, J., Zhang, Y., Sui, N., Wang, B., 2019. AtSIZ1 improves salt tolerance by maintaining ionic homeostasis and osmotic balance in Arabidopsis. Plant Sci. 55–67.
Hong, B., Tong, Z., Ma, N., Kasuga, M., Yamaguchi, K., Shinozaki, K., Gao, J.P., 2006. Expression of the Arabidopsis DREB1A gene in transgenic chrysanthemum enhances tolerance to low temperature. J. Hortic. Sci. Biotechnol. 81 (6), 1002–1008.
Hong, B., Yang, Y., Wang, T., Yamaguchi, K., Shinozaki, K., Gao, J., 2009. Over-expression of AtDREB1A in chrysanthemum enhances tolerance to heat stress. Plant Mol. Biol. 70 (3), 231–240.
Hsieh, T.H., Lee, J.T., Yang, P.T., Chiu, L.H., Charng, Y., Wang, Y.C., Chan, M.T., 2004. Heterology expression of the *Arabidopsis* C-repeat/dehydration response element binding factor1 gene confers elevated tolerance to chilling and oxidative stresses in transgenic tomato. Plant Physiol. 135 (2), 1086–1094.
Huang, J., Zhang, H.S., Cao, Y.J., Wang, D., Wang, J.F., Yang, J.S., 2002. Cloning and sequencing analysis of a novel C2H2 zinc finger protein cDNA from rice. J. Nanjing Agric. Univ. 25(3).
Ijaz, M., Qamar, S., Bukhari, S.A., Malik, K., 2019. Abiotic stress signaling in rice crop. In: Advances in Rice Research for Abiotic Stress Tolerance. Woodhead Publishing, pp. 551–569.
Imran, Q.M., Hussain, A., Lee, S.U., Mun, B.G., Falak, N., Loake, G.J., Yun, B.W., 2018. Transcriptome profile of NO-induced Arabidopsis transcription factor genes suggests their putative regulatory role in multiple biological processes. Sci. Rep. 8 (1), 771.
Joshi, R., Wani, S.H., Singh, B., Bohra, A., Dar, Z.A., Lone, A., Pareek, A., Pareek, S.L., 2016. Transcription factors and plants response to drought stress: current understanding and future directions. Front. Plant Sci. 7, 1029.
Karaba, A., Dixit, S., Greco, R., Aharoni, A., Trijatmiko, K., Marsch, R., Martinez, N., Krishnan, A., Nataraja, K.N., Udayakumar, M., Pereira, A., 2007. Improvement of water use efficiency in rice by expression of HARDY, an *Arabidopsis* drought and salt tolerance gene. Proc. Natl. Acad. Sci. 104 (39), 15270–15275.

Kasuga, M., Miura, S., Shinozaki, K., Yamaguchi, K., Shinozaki, K., 2004. A combination of the *Arabidopsis* DREB1A gene and stress-inducible rd29A promoter improved drought- and low-temperature stress tolerance in tobacco by gene transfer. Plant Cell Physiol. 45 (3), 346–350.

Klug, A., 2010. The discovery of zinc fingers and their applications in gene regulation and genome manipulation. Annu. Rev. Biochem. 79, 213–231.

Klug, A., Schwabe, J.W., 1995. Protein motifs zinc fingers. FASEB J. 9 (8), 597–604.

Kosarev, P., Mayer, K.F., Hardtke, C.S., 2002. Evaluation and classification of RING-finger domains encoded by the *Arabidopsis* genome. Genome Biol. 3 (4) 0016-1.

Kubo, K.I., Kanno, Y., Nishino, T., Takatsuji, H., 2000. Zinc-finger genes that specifically express in pistil secretory tissues of petunia. Plant Cell Physiol. 41 (3), 377–382.

Laity, J.H., Lee, B.M., Wright, P.E., 2001. Zinc finger proteins: new insights into structural and functional diversity. Curr. Opin. Struct. Biol. 11 (1), 39–46.

Latchman, D.S., 1997. Transcription factors: an overview. Int. J. Biochem. Cell Biol. 29 (12), 1305–1312.

Lim, S.D., Yim, W.C., Moon, J.C., Kim, D.S., Lee, B.M., Jang, C.S., 2010. A gene family encoding RING finger proteins in rice: their expansion, expression diversity, and co-expressed genes. Plant Mol. Biol. 72 (4–5), 369–380.

Liu, Q., Kasuga, M., Sakuma, Y., Abe, H., Miura, S., Yamaguchi, K., Shinozaki, K., 1998. Two transcription factors, DREB1 and DREB2, with an EREBP/AP2 DNA binding domain separate two cellular signal transduction pathways in drought-and low-temperature-responsive gene expression, respectively, in *Arabidopsis*. Plant Cell 10 (8), 1391–1406.

Luo, M., Bilodeau, P., Koltunow, A., Dennis, E.S., Peacock, W.J., Chaudhury, A.M., 1999. Genes controlling fertilization-independent seed development in *Arabidopsis thaliana*. Proc. Natl. Acad. Sci. 96 (1), 296–301.

Lyu, T., Cao, J., 2018. Cys2/His2 zinc-finger proteins in transcriptional regulation of flower development. Int. J. Mol. Sci. 19 (9), 2589.

Malik, K., 2019. Abiotic stress signaling in rice crop. In: Advances in Rice Research for Abiotic Stress Tolerance. Woodhead Publishing, p. 551.

Mazumdar, P., Lau, S.E., Wee, W.Y., Singh, P., Harikrishna, J.A., 2017. Genome-wide analysis of the CCCH zinc- finger gene family in banana (*Musa acuminata*): an insight into motif and gene structure arrangement, evolution and salt stress responses. Trop. Plant Biol. 10 (4), 177–193.

Mickelbart, M.V., Hasegawa, P.M., Bailey, J., 2015. Genetic mechanisms of abiotic stress tolerance that translate to crop yield stability. Nat. Rev. Genet. 16 (4), 237.

Miller, J., McLachlan, A.D., Klug, A., 1985. Repetitive zinc-binding domains in the protein transcription factor IIIA from *Xenopus oocytes*. EMBO J. 4 (6), 1609–1614.

Molnár, G., Bancoş, S., Nagy, F., Szekeres, M., 2002. Characterisation of BRH1, a brassinosteroid-responsive RING-H2 gene from *Arabidopsis thaliana*. Planta 215 (1), 127–133.

Nakano, T., Suzuki, K., Fujimura, T., Shinshi, H., 2006. Genome-wide analysis of the ERF gene family in Arabidopsis and rice. Plant Physiol. 140 (2), 411–432.

Novak, S., 2019. Plant biotechnology applications of zinc finger technology. In: Transgenic Plants. Humana Press, New York, NY, pp. 295–310.

Oh, S.J., Song, S.I., Kim, Y.S., Jang, H.J., Kim, S.Y., Kim, M., Kim, Y.K., Nahm, B.H., Kim, J.K., 2005. Arabidopsis CBF3/DREB1A and ABF3 in transgenic rice increased tolerance to abiotic stress without stunting growth. Plant Physiol. 138 (1), 341–351.

Omichinski, J.G., Clore, G.M., Schaad, O., Felsenfeld, G., Trainor, C., Appella, E., Stahl, S.J., Gronenborn, A.M., 1993. NMR structure of a specific DNA complex of Zn-containing DNA binding domain of GATA-1. Science 261 (5120), 438–446.

Pellegrineschi, A., Reynolds, M., Pacheco, M., Brito, R.M., Almeraya, R., Yamaguchi, K., Shinozaki, K., Hoisington, D., 2004. Stress-induced expression in wheat of the *Arabidopsis thaliana* DREB1A gene delays water stress symptoms under greenhouse conditions. Genome 47 (3), 493–500.
Qin, F., Kakimoto, M., Sakuma, Y., Maruyama, K., Osakabe, Y., Tran, L.S.P., Shinozaki, K., Yamaguchi, K., 2007. Regulation and functional analysis of ZmDREB2A in response to drought and heat stresses in Zea mays. Plant J. 50 (1), 54–56.
Reyes, J.L., Chua, N.H., 2007. ABA induction of miR159 controls transcript levels of two MYB factors during Arabidopsis seed germination. Plant J. 49 (4), 592–606.
Rubin, G.M., Yandell, M.D., Wortman, J.R., Gabor, G.L., Nelson, C.R., Hariharan, I.K., Fortini, M.E., Li, P.W., Apweiler, R., Fleischmann, W., Cherry, J.M., 2000. Comparative genomics of the eukaryotes. Science 287 (5461), 2204–2215.
Sakai, H., Medrano, L.J., Meyerowitz, E.M., 1995. Role of SUPERMAN in maintaining Arabidopsis floral whorl boundaries. Nature 378 (6553), 199.
Sakuma, Y., Maruyama, K., Osakabe, Y., Qin, F., Seki, M., Shinozaki, K., Yamaguchi, K., Shinozaki, K., 2006. Functional analysis of an Arabidopsis transcription factor, DREB2A, involved in drought-responsive gene expression. Plant Cell 18 (5), 1292–1309.
Shukla, V.K., Doyon, Y., Miller, J.C., DeKelver, R.C., Moehle, E.A., Worden, S.E., Mitchell, J.C., Arnold, N.L., Gopalan, S., Meng, X., Choi, V.M., 2009. Precise genome modification in the crop species *Zea mays* using zinc-finger nucleases. Nature 459 (7245), 437.
Singh, B., Bohra, A., Mishra, S., Joshi, R., Pandey, S., 2015. Embracing new-generation 'omics' tools to improve drought tolerance in cereal and food-legume crops. Biol. Plant. 59 (3), 413–428.
Singh, P.K., Srivastava, D., Tiwari, P., Tiwari, M., Verma, G., Chakrabarty, D., 2019. Drought tolerance in plants: molecular mechanism and regulation of signaling molecules. In: Plant Signaling Molecules. Woodhead Publishing, pp. 105–123.
Stone, S.L., Hauksdóttir, H., Troy, A., Herschleb, J., Kraft, E., Callis, J., 2005. Functional analysis of the RING-type ubiquitin ligase family of Arabidopsis. Plant Physiol. 137 (1), 13–30.
Sulmon, C., Van Baaren, J., Cabello-Hurtado, F., Gouesbet, G., Hennion, F., Mony, C., Renault, D., Bormans, M.E., Amrani, A., Wiegand, C., Gerard, C., 2015. Abiotic stressors and stress responses. Environ. Pollut. 202, 66–77.
Sun, B., Zhao, Y., Shi, S., Yang, M., Xiao, K., 2019. TaZFP1, a C2H2 type-ZFP gene of *T. aestivum*, mediates salt stress tolerance of plants by modulating diverse stress-defensive physiological processes. Plant Physiol. Biochem. 136, 127–142.
Suzuki, N., Rivero, R.M., Shulaev, V., Blumwald, E., Mittler, R., 2014. Abiotic and biotic stress combinations. New Phytol. 203 (1), 32–43.
Szczepek, M., Brondani, V., Buchel, J., Serrano, L., Segal, D.J., Cathomen, T., 2007. Structure-based redesign of the dimerization interface reduces the toxicity of zinc-finger nucleases. Nat. Biotechnol. 25 (7), 786.
Todaka, D., Yamaguchi, K., Shinozaki, K., 2015. Recent advances in the dissection of drought-stress regulatory networks and strategies for development of drought-tolerant transgenic rice plants. Front. Plant Sci. 6, 84.
Van DerKrol, A.R., Van Poecke, R.M., Vorst, O.F., Voogt, C., Van Leeuwen, W., Borst-Vrensen, T.W., Takatsuji, H., Van Derplas, L.H., 1999. Expression of the petunia zinc finger transcription factor gene ZPT2-2. Plant Physiol. 121, 1153–1162.
Wallace, J.S., Acreman, M.C., Sullivan, C.A., 2003. The sharing of water between society and ecosystems: from conflict to catchment-based co-management. Philos. Trans. R. Soc. Lond. B Biol. Sci. 358 (1440), 2011–2026.
Wang, D., Guo, Y., Wu, C., Yang, G., Li, Y., Zheng, C., 2008. Genome-wide analysis of CCCH zinc finger family in Arabidopsis and rice. BMC Genomics 9 (1), 44.

Wang, K., Ding, Y., Cai, C., Chen, Z., Zhu, C., 2019. The role of C2H2 zinc finger proteins in plant responses to abiotic stresses. Physiol. Plant. 165 (4), 690–700.

Wheeler, T., VonBraun, J., 2013. Climate change impacts on global food security. Science 341 (6145), 508–513.

Wright, D.A., Townsend, J.A., Irwin, R.J., Rajagopal, J., Lonosky, P.M., Hall, B.D., Jondle, M.D., Voytas, D.F., 2005. High-frequency homologous recombination in plants mediated by zinc-finger nucleases. Plant J. 44 (4), 693–705.

Xu, Z.S., Ni, Z.Y., Liu, L., Nie, L.N., Li, L.C., Chen, M., Ma, Y.Z., 2008. Characterization of the TaAIDFa gene encoding a CRT/DRE-binding factor responsive to drought, high-salt, and cold stress in wheat. Mol. Gen. Genomics 280 (6), 497–508.

Yoshida, T., Mogami, J., Yamaguchi, K., Shinozaki, K., 2014. ABA-dependent and ABA-independent signaling in response to osmotic stress in plants. Curr. Opin. Plant Biol. 21, 133–139.

Yuan, S., Li, X., Li, R., Wang, L., Zhang, C., Chen, L., Hao, Q., Zhang, X., Chen, H., Shan, Z., Yang, Z., 2018. Genome-wide identification and classification of soybean C2H2 zinc finger proteins and their expression analysis in legume-rhizobium symbiosis. Front. Microbiol. 9, 126.

Yun, J.Y., Weigel, D., Lee, I., 2002. Ectopic expression of SUPERMAN suppresses development of petals and stamens. Plant Cell Physiol. 43 (1), 52–57.

Zhu, H., Wang, G., Qian, J., 2016. Transcription factors as readers and effectors of DNA methylation. Nat. Rev. Genet. 17 (9), 551.

Further reading

Perez-Clemente, R.M., Vives, V., Zandalinas, S.I., Lopez, M.F., Munoz, V., Gomez, A., 2013. Biotechnological approaches to study plant responses to stress. BioMed Res. Int. 1–10. Article ID 654120.

CHAPTER 4

NAC transcription factor family in rice: Recent advancements in the development of stress-tolerant rice

Bharathi Raja Ramadoss[a], Manu Pratap Gangola[a], and Selvakumar Gurunathan[b]
[a]Department of Plant Sciences, College of Agriculture and Bioresources, University of Saskatchewan, Saskatoon, SK, Canada
[b]DRDO-BU-Centre for Life Sciences, Bharathiar University Campus, Coimbatore, Tamil Nadu, India

1. Introduction

Plants being sessile continuously suffer from a range of environmental stresses (Gangola and Ramadoss, 2018). Abiotic stresses such as heat and cold, drought, salinity adversely affect plant growth and development which resulted in reduction in more than 50% of yield in major food crops (Lamaoui et al., 2018). In addition to this, global warming and current climatic prediction show increasing average temperature, a changing distribution of annual rainfall, increasing sea level, and so on (Stuecker et al., 2018). This will be coexisting with increased frequency of drought, flood, heat, and salinization to trigger the yield loss due to abiotic stresses. To combat these adverse effects, plants have developed various responsive or defensive mechanisms/gene expression including stress-sensing mechanisms, antioxidant system, signaling pathways, secondary metabolites biosynthesis, and other defensive pathways (Gangola and Ramadoss, 2018). To date, a large number of stress-responsive genes have been identified in many crop plants including rice and *Arabidopsis*. These stress-responsive genes are generally classified into two types: functional genes, which codes for important enzymes and metabolic proteins and directly protect cells from stress (Shinozaki et al., 2003), and regulatory genes, which codes for various regulatory proteins including transcription factors (TFs) and protein kinases that regulates the expression of stress response genes.

Alteration in gene expression and cellular metabolism trigger biotic and abiotic stresses and produce a wide range of plant responses

Transcription Factors for Abiotic Stress Tolerance in Plants
https://doi.org/10.1016/B978-0-12-819334-1.00004-6

ultimately affecting the plant growth and development. TFs include a wide number of proteins, excluding RNA polymerase involved in the initiation and regulation (overexpression or suppression) of transcription during gene expression. TFs have DNA-binding domains that give them the ability to bind with DNA with specific regions called enhancers or promoters. Some TFs binds to a DNA promoter sequence near start codons and help to form transcription initiation. Other TFs possess ability to bind to regulatory sequences (enhancer sequence) and can influence the transcription through stimulation or repression (Williams et al., 2007). The action of TFs allows for distinctive expression of luxury genes in different cells of the plants during growth and development.

Recent development in molecular approaches have laid foundation to understand the important role of TFs in regulating various biotic and abiotic stresses through differential responses and some of them have been found to be essential for improving stress tolerance in crop plants (Nakashima et al., 2012). Genetic manipulation of TFs has been shown to be a successful procedure in altering traits like drought/salinity tolerance instead of modifying key candidate genes involved in the metabolic pathway because of the capacity of TFs in altering expression of genes involved in various stress tolerance (Turan et al., 2012). Previous studies on TFs have clearly demonstrated that DREB, bZIP, zinc finger, MYB, WRKY, and NAC play an important role in regulating plant responses under abiotic stresses (Zhu, 2002; Yadav et al., 2013; Agarwal et al., 2006; Shen et al., 2012).

NAC proteins (NAM, no apical meristem; ATAF, *Arabidopsis* transcription activation factor; and CUC, cup-shaped cotyledon) are a family of TFs which has many important regulatory roles in crop plant-containing highly conserved NAC subdomains in the N-terminal and a variable domain at the C-terminal (Olsen et al., 2005). The NAC TF is one of the largest families of TFs in plants and plays an important role in the regulation of the transcriptional reprogramming associated with various biotic and abiotic stress responses (Nakashima et al., 2012). The NAC domain consists of ~150 amino acids with five conserved regions (A to E). Structural analysis was used to identify the functional dimer formed by the NAC domain (Olsen et al., 2005). Recently, the structure of rice stress-responsive NAC protein (SNAC1, stress-responsive NAC1) was identified (Chen et al., 2011). The structure of SNAC1 NAC domain shares a similarity with *Arabidopsis* NAC domain ANACO19 (Ernst et al., 2004). This chapter presents the recent advancements in NAC family TFs and its developments in stress tolerance in rice.

2. NAC TFs in rice

Rice (*Oryza sativa* L.) is a staple crop for more than half of the world's population and a primary source of diet (Raja et al., 2017). During green revolution, population in low-income countries grew by 90% but the production of rice increased by 130% from 257 million tons in 1966 to 600 million tons in 2000 (Khush, 2005). Therefore, it is estimated to produce 40% more yield in 2030. However, global warming, mean temperature increase, prolonged drought and flood, increased soil salinity etc., poses additional burden on the researcher to develop climate-resilient rice varieties suitable for different abiotic stresses without compromising the quality of rice to achieve the projected yield to feed the growing world population in 2030 (Khush, 2005). To achieve this target, different plant breeding approaches coupled with various molecular methodologies have been used to explore/discover the structural genes which can directly or indirectly contribute toward yield increase. Recent advancements in molecular biology laid foundation to explore the role of regulatory genes toward stress tolerance in rice.

NAC TFs constitute one of the largest families, with approximately 151 genes in rice, 79 in grape, 26 in citrus, 163 in poplar, 152 genes in both soybean and tobacco, and 117 genes in *Arabidopsis thaliana* (Nuruzzaman et al., 2010, 2012a; Rushton et al., 2008; Hu et al., 2010; Le et al., 2011). Recently, Nuruzzaman et al. (2010) examined the genomic distribution of NAC genes and identified the position of NAC TFs based on MSU database. Altogether, 135 out of 151 OsNAC genes were localized on 12 chromosomes with irregular distribution. The OsNAC genes were distributed individually or in cluster throughout the chromosome. Among 12 chromosomes, third and first chromosome possess 17 and 15 NAC genes, respectively, while chromosome 4 and 9 hold OsNAC genes. Less than 10 OsNAC genes were found to be in chromosome 4, 5, 6, and 9 (Nuruzzaman et al., 2010). The variation in OsNAC genes is attributed by segmental and tandem duplication events occurred during evolution. Researcher reported that 36 OsNAc genes were involved in tandem duplication with two to four repeats (Kikuchi et al., 2000). The OsNAC genes encoding different small to larger protein with 270–329 amino acids in OsNAC3-OsNAC7, while OsNAC8 gene encodes 486 amino acids with highly conserved N-terminal segments which resembles the domain of NAM (Kikuchi et al., 2000). On the other hand, C-terminal regions were highly variable with simple sequence repeats in amino acids such as DDDDLLHHHHH in OsNAC7 or GGGGGAAA in OsNAC5. Based on NAC domain, NAC genes fall into several subfamilies, i.e., NAM, OsNAC3, and ATAF subfamilies. In addition to this, NAC domain could be divided into subdomains (A–E), which is distinct by blocks of heterogeneous amino acids or gaps (Kikuchi et al., 2000; Ooka et al., 2003). While NAC domains were rich in basic amino acid of R, K, and H, unequal distribution of positive and negative amino acids were found in each subdomain. However, subdomain B contains a higher proportion of acidic amino acids, while C and D are rich in basic amino acids and poor in acidic amino acids (Kikuchi et al., 2000; Ooka et al., 2003).

3. NAC TFs in stress tolerance

Depending on the stress, NAC TFs activate and regulate the downstream target genes through binding to the NAC recognition sequence (NACRS) in the promoter region in order to activate abiotic stress tolerance. Drought-related induction of NAC TFs was first reported in *Arabidopsis*. Subsequent transgenic studies showed that overexpression of *A. thaliana* (At) NAC genes (ANAC019, ANAC055, and ANAC072) in transgenic plants alter the expression of salinity, drought, and cold stress inducible genes (Tran et al., 2004). ATAF1, COR47, ERD10, KIN1, RD22, and RD29A TFs in *Arabidopsis* showed a positive regulation of drought tolerance (Lu et al., 2007). Soybean TFs GmNAC20 and DREB1A/CBF3 showed enhanced expression in drought and freezing tolerance (Hao et al., 2011). NAM-B1 TF in wheat played a significant role in nutrient remobilization from leaves to developing seeds (Ricachenevsky et al., 2013). TaNAC4 TFs gene functioned as an activator in wheat biotic and abiotic stress tolerance (Xia et al., 2010). In addition to this, some of the researchers have reported vital functions of NAC TFs both in biotic and abiotic stress tolerance (Xia et al., 2010; Nuruzzaman et al., 2013).

The general process of abiotic stress tolerance in plants through NAC TFs is described in Fig. 1. Describing the process of NAC TFs in abiotic stress tolerance through single cascade system is not feasible. In response to stress, plants observe and send the signal transduction to activate respective pathway. Luxury genes/NAC TFs in the pathway are activated to produce NAC mRNA through the process of transcription and translation to produce NAC protein. NAC protein may or may not undergo posttranslational modification to make

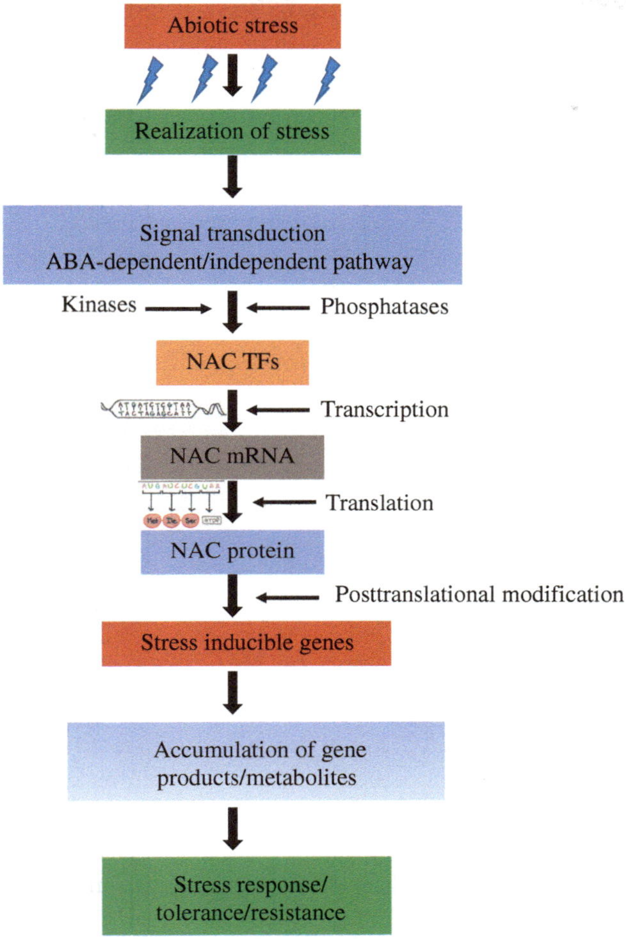

Fig. 1 Diagrammatic representation of transcriptional regulatory mechanism of NAC TFs in abiotic stress tolerance.

stress-inducible gene to accumulate gene products/metabolites thus making plants more tolerant to stress. Multiple interaction and regulatory networks were also associated to produce stress response in plants. The regulation of TFs can be modified into two different components: upstream modification through multiple interaction and regulatory networks, and posttranscription level through glycosylation, *S*-nitrosylation, methylation, *N*-acetylation, sumoylation, ubiquitination, and phosphorylation thus modifying the expression of stress responsive genes.

The recently characterized NAC gene in rice, SNAC1 and OsNAC6/SNAC2, shows significant increase in drought, salt, and cold tolerance during seedling development, respectively (Hu et al., 2006). In the field condition, transgenic plants carrying SNAC1 TFs increased the drought tolerance and produced 22%–34% of greater seed set/yield than the control plants when exposed to severe drought during flowering (Hu et al., 2006; Nakashima et al., 2007). The transgenic rice plants overexpressing OsNAC045 TFs upregulates two stress responsive genes, PM1 and LEA3, which improves salt and drought tolerance at seedling stage (Zheng et al., 2009). SNAC3 in rice provides high temperature, drought, and oxidative stress tolerance by altering ROS homeostasis (Fang et al., 2015). Similarly, different NAC TFs play different role in altering the expression of various genes in rice under different stress conditions (Table 1). In addition to abiotic stress tolerance, NAC TFs assist in several biological and physiological functions in plants and provide key regulators of various growth-related developmental processes. On the other hand, only few of them have been fully characterized for their functionality and their role in abiotic stress tolerance in rice.

3.1 Drought stress modulation by NAC TFs

Several transgenic studies have been conducted to understand the role and mechanism of individual NAC TFs in drought tolerance. Overexpression of SNAC1 significantly enhances seed set (22%–34% higher seed setting than control) under field reproductive stage drought condition without altering phenotype or yield loss (Hu et al., 2006). Transgenic rice plants overexpressing OsNAC6 showed an enhanced drought tolerance, salt stress tolerance, and displayed less susceptible to blast *Magnaporthe grisea* (Nakashima et al., 2007). In another study, transgenic plants carrying OsNAC6 TFs were shown to upregulate the genes involved in nicotianamine (NA) biosynthesis, membrane modification, glutathione relocation, glycosylation, and 3′-phophoadenosine 5′-phosphosufhate accumulation, which is involved in multiple drought-tolerant pathways to mediate the structural adaptation of roots, increase the number of roots and root diameter to enhance the drought tolerance under water-limiting conditions (Lee et al., 2017).

Another NAC TFs gene, SNAC3, was characterized for various abiotic stress tolerance. Overexpression of SNAC3 in rice enhance the tolerance of TFs transgenic plants to drought, heat, and oxidative stress through H_2O_2 homeostasis and controlling ROS gene expression (Fang et al., 2015). The field performance of transgenic plants overexpressing OsNAC5 under

Table 1 Abiotic stress response of different NAC transcription factors.

Genes/TFs	Abiotic stress	Method	References
SNAC2/OsNAC6/ Osg66120	Drought, salinity, and cold tolerance	Overexpression	Hu et al. (2008)
OsNAC5/ Os11g08210	Salt and cold tolerance	Overexpression	Sperotto et al. (2009)
ONAC04/ Os11g033005	Drought, salt, and cold tolerance	Overexpression	Zheng et al. (2009)
OsNAC10/ Os11g03300	Drought and salt tolerance	Overexpression	Jeong et al. (2010)
Os01g28050, Os01g29840, Os0704560, Os10g38834, Os03g12120, Os03g59730, Os06g15690, Os08g06140, Os08g33670	Drought	Microarray	Nuruzzaman et al. (2012b)
Os12g41680, Os07g48550, Os11g03300, Os12g03040, Os01g66120, Os05g34830	Drought, cold, submergence, laid down submergence tolerance	Microarray	Nuruzzaman et al. (2010)
Os02g34970, Os07g48450, Os01g01430, Os01g48460	Drought, submergence, laid down submergence tolerance	Microarray	Nuruzzaman et al. (2010)
OsOAT, SNAC2	Drought and oxidative stress tolerance	Overexpression	You et al. (2013)
SNAC1, OsSR01c	Oxidative stress tolerance	Overexpression	You et al. (2013)
ONAC063	High-temperature and high-salinity tolerance	Transactivation	Yokotani et al. (2009)

Continued

Table 1 Abiotic stress response of different NAC transcription factors—cont'd

Genes/TFs	Abiotic stress	Method	References
SNAC1	Drought and salt tolerance	Overexpression	Hu et al. (2006)
OsNAC2	Cold tolerance	Overexpression	Hu et al. (2008)
OsNAC3	High-temperature and drought tolerance	Overexpression	Fang et al. (2015)
OsNAC6	Drought and salt tolerance	Overexpression	Nakashima et al. (2007)
OsNAC5	Drought and salt tolerance	Overexpression	Song et al. (2011)
EcNAC67 (*Eleusine coracana* L.)	Salinity and drought tolerance in rice	Overexpression	Rahman et al. (2016)
OsNAC041	Salt stress	Overexpression and mutation using CRISPR/Cas9	Bo et al. (2019)
SNAC1	Drought tolerance	Overexpression	Li et al. (2019)
OsNAC14	Drought tolerance	Overexpression	Shim et al. (2018)
ONAC022	Drought and salt tolerance	Overexpression	Hong et al. (2016)
ONAC066	Drought and oxidative stress	Transactivation	Yuan et al. (2019)

the control of two different promoters (root-specific and constitutive) revealed that the grain yield of overexpressing OsNAC5 (root) and OsNAC5 (constitutive) plants were increased by 9%–23% and 9%–26%, respectively. In drought condition, plants carrying OsNAC5 TFs with root-specific promoters produced grain yield 22%–63% higher than control plants (Jeong et al., 2013). Further, microarray-based expression analysis revealed that 19 upregulated genes in overexpressing (root-specific promoters) OsNAC5 TFs transgenic roots, which includes MERI5, GLP, PDX, and *O*-methyltransferase, were involved in root growth and development. Similarly, root-specific and constitutive promoters were used to overexpress OsNAC9 TFs in rice and produce transgenic plant (root-specific promoters) with capacity to produce a yield increase of 28%–72% and 13%–18% under drought condition (Redillas et al., 2012). In addition to field performance, OsNAC9 TFs are also involved in threefold upregulation of 40 stress responsive genes, which include ABA biosynthesis gene, calcium-transporting ATPase, 9-*cis*-epoxycarotenoid dioxygenase, O-methyltransferase, wall-associated kinases, genes involved in cell elongation and morphogenesis, and gene involved in lignin biosynthesis (Redillas et al., 2012). These upregulated genes in overexpressed OsNAC9 transgenic roots may play important role in root modification to confer the drought-tolerant phenotype. Another transgenic study revealed that root-specific expression of OsNAC10 TFs enlarges roots and significantly increases drought tolerance at the reproductive stage, increasing grain yield by 25%–42% and by 5%–14% over controls in the field under drought and normal conditions, respectively (Jeong et al., 2010). Root diameter of root-specific overexpressed OsNAC10 transgenic plants showed 1.25-fold increased than that of control plants. Thus, root-specific overexpression of OsNAC10 enlarges root diameter and increases the yield and drought tolerance under field drought conditions (Jeong et al., 2010).

Recently, Shim et al. (2018) have functionally characterized the rice drought responsive transcription factor OsNAC14. Overexpression of OsNAC14 in rice resulted in enhanced drought tolerance during vegetative and reproductive stage. Overexpressed OsNAC14 TFs transgenic plants demonstrated higher number of productive tillers and grain filling rate compared to control plants under drought conditions. Further, RNA-sequencing analysis revealed that OsNAC14 overexpression enhances the expression of stress responsive, defense-related, strigolactone biosynthesis, and DNA damage repair genes to provide drought tolerance through DNA damage repair and defense response.

3.2 High-/low-temperature stress modulation by NAC TFs

Agricultural production is highly influenced by high- and low-temperature stress as a result of global warming. In the last two decades, tremendous studies have been performed to understand the role of NAC TFs in both heat and cold stress tolerance. NAC TF gene ONAC063 in rice roots responds to a different combination of high-temperature stress (Yokotani et al., 2009).

Different NAC TFs SNAC2/OsNAC6, OsNAC10, TaNAC4, NTL6, TaNAC4a, TaNAC6, TaNAC7, TaNAC13, and TaNTL5 are highly induced when plants are exposed to cold stress (Jeong et al., 2010; Xia et al., 2010; Tang et al., 2012). Analysis of SNAC2 TF expression and promoter activity suggests that the SNAC2 TFs gene is induced by low temperature and 50% transgenic plants overexpressing SNAC2 survived when exposed to 4–8°C for 5 days. These plants had higher cell membrane integrity and stability than wild-type plants when exposed to cold stress. Further, microarray-based expression studies revealed that 8 out of 14 OsNAC genes were induced during various drought conditions (Nuruzzaman et al., 2012b).

3.3 Salinity stress modulation by NAC TFs

Salinity stress causes a wide range of plant responses through gene expression and cellular metabolism to alter physiological, morphological, and ultrastructural/functional changes in the cell to produce normal plant growth and development. Transgenic studies revealed that overexpression of Os01g66120/OsNAC2/6, Os11g03300/OsNAC10, OsNAC5, and OsNAC6 TFs increases tolerance to salinity stress (Nakashima et al., 2007, 2009; Takasaki et al., 2010; Jeong et al., 2010, 2013). Further, overexpression of other NAC TFs ONAC04/Os11g033005, SNAC2/OsNAC6/Os01g66120, and ONAC063 also showed significant amount of salinity tolerance in rice (Zheng et al., 2009; Hu et al., 2008; Yokotani et al., 2009).

In additional to overexpressing same species NAC TFs, cross species gene expression analysis was also carried out to understand the role of NAC TFs in salinity tolerance. Rice NAC TFs SNAC1 expressed in transgenic cotton plants exhibited enhanced salinity and drought tolerance through better root development and decreased transpiration rate (Liu et al., 2014). In another study, overexpression of wheat NAC TF TaNAC67 in *Arabidopsis* was found to produce plants with enhanced salinity, drought, and low-temperature stress (Mao et al., 2014). Recently, finger millet NAC TF EcNAC67 was isolated from salinity-tolerant finger millet genotype Trichy 1 and engineered into rice to study the salinity tolerance. Overexpression of EcNAC67 TFs in rice plants showed an enhanced salinity and drought tolerance (Rahman et al., 2016). In addition, overexpression of another finger millet TFs EcNAC1 in tobacco showed an increased salinity tolerance (Ramegowda et al., 2012).

4. Pleiotropic function of NAC TFs

Functional analysis of several NAC TFs has demonstrated to produce multiple roles when exposed to various stress factors in different signaling pathways, as it can be deduced from their induced expression profiles by various stress factors. Overexpression of rice NAC TFs Os04g0477300 showed different functions of pathogen defense, senescence, and responses to boron and phosphate deficiency (Uauy et al., 2006; Waters et al., 2009; Nilsson et al., 2010; Ochiai et al., 2011). Both drought and salinity tolerance were observed in plants overexpressed with SNAC1, SNAC2/OsNAC6/Osg66120, OsNAC6, OsNAC045, EcNAC67, and ONAC022 (Table 1). Similarly, different combination of abiotic and biotic stresses has been observed in different transgenic plants overexpressed with NAC TFs. Different NAC genes in rice and other plants are activated/affected in response to abiotic/biotic stress. Accumulation of SiNAC transcripts in young spikes were strongly induced by salinity, dehydration, ethephon, and methyl jasmonate (Distelfeld et al., 2012). This study revealed that multiple functions of single NAC TFs gene are involved in regulation/cross talk between different signaling pathways.

5. Conclusion

Most of the functional studies on NAC TFs have been carried out to improve the biotic stress tolerance in crop plants. Limited effort has been taken to understand the role of individual NAC TFs in abiotic stress tolerance. However, most of the previous studies were carried through laboratory level transgenic approach. Most of the NAC TFs contribute directly or indirectly to more than one abiotic stress tolerance. Therefore, a different approach is needed to study the role of individual NAC TFs in specific abiotic stress and the way that mediate the cross talk between different signaling pathways will help to achieve better understanding of their role during stress. Further, only few field experiments were carried out to understand the degree of tolerance to various abiotic stresses. On the other hand, commercial cultivation of transgenics is still under consideration in most of the developing/middle-income countries. To develop better varieties, understanding the role of abiotic stress is essential to study the NAC TFs that is involved in regulating downstream target genes under specific condition. In addition, it is also crucial to identify the key components of the signal transduction pathways with which these factors interact.

References

Agarwal, P., Agarwal, P., Reddy, M.K., Sopory, S., 2006. Role of DREB transcription factors in abiotic and biotic stress tolerance in plants. Plant Cell Rep. 25, 1263–1274.

Bo, W., Zhaohui, Z., Huanhuan, Z., Xia, W., Binglin, L., Lijia, Y., Xiangyan, H., Deshui, Y., Xuelian, Z., Chunguo, W., Wenqin, S., 2019. Targeted mutagenesis of NAC transcription factor gene, OsNAC041, leading to salt sensitivity in rice. Rice Sci. 26, 98–108.

Chen, Q., Wang, Q., Xiong, L., Lou, Z., 2011. A structural view of the conserved domain of rice stress-responsive NAC1. Protein Cell 2, 55–63.

Distelfeld, A., Pearce, S.P., Avni, R., Scherer, B., Uauy, C., Piston, F., et al., 2012. Divergent functions of orthologous NAC transcription factors in wheat and rice. Plant Mol. Biol. 78, 515–524.

Ernst, H.A., Olsen, A.N., Larsen, S., Lo-Leggio, L., 2004. Structure of the conserved domain of ANAC, a member of the NAC family of transcription factors. EMBO Rep. 5, 297–303.

Fang, Y., Liao, K., Du, H., Xu, Y., Song, H., Li, X., Xiong, L., 2015. A stress-responsive NAC transcription factor SNAC3 confers heat and drought tolerance through modulation of reactive oxygen species in rice. J. Exp. Bot. 66, 6803–6817.

Gangola, M.P., Ramadoss, B.R., 2018. Sugars play a critical role in abiotic stress tolerance in plants. In: Biochemical, Physiological and Molecular Avenues for Combating Abiotic Stress Tolerance in Plants. Academic Press, pp. 17–38.

Hao, Y.J., Wei, W., Song, Q.X., Chen, H.W., Zhang, Y.Q., Wang, F., Zou, H.F., Lei, G., Tian, A.G., Zhang, W.K., Ma, B., 2011. Soybean NAC transcription factors promote abiotic stress tolerance and lateral root formation in transgenic plants. Plant J. 68, 302–313.

Hong, Y., Zhang, H., Huang, L., Li, D., Song, F., 2016. Overexpression of a stress-responsive NAC transcription factor gene ONAC022 improves drought and salt tolerance in rice. Front. Plant Sci. 7, 4.

Hu, H., Dai, M., Yao, J., Xiao, B., Li, X., Zhang, Q., Xiong, L., 2006. Overexpressing a NAM, ATAF, and CUC (NAC) transcription factor enhances drought resistance and salt tolerance in rice. Proc. Natl. Acad. Sci. 103, 12987–12992.

Hu, H., You, J., Fang, Y., Zhu, X., Qi, Z., Xiong, L., 2008. Characterization of transcription factor gene SNAC2 conferring cold and salt tolerance in rice. Plant Mol. Biol. 67, 169–181.

Hu, R., Qi, G., Kong, Y., Kong, D., Gao, Q., Zhou, G., 2010. Comprehensive analysis of NAC domain transcription factor gene family in *Populus trichocarpa*. BMC Plant Biol. 10, 145.

Jeong, J.S., Kim, Y.S., Baek, K.H., Jung, H., Ha, S.H., Do Choi, Y., Kim, M., et al., 2010. Root-specific expression of OsNAC10 improves drought tolerance and grain yield in rice under field drought conditions. Plant Physiol. 153, 185–197.

Jeong, J.S., Kim, Y.S., Redillas, M.C., Jang, G., Jung, H., Bang, S.W., et al., 2013. OsNAC5 overexpression enlarges root diameter in rice plants leading to enhanced drought tolerance and increased grain yield in the field. Plant Biotechnol. J. 11, 101–114.

Khush, G.S., 2005. What it will take to feed 5.0 billion rice consumers in 2030. Plant Mol. Biol. 59, 1–6.

Kikuchi, K., Ueguchi-Tanaka, M., Yoshida, K.T., Nagato, Y., Matsusoka, M., Hirano, H.Y., 2000. Molecular analysis of the NAC gene family in rice. Mol. Gen. Genet. 262, 1047–1051.

Lamaoui, M., Jemo, M., Datla, R., Bekkaoui, F., 2018. Heat and drought stresses in crops and approaches for their mitigation. Front. Chem. 6, 26. https://doi.org/10.3389/fchem.2018.00026.

Le, D.T., Nishiyama, R., Watanabe, Y., Mochida, K., Yamaguchi-Shinozaki, K., Shinozaki, K., et al., 2011. Genome-wide survey and expression analysis of the plant-specific NAC transcription factor family in soybean during development and dehydration stress. DNA Res. 18, 263–276.

Lee, D.K., Chung, P.J., Jeong, J.S., Jang, G., Bang, S.W., Jung, H., et al., 2017. The rice OsNAC6 transcription factor orchestrates multiple molecular mechanisms involving root structural adaptions and nicotianamine biosynthesis for drought tolerance. Plant Biotechnol. J. 15, 754–764.

Li, X., Chang, Y., Ma, S., Shen, J., Hu, H., Xiong, L., 2019. Genome-wide identification of SNAC1-targeted genes involved in drought response in rice. Front. Plant Sci. 10, 982.

Liu, G., Li, X., Jin, S., Liu, X., Zhu, L., Nie, Y., et al., 2014. Overexpression of Rice NAC gene SNAC1 improves drought and salt tolerance by enhancing root development and reducing transpiration rate in transgenic cotton. PLoS One 9, 1–10.

Lu, P.L., Chen, N.Z., An, R., Su, Z., Qi, B.S., Ren, F., Chen, J., Wang, X.C., 2007. A novel drought-inducible gene, ATAF1, encodes a NAC family protein that negatively regulates the expression of stress-responsive genes in Arabidopsis. Plant Mol. Biol. 63, 289–305.

Mao, X., Chen, S., Li, A., Zhai, C., Jing, R., 2014. Novel NAC transcription factor TaNAC67 confers enhanced multi-abiotic stress tolerances in Arabidopsis. PLoS One 9, 1–10.

Nakashima, K., Tran, L.P., Nguyen, D.V., Fujita, M., Maruyama, K., Todaka, D., et al., 2007. Functional analysis of a NAC-type transcription factor OsNAC6 involved in abiotic and biotic stress-responsive gene expression in rice. Plant J. 51, 617–630.

Nakashima, K., Ito, Y., Yamaguchi-Shinozaki, K., 2009. Transcriptional regulatory networks in response to abiotic stresses in Arabidopsis and grasses. Plant Physiol. 149, 88–95.

Nakashima, K., Takasaki, H., Mizoi, J., Shinozaki, K., Yamaguchi-Shinozaki, K., 2012. NAC transcription factors in plant abiotic stress responses. Biochim. Biophys. Acta 1819, 97–103.

Nilsson, L., Müller, R., Nielsen, T.H., 2010. Dissecting the plant transcriptome and the regulatory responses to phosphate deprivation. Physiol. Plant. 139, 129–143.

Nuruzzaman, M., Manimekalai, R., Sharoni, A.M., Satoh, K., Kondoh, H., Ooka, H., et al., 2010. Genome-wide analysis of NAC transcription factor family in rice. Gene 465, 30–44.

Nuruzzaman, M., Sharoni, A.M., Satoh, K., Kondoh, H., Hosaka, A., Kikuchi, S., 2012a. A genome-wide survey of the NAC transcription factor family in monocots and eudicots. In: Introduction to Genetics—DNA Methylation, Histone Modification and Gene Regulation. iConcept Press, Hong Kong.

Nuruzzaman, M., Sharoni, A.M., Satoh, K., Moumeni, A., Venuprasad, R., Serraj, R., et al., 2012b. Comprehensive gene expression analysis of the NAC gene family under normal growth conditions, hormone treatment, and drought stress conditions in rice using near-isogenic lines (NILs) generated from crossing Aday Selection (drought tolerant) and IR64. Mol. Gen. Genomics 287, 389–410.

Nuruzzaman, M., Sharoni, A.M., Kikuchi, S., 2013. Roles of NAC transcription factors in the regulation of biotic and abiotic stress responses in plants. Front. Microbiol. 4, 248.

Ochiai, K., Shimizu, A., Okumoto, Y., Fujiwara, T., Matoh, T., 2011. Suppression of a NAC-like transcription factor gene improves boron-toxicity tolerance in rice. Plant Physiol. 156, 1457–1463.

Olsen, A.N., Ernst, H.A., Leggio, L.L., Skriver, K., 2005. NAC transcription factors: structurally distinct, functionally diverse. Trends Plant Sci. 10, 79–87.

Ooka, H., Satoh, K., Doi, K., Nagata, T., Otomo, Y., Murakami, K., et al., 2003. Comprehensive analysis of NAC family genes in *Oryza sativa* and *Arabidopsis thaliana*. DNA Res. 10, 239–247.

Rahman, H., Ramanathan, V., Nallathambi, J., Duraialagaraja, S., Muthurajan, R., 2016. Over-expression of a NAC 67 transcription factor from finger millet (*Eleusine coracana* L.) confers tolerance against salinity and drought stress in rice. BMC Biotechnol. 16, 35.

Raja, R.B., Agasimani, S., Jaiswal, S., Thiruvengadam, V., Sabariappan, R., Chibbar, R.N., Ram, S.G., 2017. EcoTILLING by sequencing reveals polymorphisms in genes encoding starch synthases that are associated with low glycemic response in rice. BMC Plant Biol. 17, 13.

Ramegowda, V., Senthil-Kumar, M., Nataraja, K.N., Reddy, M.K., Mysore, K.S., Udayakumar, M., 2012. Expression of a finger millet transcription factor, EcNAC1, in tobacco confers abiotic stress-tolerance. PLoS One 7, e40397.

Redillas, M.C., Jeong, J.S., Kim, Y.S., Jung, H., Bang, S.W., Choi, Y.D., et al., 2012. The overexpression of OsNAC9 alters the root architecture of rice plants enhancing drought resistance and grain yield under field conditions. Plant Biotechnol. J. 10, 792–805.

Ricachenevsky, F.K., Menguer, P.K., Sperotto, R.A., 2013. kNACking on heaven's door: how important are NAC transcription factors for leaf senescence and Fe/Zn remobilization to seeds? Front. Plant Sci. 4, 226.

Rushton, P.J., Bokowiec, M.T., Han, S., Zhang, H., Brannock, J.F., Chen, X., et al., 2008. Tobacco transcription factors: novel insights into transcriptional regulation in the Solanaceae. Plant Physiol. 147, 280–295.

Shen, H., Liu, C., Zhang, Y., Meng, X., Zhou, X., Chu, C., et al., 2012. OsWRKY30 is activated by MAP kinases to confer drought tolerance in rice. Plant Mol. Biol. 80, 241–253.

Shim, J.S., Oh, N., Chung, P.J., Kim, Y.S., Choi, Y.D., Kim, J.K., 2018. Overexpression of OsNAC14 improves drought tolerance in rice. Front. Plant Sci. 9, 310.

Shinozaki, K., Yamaguchi-Shinozaki, K., Seki, M., 2003. Regulatory network of gene expression in the drought and cold stress responses. Curr. Opin. Plant Biol. 6 (5), 410–417.

Song, S.Y., Chen, Y., Chen, J., Dai, X.Y., Zhang, W.H., 2011. Physiological mechanisms underlying OsNAC5-dependent tolerance of rice plants to abiotic stress. Planta 234, 331–345.

Sperotto, R.A., Ricachenevsky, F.K., Duarte, G.L., Boff, T., Lopes, K.L., Sperb, E.R., et al., 2009. Identification of up-regulated genes in flag leaves during rice grain filling and characterization of OsNAC5, a new ABA-dependent transcription factor. Planta 230, 985–1002.

Stuecker, M.F., Tigchelaar, M., Kantar, M.B., 2018. Climate variability impacts on rice production in the Philippines. PLoS One 13, e0201426.

Takasaki, H., Maruyama, K., Kidokoro, S., Ito, Y., Fujita, Y., Shinozaki, K., et al., 2010. The abiotic stress-responsive NAC-type transcription factor OsNAC5 regulates stress-inducible genes and stress tolerance in rice. Mol. Gen. Genomics 284, 173–183.

Tang, Y., Liu, M., Gao, S., Zhang, Z., Zhao, X., Zhao, C., et al., 2012. Molecular characterization of novel TaNAC genes in wheat and overexpression of TaNAC2a confers drought tolerance in tobacco. Physiol. Plant. 144, 210–224.

Tran, L.S., Nakashima, K., Sakuma, Y., Simpson, S.D., Fujita, Y., Maruyama, K., Fujita, M., et al., 2004. Isolation and functional analysis of Arabidopsis stress-inducible NAC transcription factors that bind to a drought-responsive cis-element in the early responsive to dehydration stress 1 promoter. Plant Cell 16, 2481–2498.

Turan, S., Cornish, K., Kumar, S., 2012. Salinity tolerance in plants: breeding and genetic engineering. Aust. J. Crop. Sci. 6, 1337–1348.

Uauy, C., Distelfeld, A., Fahima, T., Blechl, A., Dubcovsky, J., 2006. A NAC gene regulating senescence improves grain protein, zinc, and iron content in wheat. Science 314, 1298–1301.

Waters, B.M., Uauy, C., Dubcovsky, J., Grusak, M.A., 2009. Wheat (*Triticum aestivum*) NAM proteins regulate the translocation of iron, zinc, and nitrogen compounds from vegetative tissues to grain. J. Exp. Bot. 60, 4263–4274.

Williams, R.B., Chan, E.K., Cowley, M.J., Little, P.F., 2007. The influence of genetic variation on gene expression. Genome Res. 17, 1707–1716.

Xia, N., Zhang, G., Liu, X.Y., Deng, L., Cai, G.L., Zhang, Y., Wang, X.J., Zhao, J., Huang, L.L., Kang, Z.S., 2010. Characterization of a novel wheat NAC transcription factor gene involved in defense response against stripe rust pathogen infection and abiotic stresses. Mol. Biol. Reprod. 37, 3703–3712.

Yadav, N.R., Taunk, J., Rani, A., Aneja, B., Yadav, R.C., 2013. Role of transcription factors in abiotic stress tolerance in crop plants. In: Tuteja, N., Gill, S.S. (Eds.), Climate Change and Plant Abiotic Stress Tolerance. Wiley-VCH Verlag GmbH & Co, Germany, pp. 605–640.

Yokotani, N., Ichikawa, T., Kondou, Y., Matsui, M., Hirochika, H., Iwabuchi, M., et al., 2009. Tolerance to various environmental stresses conferred by the salt-responsive rice gene ONAC063 in transgenic Arabidopsis. Planta 229, 1065–1075.

You, J., Zong, W., Li, X., Ning, J., Hu, H., Li, X., et al., 2013. The SNAC1-targeted gene OsSRO1c modulates stomatal closure and oxidative stress tolerance by regulating hydrogen peroxide in rice. J. Exp. Bot. 64, 569–583.

Yuan, X., Wang, H., Cai, J., Bi, Y., Li, D., Song, F., 2019. Rice NAC transcription factor ONAC066 functions as a positive regulator of drought and oxidative stress response. BMC Plant Biol. 19, 278.

Zheng, X., Chen, B., Lu, G., Han, B., 2009. Overexpression of a NAC transcription factor enhances rice drought and salt tolerance. Biochem. Biophys. Res. Commun. 379, 985–989.

Zhu, J.K., 2002. Salt and drought stress signal transduction in plants. Annu. Rev. Plant Biol. 53, 247–273.

CHAPTER 5

CRISPR/Cas9-based genome editing, with focus on transcription factors, for plant improvement

Seema Pradhan[a], Chandra Kant[b], and Vimal Pandey[c]
[a]Institute of Life Sciences, Bhubaneswar, India
[b]Department of Botany, D S College, Aligarh, India
[c]National Institute of Plant Genome Research, New Delhi, India

1. Introduction

The earth came into being about 4.6 billion years ago and according to paleontological evidence, life originated around 3.5 billion years ago (Yin et al., 2002; Pearce et al., 2018), only after the harsh environment of our planet became conducive to support life forms. Along with the arrival of life forms came the process of evolution which ensured that only those organisms that adapted to their environment survived to continue their lineage (Smith, 1978). Modern evolution biologists claim that the act of adaptation alone cannot ensure the continuity of an organisms' pedigree. In order to survive as well as thrive, changes in an organism must occur rapidly and stably (Smith, 1978). The process of evolution is continuous and affects all life forms according to their environment. The millions of years of this phenomenon have resulted in all living things that we see today, which can be broadly categorized into flora (plants) and fauna (animals), except for few like viruses. One very evident difference between the plant and the animal kingdom is mobility. The ability of animals to move away from unfavorable or toward favorable conditions allows them a distinct advantage over the sessile plants. Therefore, plants developed extensive and intricate mechanisms to combat conditions of stress. Plants constantly face a number of biotic and abiotic stresses which may hamper their growth and lead to plant death depending on the severity of the stress.

Scientists now estimate that the first land plants appeared on the Earth about 500 million years ago and that they were simple, without very

Transcription Factors for Abiotic Stress Tolerance in Plants
https://doi.org/10.1016/B978-0-12-819334-1.00005-8

well-defined root and shoot systems, e.g., liverworts (Morris et al., 2018). Over time, they accumulated changes to develop into the complex organisms they are today. Their journey of diversification has been widely studied by paleobotanists to predict future adaptations and how they might be beneficial to the survival of mankind. Although early biologists considered only the phenotypic characters to define the characteristics of a plant, the techniques of molecular biology are the most widely used tools to study these changes at molecular level. The term "molecular" is inclusive of the three molecules defined in the central dogma for biology, i.e., DNA, RNA, and proteins. Although the proteins are responsible for the phenotype of plants, it is the DNA that is transcribed into these proteins. The process of transcription leads to the synthesis of mRNA from DNA and this mRNA is then translated into protein. Therefore, any change in the DNA may also affect the consequent protein, thereby leading to a change in the appearance of the plant. Hence, it is necessary to understand the various mechanisms by which the heritable changes occur in the DNA and how they can be manipulated to produce improved varieties of useful plants.

2. Abiotic stresses affecting plant growth and development

Abiotic stress refers to the environmental conditions that adversely affect growth and development of organisms. The effect is more pronounced in plants as being sessile in nature, they cannot move away from the source of stress. A report by Boyer (1982) indicated that hostile environmental conditions may limit crop production by as much as 70%. Various types of environmental conditions can prove to be detrimental to plant survival. One of the most widespread abiotic stresses encountered by plants world over is drought or unavailability of water (Boyer, 1982; Araus et al., 2002). It leads to stomatal closure and limits gas exchange. A more extensive loss of water leads to desiccation, which can potentially lead to the disruption of metabolism and cell structure and eventually to the termination of enzyme catalyzed reactions (Jaleel et al., 2007). Many plants have developed mechanisms to evade, resist, or tolerate drought stress. Many plants escape drought by flowering early and completing their life cycle faster (Shavrukov et al., 2017). A number of mechanisms for drought stress tolerance involve hormones like ABA and ethylene (Sah et al., 2016; Goda et al., 2008). Salinity stress affects plants in a similar way where the high concentration of salt in the soil creates an osmotic imbalance causing loss of water from the plant through roots, thereby leading to drought stress-like conditions (Hu and Schmidhalter, 2005). A number of biochemical pathways work in a concerted effort to facilitate retention and/or acquisition of water, protect chloroplast functions, and maintain ion homeostasis (Parida and Das, 2005). Over the years, a number of studies have reported various molecules that can help maintain the osmolality to protect plants against severely saline conditions. These include glycine-betaine and proline (Ashraf and Foolad, 2007), nitrogen-containing compounds such as polyamines (Mansour, 2000), and steroidal hormones like brassinosteroids (Bajguz and Hayat, 2009). In addition, roles of various ion pumps, calcium, SOS pathways, transcription factors, mitogen-activated protein kinases, reactive oxygen species, and DEAD-box helicases have also been enumerated in salinity stress tolerance (Tuteja, 2007). Similar responses have also been recorded in response to cold stress (Eremina et al., 2016) where hormone like ABA and protein molecules like glycine betaine and proline are known to play a key role in plant protection. The presence of high concentration of heavy metals like cadmium, zinc, mercury, etc., in soil can lead to hyperaccumulation of these ions in plants, consequently affecting their nutrient value (Khan et al., 2015).

The mechanisms of overcoming abiotic stress conditions in plants often overlap with each other. For example, it is often the case that conditions of high temperature or heat stress lead to excessive loss of water from the plant, thus creating drought-like conditions. Therefore, plants adopt similar mechanisms to alleviate both these stresses (Zandalinas et al., 2018). Conditions of salinity, heavy metal, and cold stress involve hormones like ABA, osmo-protectants, various transcription factors, transporters, and signal transduction pathways (Sah et al., 2016; Ashraf and Foolad, 2007; Eremina et al., 2016; Khan et al., 2015). Thus, it is natural that there is extensive cross talk between the mechanisms of response to various abiotic stresses (Gupta et al., 2016; Nguyen et al., 2016). The most widely encountered effect of any stress on a plant is the production of reactive oxygen species (ROS) that not only function as signal transduction molecules, but can also cause extensive cellular damage and inhibit photosynthesis (Zandalinas et al., 2018). To prevent damage, ROS are normally removed by antioxidant machinery, which can be, in turn, impaired by the effect of stresses themselves (Baxter et al., 2013). Plants have developed a number of mechanisms to tolerate high levels of ROS, which include ROS-scavenging enzymes of plants such as superoxide dismutase (SOD), ascorbate peroxidase (APX), catalase (CAT), glutathione peroxidase (GPX), monodehydroascorbate reductase (MDHAR), dehydroascorbate reductase

(DHAR), glutathione reductase (GR), glutathione *S*-transferase (GST), and peroxiredoxin (PRX). These antioxidant enzymes are located in different sites of plant cells and work together to detoxify ROS (You and Chan, 2015). However, it is necessary to understand the mechanism of stress tolerance/resistance at the molecular level to develop improved crop plants and combat the imminent global food crisis.

3. Transcription factors and their role in combating abiotic stress

The term "gene expression" refers to the process of synthesis of proteins based on the genetic code contained in the DNA sequence. The process begins with transcription, which is the first step of gene expression. Transcription starts with the separation of double strand of DNA and binding of RNA polymerase to the promoter region of now single-stranded DNA to synthesize the pre-mRNA sequence and ends at the terminator sequence. The rate of transcription is regulated by proteins called transcription factors (TFs). Transcription factors include a wide number of proteins, excluding RNA polymerase, that initiate and regulate the transcription of genes. One distinct feature of transcription factors is that they have DNA-binding domains that give them the ability to bind to specific sequences of DNA called enhancer or promoter sequences. Some transcription factors bind to a DNA promoter sequence near the transcription start site and help form the transcription initiation complex. Other transcription factors bind to regulatory sequences, such as enhancer sequences, and can either stimulate or repress transcription of the related gene. These regulatory sequences can be thousands of base pairs upstream or downstream from the gene being transcribed. Regulation of transcription is the most common form of gene control. The action of transcription factors allows for unique expression of each gene in different cell types and during development. They are often considered to be the master regulators of gene expression and a single transcription factor may regulate the expression of multiple genes (Gujjar et al., 2014). A number of transcription factors have been implicated in a variety of abiotic stress responses (Table 1).

Transcription factors work by binding to the *cis*-acting elements in the promoter regions of genes leading to the initiation of transcription. But the mechanism of action of TFs in abiotic stress response is quite complex. Cross talk between the different types of stresses leads to a more complicated network of concerted regulation of various pathways that lead to an effective

Table 1 TFs reported to play crucial roles in abiotic stress tolerance.

Transcription factor	Abiotic stress response regulated by the TF	References
AP2/ERF	Submergence tolerance, drought, dehydration, ROS, cold	Jisha et al. (2015), Mishra et al. (2015), Shu et al. (2016)
DREB	Drought, cold, salt, heat	Kidokoro et al. (2015)
MYB	Drought, salt	Wei et al. (2017), Roy (2016)
NAC	Drought, salt	Shen et al. (2017)
bZIP	Drought, dehydration, salt, cold	Hu et al. (2016), Agarwal et al. (2019)
bHLH	Drought, salt	Mao et al. (2017)
WRKY	Drought, heat, salt, oxidative stress	Jiang et al. (2017)
Zinc finger TFs	Drought, heat, salt, oxidative stress	Yin et al. (2017), Liu et al. (2015)

response from the plant. A more simplified idea has been succinctly reported by Khan et al. (2018) (Fig. 1). Therefore, attempts have been made to upregulate or downregulate TFs in order to improve plant response to environmental stress. The arrival of genomics era brought a number of molecular techniques which made the process of making calculated changes at the level of genome more feasible.

Conventionally, people have manipulated TFs using a number of techniques such as overexpression and silencing in order to either escalate or diminish their expression. Overexpression of various TFs in different plants has produced plants with enhanced tolerance to abiotic stresses. Overexpression of TFs like MYB in *Arabidopsis* (Jung et al., 2008), DREB in *Pennisetum glaucum* (Agarwal et al., 2010), AP2/ERF and NAC TF in rice (Jisha et al., 2015; Hu et al., 2006), WRKY in wheat (Jiang et al., 2017) are a few examples led to an improved tolerance to salt, cold, and drought stress. The process involved isolation of the desired TF gene from a suitable source (i.e., a resistant plant species), cloning it into a suitable vector and introducing it into the plant of interest using *Agrobacterium*-mediated transformation, protoplast transformation, or gene gun. Similarly, silencing of a number of TF genes has also been carried out in order to establish the role of the TFs in abiotic stress tolerance. This approach was used to establish the role of a zinc finger homeodomain (ZF-HD) TF in salt stress tolerance in tomato (Zhao et al., 2018). A similar study also established that silencing of certain TFs may enhance stress tolerance in plants (Hu et al., 2017). Viral vector-based gene silencing

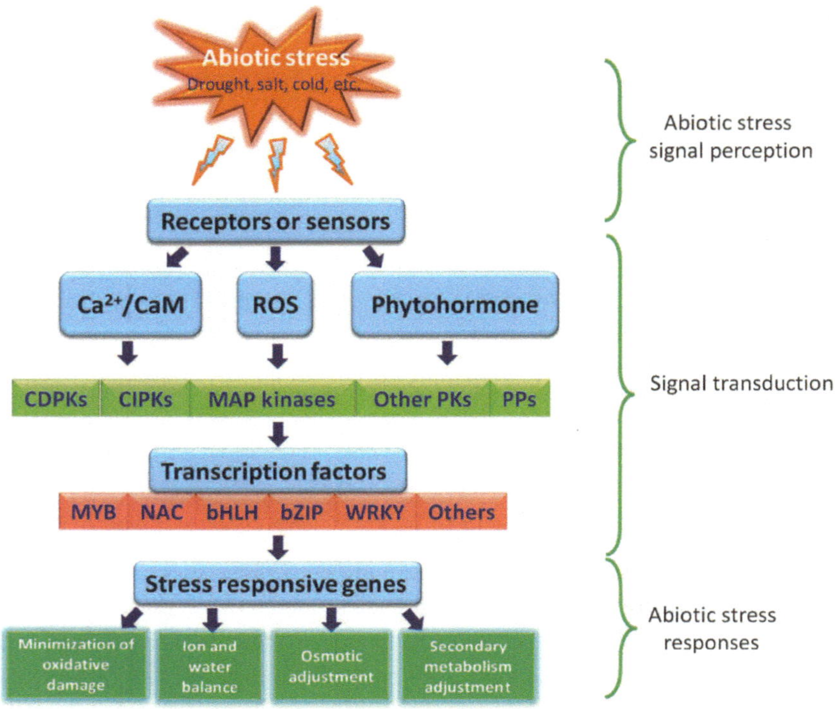

Fig. 1 A general representation of the transcription factors regulating abiotic stress response in plants as depicted by Khan et al. (2018).

technique (VIGS), and siRNA- and miRNA-induced gene silencing have proved to be an elegant tool of choice for such studies (Singh et al., 2018; Filipowicz et al., 2005). Although widely used, these approaches come with their fair share of disadvantages. Gene transfer using conventional binary vector, although very popular, is also very erratic and may not always lead to stable transformations (Mendel and Hänsch, 2017). Gene silencing using VIGS has wide applications but it is transient and cannot be stably inherited. All viral vectors may not be compatible with plants of choice and the method also does not guarantee a complete silencing of target gene (Senthil-Kumar et al., 2008). Thus there is a need for finding more efficient alternatives.

4. Genome editing: What is it and how does it work?

As discussed earlier, the genome carries the instructions for the biological activities of an organism. Therefore, stable changes in genome can bring about desired changes in the phenotype. The introduction of such stable, predetermined changes into specific sites of the genome of organisms is known as genome editing. This may involve addition, deletion, or exchange of DNA segments to bring about the desired effect. Until recently, introduction of foreign genes using *Agrobacterium*, gene gun, etc., and gene knockdown using small RNA molecules such as siRNA and miRNA were tools of choice for this purpose, but the discovery of nucleases that can be directed to cut at specific sites within the genome has revolutionized the area of gene editing.

The process of genome editing enables efficient and precise genetic modification by introducing a double-stranded break in the target sequence, which then accumulates the desired modification during the process of DNA repair. Therefore, the requisites for a successful genome-editing event are (1) recognition of the specific sequence of DNA that needs to be edited and (2) a suitable nuclease enzyme that can cut at the specified sites. The synergistic mechanism of action whereby cleavage of DNA only occurs after the recognition of specific sequence forms the backbone of modern genome-editing methods. Based on these requirements, a number of systems have been engineered to be used as programmable genome-editing machinery and are discussed here in brief (summarized in Table 2 and Fig. 2).

Table 2 Comparison of key features of the programmable genome editing methods.

Characteristics	ZFNs	TALENs	CRISPR/Cas
Type of molecule that is responsible for specificity	Zinc finger containing domains of DNA-binding proteins	TALE proteins that bind to specific promoter regions	Short complementary guide RNA sequence
Nuclease involved	FokI	FokI	CRISPR associated proteins (Cas)
Effect on nontarget site	Moderate	Low	Moderate
Ease of multiplexing	Low	Low	High
Feasibility of experiment setup	Labor-intensive and complicated	Labor-intensive and requires substantial amount of cloning	Comparatively easier and requires simple oligodesign and cloning

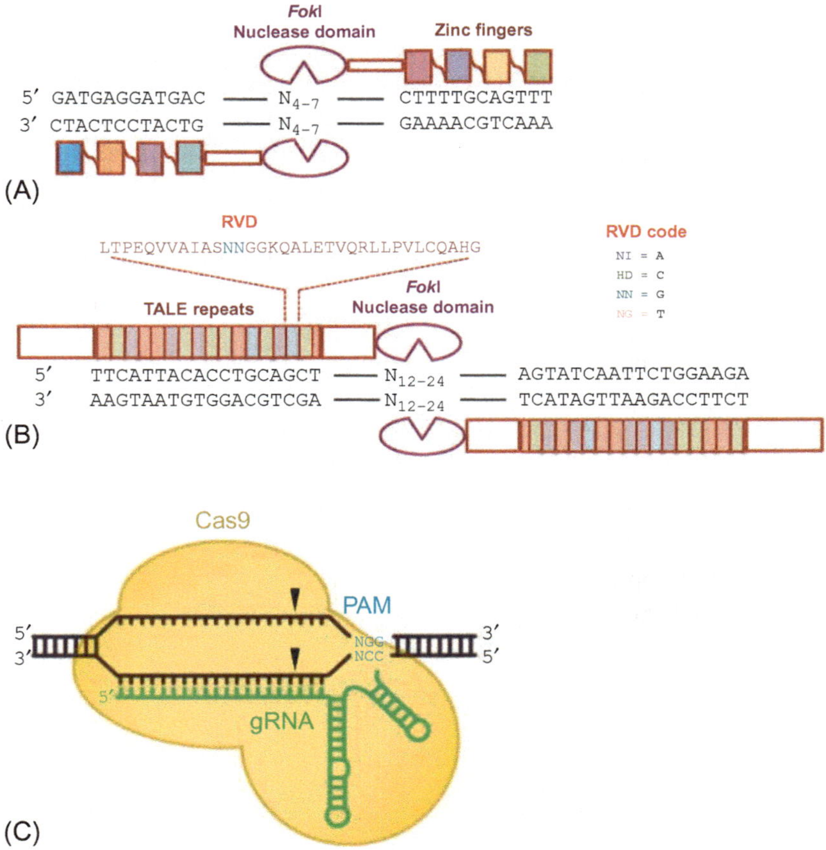

Fig. 2 The figure has been sourced as it is from the publication by Pattanayak et al. (2014). (A) A ZFN monomer is a fusion of a *FokI* nuclease cleavage domain *(purple)* to (typically) four adjoined zinc fingers each targeting three base pairs for a total of 12 base pairs recognized. Two different ZFNs bind their corresponding half-sites, allowing *FokI* dimerization and DNA cleavage between the half-sites. (B) A TALEN monomer contains an N-terminal domain followed by an array of TALE repeats *(filled boxes)*, a C-terminal domain, and a *FokI* nuclease cleavage domain *(purple)*. The 12th and 13th amino acids (the RVD, *red*) of each TALE repeat recognize a specific DNA base pair. Two different TALENs bind their corresponding half-sites, allowing *FokI* dimerization and DNA cleavage between the half-sites. (C) Cas9 protein *(yellow)* binds to target DNA in complex with a single guide RNA (sgRNA, *green*). The *Streptococcus pyogenes* Cas9 protein and sgRNA complex recognizes the PAM sequence NGG *(blue)*. *Black triangles* indicate the cleavage points in the target DNA three bases from the PAM on both DNA strands.

4.1 Zinc finger nucleases

The zinc finger nucleases (ZFN) combine the DNA binding domain of zinc finger proteins, one of the largest classes of eukaryotic transcription factor, with the nuclease domain of *FokI* restriction enzyme (Urnov et al., 2010). The zinc finger domain of the complex consists of a tandem array of C2H2 Zn fingers, each domain consisting of about 30 amino acids and each recognizing about 3 bp of DNA (Wolfe et al., 2000). The enzyme *FokI* needs to dimerize to cleave DNA (Vanamee et al., 2001). Therefore, for ZFN to effectively restructure the genome, it is necessary that two ZFN complexes must bind in the correct orientation and with enough spacing for the dimerization to occur. The prerequisite for dimerization of *FokI* creates the necessity for two DNA binding events which in turn increases the specificity of the process. This has stimulated the development of suitable variants of ZFNs that are more specific and reduce the chances of erroneous targeting to a minimum (Miller et al., 2007). Nevertheless, designing the multifinger domains of a zinc finger nuclease for gene editing purposes often requires selection or computational approaches (Pattanayak et al., 2014). Initial studies favored the use of heterodimeric ZFNs as they were reported to be highly specific (Szczepek et al., 2007). The *FokI* cleavage domain of these ZFNs is mutated such that they only allow dimerization between different ZFN monomers (Szczepek et al., 2007). This approach has been successfully applied to mutate the *ABI4* gene in *Arabidopsis* and produce abiotic stress-tolerant plants (Osakabe et al., 2010).

4.2 Transcription activator-like effector nucleases

Transcription activator-like effector (TALE) proteins can recognize and activate specific plant promoters through a set of tandem repeats. This formed the basis for the conception of a genome-editing system consisting of chimeric nucleases called TALE nucleases (TALENs) (Jankele and Svoboda, 2014). TALENs are similar to ZFNs and comprise a nonspecific *Fok*I nuclease domain fused to a customizable DNA-binding domain. TALE repeats are naturally found in the plant pathogen *Xanthomonas* and are part of transcriptional activator proteins that lead to gene expression upon binding to specific promoter elements in the plant host cell (Kay et al., 2007). Based on the studies conducted so far, TALE repeats have 34 amino acid sequences, with each repeat being able to recognize one base pair of DNA (Deng et al., 2012). The DNA-binding specificity of each repeat is determined by two amino acids referred to as the repeat-variable di-residue (RVD) (Moscou and Bogdanove, 2009).

While TALE proteins, in general, can be designed to bind any DNA sequence of interest, studies have demonstrated that the 5′-most nucleotide base of the DNA sequence bound by a TALE protein should always be a thymidine and that a deviation from this requirement can affect the efficacy of TALENs (Lamb et al., 2013; Joung and Sander, 2013).

4.3 CRISPR/Cas

Unlike the ZFN and TALEN systems, RNA-guided Cas nucleases do not require a DNA-binding protein but rather depend on a stretch of RNA sequence complementary to the target site for specificity. The clustered regularly interspaced short palindromic repeats (CRISPR) regions and their associated nucleases (CRISPR-associated proteins or Cas) form the basis of this popular system that has emerged in the field of targeted genome editing. The CRISPR/Cas system consists of two elements: single guide RNA or sgRNA, which confers specificity, and Cas nuclease, which cleaves the target DNA (Wang et al., 2016). It is an adaptive immunity machinery present in many bacteria and works to defend the bacterial genomes through endonuclease activity against foreign DNA sequences. Generally, the process takes place in three steps. First, the fragments of invading DNA (termed protospacers) get incorporated into host CRISPR locus as spacers between the transcribed CRISPR RNA (crRNA) repeats. Second, Cas proteins are expressed, the CRISPR array containing acquired spacers is transcribed into pre-crRNA, and the pre-crRNA is cleaved and processed into mature crRNAs by Cas proteins and host factors (Deltcheva et al., 2011). Third, based on the target-specific spacer sequence present in the processed crRNA, the appropriate Cas protein recognizes the target and leads to the cleavage of foreign invading genome, thereby protecting the bacterial cell from infection. In addition, many CRISPR/Cas systems require the presence of a sequence-specific protospacer adjacent motif (PAM) that is adjacent to the crRNA target site in the invading genome (Mojica et al., 2009). The absence of PAM in host genome is a mechanism which protects the host from its endogenous CRISPR/Cas system. This system has become widely popular for the purpose of genome editing and will be discussed in details.

4.4 DNA repair to introduce the mutation in desired target

After being cleaved by any of the abovementioned mechanisms, there are methods by which the organism's system tries to repair the damage caused due to the breaking of double-stranded DNA (known as double-stranded breaks or DSBs). It is only during this process of mending the DNA that a number of mutations are introduced into the repaired site on the genome, thus bringing about the desired change or edit. Plants being sessile organisms, these processes need to be flexible and yet under very stringent regulations so as to allow them to adapt to environmental stresses. There are two major categories of DNA repair mechanisms that can restore the double-stranded structure: nonhomologous end joining (NHEJ) and homology directed repair (HDR), which mostly occurs through homologous recombination (HR). NHEJ of a nuclease-induced double-stranded break leads to the efficient introduction of variable length insertion/deletion mutations that originate at the site of the break. Thus, NHEJ-mediated repair of DSBs introduced into gene-coding sequences will often yield frameshift mutations that can lead to knockout of gene function. Alternatively, in the event of availability of a double-stranded DNA "donor template," homology-directed recombination (HDR) can be used to introduce precise nucleotide substitutions or for insertions of up to 7.6 kb at or near the site of the break (Moehle et al., 2007). The most common example of NHEJ in plants is the integration of T-DNA into plant genome (Gorbunova and Levy, 1999).

The process of DSB repair begins with DNA damage recognition by a complex network of DNA damage response mechanism and components of the cell-cycle regulation and requires modification of the nucleosomes at the damaged site (Van Attikum and Gasser, 2009). First, the signal transduction machinery of the plant relays the information about the site of DNA damage. Two phosphatidylinositol 3 kinase-like (PI3K) protein kinases, ATM and ATR, have been recognized to be primary signal transducers of DNA damage, which initiate a phosphorylation-mediated signal transduction cascade that leads to cell-cycle arrest and repair of DSBs (Manova and Gruszka, 2015). Then, the heterodimer Ku70/Ku80 recognizes this signal and forms a ring structure, which binds both DNA ends, brings them together, and prevents their degradation (Mannuss et al., 2012). This is followed by the recruitment of DNA-PKcs kinase, which promotes or restrains the access to damaged DNA ends. However, the DNA-PKcs kinase has not been identified in plants (Nishizawa-Yokoi et al., 2012). The DNA ends at the site of damage are then processed by several proteins to restore the correct phospho and hydroxyl groups at both the 5′ and 3′ DNA ends, respectively. The NHEJ pathway ends in DNA ligation, which is mediated by various proteins, such as the LigIV/XRCC4, XLF, PARP3 [Poly (ADP-ribose) polymerase 3], and the APLF protein (Rulten et al., 2011).

On the other hand, HR constitutes a minor DSB repair pathway. The most common intermediate step of the DSB repair process is resection of DNA ends and exposure of single-stranded 3′ overhangs. This was proposed as the first step for the majority of DSB repair pathways in plants and seems to be mediated by the MRN complex. Plant homologs of MRE11, RAD50, and NBS1, which form the MRN complex, have been identified (Akutsu et al., 2007).

5. Why choose CRISPR/Cas9?

In a comprehensive review about the CRISPR/Cas9 genome-editing system, Wang et al. (2016) described the current models of Cas systems that are being used for targeted genome editing. The CRISPR system was discovered in bacteria and has been modified to suit the needs of researchers. It is evident that this immune system works through the cooperation of many diverse Cas proteins. However, Cas9 dominates the genome-editing arena because of certain unique characteristics which make it the tool of choice.

5.1 It is simple

Based on differences in their components and mechanisms of action, CRISPR systems have been divided into two major classes: class 1 and class 2 (Makarova et al., 2015). The class 1 systems require a large complex of several effector proteins, but in the class 2 systems only one RNA-guided endonuclease is required to mediate cleavage of invading genetic material. The CRISPR/Cas9 system is a type II, class 2 CRISPR system and is the most widely used genome-editing system today (Wang et al., 2016). It consists of a number of components that work in synchronization to bring about desired edits. The general working mechanism of CRISPR/Cas-mediated genome editing has been discussed earlier. Type II CRISPR systems such as CRISPR/Cas9 require a noncoding trans-activating CRISPR RNA (tracrRNA), which hybridizes with the crRNA repeat sequence. This is essential for crRNA crRNA processing, Cas9 binding, and Cas9-mediated target cleavage (Jinek et al., 2012). Thus, Cas9-guided cleavage depends on two hybridization events where the crRNA recognizes the target DNA through an approximately 20-base pair (bp) Watson-Crick base-pairing region and the tracrRNA that hybridizes with crRNA (Jinek et al., 2012). For practical application, this crRNA-tracrRNA duplex can be fused into a chimeric single guide RNA (sgRNA). The sgRNA-Cas9 complex can bind to the target DNA (complementary to the sgRNA sequence) and adjacent to PAM sequence (usually having the consensus sequence NGG, where N is any nucleotide). Therefore, the Cas9 system provides ease of customization to suit the varied requirements of researchers.

5.2 It is easy to customize

The Cas9 protein contains two nuclease domains: an HNH nuclease domain that cleaves the target strand ofDNA (complementary to the guide RNA) and a RuvC-like nuclease domain that cleaves the nontarget strand, thus creating a double-strand break (Jinek et al., 2012). However, mutation in any one of these domains turns it into a nickase Cas9 (nCas9), which cleaves only one strand of DNA. So, a pair of nCas9 can be used like TALENs to create DSBs (Ran et al., 2013). This strategy increases the specificity of Cas9 along with providing the ease of using sgRNA. Mutating both nuclease domains of Cas9 generates dCas9, which lacks nuclease activity but retains its RNA-guided DNA-binding activity. This allows dCas9 to be fused to other effectors to mediate site-specific genetic and epigenetic regulation without cleaving the target DNA, as well as specific DNA binding for several other applications (Gilbert et al., 2013; Hilton et al., 2015).

5.3 It is highly specific

The interaction of Cas9-sgRNA complex with the target DNA requires specific Watson-Crick pairing between regions of complementarity along with the added criteria that Cas9 must also interact with the PAM sequence adjacent to sgRNA target site (Wang et al., 2016). Also, the activation of Cas9 requires a conformational change to occur in the HNH domain which depends on proper base pairing between guide RNA and target DNA (Sternberg et al., 2015). The PAM region also contributes to increasing specificity of this system. The PAM-proximal region or the seed region is the site where the double-stranded DNA starts to separate and facilitates the formation of sgRNA-DNA heteroduplex and has a decisive role in influencing the binding and cleavage specificity of Cas9 (Sternberg et al., 2014). In addition to sequence-specific interactions, there are also a number of structure-specific interactions that play a role in the binding of Cas9 to its target sgRNA scaffold (Anders et al., 2014). All these mechanisms ensure high fidelity of the Cas9 system.

6. CRISPR/Cas9 for editing transcription factor genes in plants

Thus far, we have already established that TFs play a critical role in plant response to abiotic stress and that CRISPR/Cas9 is an efficient and feasible system for genome editing. So, applying Cas9-based DNA editing to target genes encoding relevant TFs would be a very effective method for their functional characterization. Therefore, this method has been successfully employed in animals and plants to study gene regulation. *CBF* genes are known to encode AP2/ERF transcription factors which are involved in cold acclimation (Medina et al., 2011). When plants are exposed to low temperatures, CBF genes are rapidly upregulated and the encoded proteins promote the transcription of downstream cold-responsive (COR) genes, which in turn increase the freezing tolerance of plants (Zhao et al., 2016). CRISPR/Cas9-mediated genome editing was used to generate *cbf* triple mutants in wheat where they targeted the CDS as well as intergenic regions of three CBF-coding genes (Zhao and Zhu, 2016). The study led to some important conclusions. It revealed that although the triple *cbf* mutants were very sensitive to cold stress, they did not completely lose the ability to acclimate to cold, implying that other components or pathways are also involved in cold acclimation. In another study, sgRNA sequences were designed to target *TaDREB2* and *TaERF3* genes in wheat and administered through protoplast transformation (Kim et al., 2018). This study reiterated the specificity of CRISPR/Cas9 system and also pointed out the importance of codon optimization (Zhou et al., 2016) when designing guide RNA sequences. Rice is a staple crop for many countries in the world and is regularly subjected to water and salinity stress. The CRSPR/Cas9 system was used to target salt-sensitive *OsERF922* gene (Liu et al., 2012) and a drought-response gene *OsDREB* (Hoang et al., 2016) in rice. The *CBF* gene has also been edited in tomato plants to study the effects of chilling stress (Li et al., 2018). CRISPR/Cas9 system has been used to successfully edit *SlMAPK3* gene in tomato which led to reduced expression of TFs like *SlDREB* and resulted in resistance to salt stress (Wang et al., 2017). Similarly, the *StERF* gene regulated by drought and salt stress in *Solanum tuberosum* was also edited by CRISPR/Cas9 endonuclease (Wang et al., 2015).

7. Future perspectives

CRISPR/Cas9 is a powerful machinery that has a wide range of applications due to its versatile nature. Since its conception, researchers have found different ways to use it to achieve the desired results. Continued efforts to get more varied functions by "tweaking" this machinery have produced promising results.

7.1 CRISPR interference and CRISPR activation

The process of gene regulation in higher organism is not only complex but also key to its survival. Abiotic stress tolerance in plants is an especially complicated system involving cross talk, redundancy, and layers of gene regulatory elements. Gene knockdown and gene activation can prove to be valuable methods to study gene regulation. The deactivated Cas9 (dCas9) lacks nuclease activity while retaining its ability to bind both sgRNA and targeted DNA (Wang et al., 2016). Binding of the dCas9 protein to target DNA can prevent binding of TFs to the promoter region, thereby leading to gene silencing. This method is termed as CRISPR interference and has been employed successfully in the bacterial system (Bikard et al., 2013). In a similar manner, transcriptional activators could be fused to dCas9 to specifically activate the expression of certain genes in a method termed CRISPR activation. In addition, the sgRNA can be modified and turned into a scaffold to recruit transcriptional regulators (Zalatan et al., 2015; Konermann et al., 2015). A combination of both these techniques can be employed in plants to study the effect of TFs in abiotic stress tolerance.

8. Practical considerations for designing genome-editing experiments using CRISPR/Cas9

Although a very efficient and powerful technique, CRISPR/Cas9 does come with its set of drawbacks. However, careful planning and execution of experiments can ensure that these "pitfalls" do not affect the outcome.

8.1 Optimizing the activity of Cas9 and sgRNA

The binding of Cas9 to sgRNA and target DNA creates a conformational rearrangement in the Cas9 nuclease domain. Large-scale conformational change can affect the nuclease activity of Cas9. Thus, it is very important to design the sgRNA sequence very carefully so as to get optimal activity of Cas9 nuclease. Although a 20-nt long complementary sgRNA can guide Cas9 to target gene, one study reported that sgRNAs with +85 nucleotide tracrRNA tails increased the activity of Cas9 and induced higher level of indels in vivo (Hsu et al., 2013). The same study also reported that both concatenated and interspaced two base mismatches, which occurred in the proximal region of PAM, greatly reduced Cas9 activity. The concentration of sgRNA and Cas9 are also known to play a role in Cas9 activity (Kim et al., 2014). In addition, not all sgRNAs are effective in targeting specified DNA. There are a number of criteria that must be fulfilled while designing sgRNA so as to increase the effectiveness of the system (Peng et al., 2016).

8.2 Choosing the delivery method

There are a number of approaches that can be used to deliver the sgRNA and Cas9 constructs into the plant system. Plasmids are the most commonly used vectors for the delivery of these chimeric guide RNA and Cas9 and have proved to be quite efficient in plants (Ma et al., 2015). However, some instances have been reported where the constructs were randomly inserted in the genome (Kim et al., 2014). Thus, methods like microinjection and viral vectors may also be explored.

8.3 Avoid off-target sites

The incidences of targeting off-target sites in CRISPR/Cas9 system are low to moderate, which means that such occurrences cannot be completely ruled out. So, it is advisable to use computational methods to predict off-target sites, which can be deduced from alignment algorithms using a "seed sequence" adjacent to the NGG PAM (Hsu et al., 2013; Pattanayak et al., 2013). However, this requires prior knowledge of the genome sequence of the organism. The use of paired Cas9 nickases can also reduce the incidence of directing the nuclease to off-target site (Shen et al., 2014) and so can use of truncated sgRNAs with a length of 17–18 nucleotides (Fu et al., 2014).

8.4 The method of DNA repair affects outcome

DNA repair after DSBs cut by Cas9 nuclease at desired target sites can occur by two mechanisms, NHEJ and HDR, to introduce the desired modification. Of these, NHEJ is the more frequently used repair pathway. Although NHEJ is suitable for generating indels, it cannot edit the genome precisely or in the same way as HDR. As a result of the low incidence of HDR, improving the incidence of precise insertional mutagenesis using CRISPR/Cas9 is a challenge. One way to tilt the DNA repair mechanism in favor of HDR would be to suppress the key components of NHEJ such as KU70, KU80, and DNA ligase IV (Chu et al., 2015; Basu et al., 2015).

CRISPR/Cas9 is a powerful method that has revolutionized the area of genetic manipulation. It has widespread applications and, if used correctly, can lead to vast improvements in the existing gene pool of many underutilized crops. Genome editing allows us to study gene regulation and generate improved varieties without introducing foreign genes into the plant, thereby largely avoiding the controversy surrounding transgenic plants. Like all techniques, it comes with its set of pros and cons. But a clear understanding of the mechanisms involved in plant response to abiotic stress and the mechanism of action of CRISPR/Cas9 system can allow us to apply this method successfully to achieve improvement in the area of abiotic stress tolerance by plants.

References

Agarwal, P., Agarwal, P.K., Joshi, A.J., Sopory, S.K., Reddy, M.K., 2010. Overexpression of PgDREB2A transcription factor enhances abiotic stress tolerance and activates downstream stress-responsive genes. Mol. Biol. Rep. 37 (2), 1125.

Agarwal, P., Baranwal, V.K., Khurana, P., 2019. Genome-wide analysis of bZIP transcription factors in wheat and functional characterization of a TabZIP under abiotic stress. Sci. Rep. 9 (1), 4608.

Akutsu, N., Iijima, K., Hinata, T., Tauchi, H., 2007. Characterization of the plant homolog of Nijmegen breakage syndrome 1: involvement in DNA repair and recombination. Biochem. Biophys. Res. Commun. 353 (2), 394–398.

Anders, C., Niewoehner, O., Duerst, A., Jinek, M., 2014. Structural basis of PAM-dependent target DNA recognition by the Cas9 endonuclease. Nature 513 (7519), 569.

Araus, J.L., Slafer, G.A., Reynolds, M.P., Royo, C., 2002. Plant breeding and drought in C3 cereals: what should we breed for? Ann. Bot. 89 (7), 925–940.

Ashraf, M.F.M.R., Foolad, M., 2007. Roles of glycine betaine and proline in improving plant abiotic stress resistance. Environ. Exp. Bot. 59 (2), 206–216.

Bajguz, A., Hayat, S., 2009. Effects of brassinosteroids on the plant responses to environmental stresses. Plant Physiol. Biochem. 47 (1), 1–8.

Basu, S., Aryan, A., Overcash, J.M., Samuel, G.H., Anderson, M.A., Dahlem, T.J., … Adelman, Z.N., 2015. Silencing of end-joining repair for efficient site-specific gene insertion after TALEN/CRISPR mutagenesis in Aedes aegypti. Proc. Natl. Acad. Sci. 112 (13), 4038–4043.

Baxter, A., Mittler, R., Suzuki, N., 2013. ROS as key players in plant stress signalling. J. Exp. Bot. 65 (5), 1229–1240.

Bikard, D., Jiang, W., Samai, P., Hochschild, A., Zhang, F., Marraffini, L.A., 2013. Programmable repression and activation of bacterial gene expression using an engineered CRISPR-Cas system. Nucleic Acids Res. 41 (15), 7429–7437.

Boyer, J.S., 1982. Plant productivity and environment. Science 218 (4571), 443–448.

Chu, V.T., Weber, T., Wefers, B., Wurst, W., Sander, S., Rajewsky, K., Kühn, R., 2015. Increasing the efficiency of homology-directed repair for CRISPR-Cas9-induced precise gene editing in mammalian cells. Nat. Biotechnol. 33 (5), 543.

Deltcheva, E., Chylinski, K., Sharma, C.M., Gonzales, K., Chao, Y., Pirzada, Z.A., … Charpentier, E., 2011. CRISPR RNA maturation by trans-encoded small RNA and host factor RNase III. Nature 471 (7340), 602.

Deng, D., Yan, C., Pan, X., Mahfouz, M., Wang, J., Zhu, J.K., Shi, Y., Yan, N., 2012. Structural basis for sequence-specific recognition of DNA by TAL effectors. Science 335 (6069), 720–723.

Eremina, M., Rozhon, W., Poppenberger, B., 2016. Hormonal control of cold stress responses in plants. Cell. Mol. Life Sci. 73 (4), 797–810.
Filipowicz, W., Jaskiewicz, L., Kolb, F.A., Pillai, R.S., 2005. Post-transcriptional gene silencing by siRNAs and miRNAs. Curr. Opin. Struct. Biol. 15 (3), 331–341.
Fu, Y., Sander, J.D., Reyon, D., Cascio, V.M., Joung, J.K., 2014. Improving CRISPR-Cas nuclease specificity using truncated guide RNAs. Nat. Biotechnol. 32 (3), 279.
Gilbert, L.A., Larson, M.H., Morsut, L., Liu, Z., Brar, G.A., Torres, S.E., ... Lim, W.A., 2013. CRISPR-mediated modular RNA-guided regulation of transcription in eukaryotes. Cell 154 (2), 442–451.
Goda, H., Sasaki, E., Akiyama, K., Maruyama-Nakashita, A., Nakabayashi, K., Li, W., ... Kiba, T., 2008. The AtGenExpress hormone and chemical treatment data set: experimental design, data evaluation, model data analysis and data access. Plant J. 55 (3), 526–542.
Gorbunova, V., Levy, A.A., 1999. How plants make ends meet: DNA double-strand break repair. Trends Plant Sci. 4 (7), 263–269.
Gujjar, R.S., Akhtar, M., Singh, M., 2014. Transcription factors in abiotic stress tolerance. Indian J. Plant Physiol. 19 (4), 306–316.
Gupta, P., Sharma, R., Sharma, M.K., Sharma, M.P., Satpute, G.K., Garg, S., ... Pareek, A., 2016. Signaling cross talk between biotic and abiotic stress responses in soybean. In: Abiotic and Biotic Stresses in Soybean Production. Academic Press, pp. 27–52.
Hilton, I.B., D'ippolito, A.M., Vockley, C.M., Thakore, P.I., Crawford, G.E., Reddy, T.E., Gersbach, C.A., 2015. Epigenome editing by a CRISPR-Cas9-based acetyltransferase activates genes from promoters and enhancers. Nat. Biotechnol. 33 (5), 510.
Hoang, T., Tran, T., Nguyen, T., Williams, B., Wurm, P., Bellairs, S., Mundree, S., 2016. Improvement of salinity stress tolerance in rice: challenges and opportunities. Agronomy 6 (4), 54.
Hsu, P.D., Scott, D.A., Weinstein, J.A., Ran, F.A., Konermann, S., Agarwala, V., ... Cradick, T.J., 2013. DNA targeting specificity of RNA-guided Cas9 nucleases. Nat. Biotechnol. 31 (9), 827.
Hu, Y., Schmidhalter, U., 2005. Drought and salinity: a comparison of their effects on mineral nutrition of plants. J. Plant Nutr. Soil Sci. 168 (4), 541–549.
Hu, H., Dai, M., Yao, J., Xiao, B., Li, X., Zhang, Q., Xiong, L., 2006. Overexpressing a NAM, ATAF, and CUC (NAC) transcription factor enhances drought resistance and salt tolerance in rice. Proc. Natl. Acad. Sci. 103 (35), 12987–12992.
Hu, W., Yang, H., Yan, Y., Wei, Y., Tie, W., Ding, Z., ... Li, K., 2016. Genome-wide characterization and analysis of bZIP transcription factor gene family related to abiotic stress in cassava. Sci. Rep. 6, 22783.
Hu, J., Chen, G., Yin, W., Cui, B., Yu, X., Lu, Y., Hu, Z., 2017. Silencing of SlHB2 improves drought, salt stress tolerance, and induces stress-related gene expression in tomato. J. Plant Growth Regul. 36 (3), 578–589.
Jaleel, C.A., Manivannan, P., Sankar, B., Kishorekumar, A., Gopi, R., Somasundaram, R., Panneerselvam, R., 2007. Induction of drought stress tolerance by ketoconazole in Catharanthus roseus is mediated by enhanced antioxidant potentials and secondary metabolite accumulation. Colloids Surf. B Biointerfaces 60, 201–206.
Jankele, R., Svoboda, P., 2014. TAL effectors: tools for DNA targeting. Brief Funct. Genomics 13 (5), 409–419.
Jiang, J., Ma, S., Ye, N., Jiang, M., Cao, J., Zhang, J., 2017. WRKY transcription factors in plant responses to stresses. J. Integr. Plant Biol. 59 (2), 86–101.
Jinek, M., Chylinski, K., Fonfara, I., Hauer, M., Doudna, J.A., Charpentier, E., 2012. A programmable dual-RNA-guided DNA endonuclease in adaptive bacterial immunity. Science 337 (6096), 816–821.

Jisha, V., Dampanaboina, L., Vadassery, J., Mithöfer, A., Kappara, S., Ramanan, R., 2015. Overexpression of an AP2/ERF type transcription factor OsEREBP1 confers biotic and abiotic stress tolerance in rice. PLoS One 10 (6), e0127831.
Joung, J.K., Sander, J.D., 2013. TALENs: a widely applicable technology for targeted genome editing. Nat. Rev. Mol. Cell Biol. 14 (1), 49.
Jung, C., Seo, J.S., Han, S.W., Koo, Y.J., Kim, C.H., Song, S.I., … Cheong, J.J., 2008. Overexpression of AtMYB44 enhances stomatal closure to confer abiotic stress tolerance in transgenic Arabidopsis. Plant Physiol. 146 (2), 623–635.
Kay, S., Hahn, S., Marois, E., Hause, G., Bonas, U., 2007. A bacterial effector acts as a plant transcription factor and induces a cell size regulator. Science 318 (5850), 648–651.
Khan, A., Khan, S., Khan, M.A., Qamar, Z., Waqas, M., 2015. The uptake and bioaccumulation of heavy metals by food plants, their effects on plants nutrients, and associated health risk: a review. Environ. Sci. Pollut. Res. 22 (18), 13772–13799.
Khan, S.A., Li, M.Z., Wang, S.M., Yin, H.J., 2018. Revisiting the role of plant transcription factors in the battle against abiotic stress. Int. J. Mol. Sci. 19 (6), 1634.
Kidokoro, S., Watanabe, K., Ohori, T., Moriwaki, T., Maruyama, K., Mizoi, J., … Yamaguchi-Shinozaki, K., 2015. Soybean DREB 1/CBF-type transcription factors function in heat and drought as well as cold stress-responsive gene expression. Plant J. 81 (3), 505–518.
Kim, S., Kim, D., Cho, S.W., Kim, J., Kim, J.S., 2014. Highly efficient RNA-guided genome editing in human cells via delivery of purified Cas9 ribonucleoproteins. Genome Res. 24 (6), 1012–1019.
Kim, D., Alptekin, B., Budak, H., 2018. CRISPR/Cas9 genome editing in wheat. Funct. Integr. Genomics 18 (1), 31–41.
Konermann, S., Brigham, M.D., Trevino, A.E., Joung, J., Abudayyeh, O.O., Barcena, C., … Nureki, O., 2015. Genome-scale transcriptional activation by an engineered CRISPR-Cas9 complex. Nature 517 (7536), 583.
Lamb, B.M., Mercer, A.C., Barbas III, C.F., 2013. Directed evolution of the TALE N-terminal domain for recognition of all 5′ bases. Nucleic Acids Res. 41 (21), 9779–9785.
Li, R., Zhang, L., Wang, L., Chen, L., Zhao, R., Sheng, J., Shen, L., 2018. Reduction of tomato-plant chilling tolerance by CRISPR–Cas9-mediated SlCBF1 mutagenesis. J. Agric. Food Chem. 66 (34), 9042–9051.
Liu, D., Chen, X., Liu, J., Ye, J., Guo, Z., 2012. The rice ERF transcription factor OsERF922 negatively regulates resistance to Magnaporthe oryzae and salt tolerance. J. Exp. Bot. 63 (10), 3899–3911.
Liu, Q., Wang, Z., Xu, X., Zhang, H., Li, C., 2015. Genome-wide analysis of C2H2 zinc-finger family transcription factors and their responses to abiotic stresses in poplar (*Populus trichocarpa*). PLoS One 10 (8), e0134753.
Ma, X., Zhang, Q., Zhu, Q., Liu, W., Chen, Y., Qiu, R., … Xie, Y., 2015. A robust CRISPR/Cas9 system for convenient, high-efficiency multiplex genome editing in monocot and dicot plants. Mol. Plant 8 (8), 1274–1284.
Makarova, K.S., Wolf, Y.I., Alkhnbashi, O.S., Costa, F., Shah, S.A., Saunders, S.J., … Horvath, P., 2015. An updated evolutionary classification of CRISPR–Cas systems. Nat. Rev. Microbiol. 13 (11), 722.
Mannuss, A., Trapp, O., Puchta, H., 2012. Gene regulation in response to DNA damage. Biochim. Biophys. Acta Gene Regul. Mech. 1819 (2), 154–165.
Manova, V., Gruszka, D., 2015. DNA damage and repair in plants—from models to crops. Front. Plant Sci. 6, 885.
Mansour, M.M.F., 2000. Nitrogen containing compounds and adaptation of plants to salinity stress. Biol. Plant. 43 (4), 491–500.

Mao, K., Dong, Q., Li, C., Liu, C., Ma, F., 2017. Genome wide identification and characterization of apple bHLH transcription factors and expression analysis in response to drought and salt stress. Front. Plant Sci. 8, 480.

Medina, J., Catalá, R., Salinas, J., 2011. The CBFs: three Arabidopsis transcription factors to cold acclimate. Plant Sci. 180 (1), 3–11.

Mendel, R.R., Hänsch, R., 2017. Gene transfer to higher plants. In: Molecular Methods in Plant Pathology. CRC Press (Taylor & Francis Group), pp. 189–202. https://doi.org/10.1201/9780203746523.

Miller, J.C., Holmes, M.C., Wang, J., Guschin, D.Y., Lee, Y.L., Rupniewski, I., … Gregory, P.D., 2007. An improved zinc-finger nuclease architecture for highly specific genome editing. Nat. Biotechnol. 25 (7), 778.

Mishra, S., Phukan, U.J., Tripathi, V., Singh, D.K., Luqman, S., Shukla, R.K., 2015. PsAP2 an AP2/ERF family transcription factor from *Papaver somniferum* enhances abiotic and biotic stress tolerance in transgenic tobacco. Plant Mol. Biol. 89 (1–2), 173–186.

Moehle, E.A., Rock, J.M., Lee, Y.L., Jouvenot, Y., DeKelver, R.C., Gregory, P.D., … Holmes, M.C., 2007. Targeted gene addition into a specified location in the human genome using designed zinc finger nucleases. Proc. Natl. Acad. Sci. 104 (9), 3055–3060.

Mojica, F.J., Díez-Villaseñor, C., García-Martínez, J., Almendros, C., 2009. Short motif sequences determine the targets of the prokaryotic CRISPR defence system. Microbiology 155 (3), 733–740.

Morris, J.L., Puttick, M.N., Clark, J.W., Edwards, D., Kenrick, P., Pressel, S., … Donoghue, P.C., 2018. The timescale of early land plant evolution. Proc. Natl. Acad. Sci. 115 (10), E2274–E2283.

Moscou, M.J., Bogdanove, A.J., 2009. A simple cipher governs DNA recognition by TAL effectors. Science 326 (5959), 1501.

Nguyen, D., Rieu, I., Mariani, C., van Dam, N.M., 2016. How plants handle multiple stresses: hormonal interactions underlying responses to abiotic stress and insect herbivory. Plant Mol. Biol. 91 (6), 727–740.

Nishizawa-Yokoi, A., Nonaka, S., Saika, H., Kwon, Y.I., Osakabe, K., Toki, S., 2012. Suppression of Ku70/80 or Lig4 leads to decreased stable transformation and enhanced homologous recombination in rice. New Phytol. 196 (4), 1048–1059.

Osakabe, K., Osakabe, Y., Toki, S., 2010. Site-directed mutagenesis in Arabidopsis using custom-designed zinc finger nucleases. Proc. Natl. Acad. Sci. 107 (26), 12034–12039.

Parida, A.K., Das, A.B., 2005. Salt tolerance and salinity effects on plants: a review. Ecotoxicol. Environ. Saf. 60 (3), 324–349.

Pattanayak, V., Lin, S., Guilinger, J.P., Ma, E., Doudna, J.A., Liu, D.R., 2013. High-throughput profiling of off-target DNA cleavage reveals RNA-programmed Cas9 nuclease specificity. Nat. Biotechnol. 31 (9), 839.

Pattanayak, V., Guilinger, J.P., Liu, D.R., 2014. Determining the specificities of TALENs, Cas9, and other genome-editing enzymes. In: Methods in Enzymology. vol. 546. Academic Press, pp. 47–78.

Pearce, B.K., Tupper, A.S., Pudritz, R.E., Higgs, P.G., 2018. Constraining the time interval for the origin of life on Earth. Astrobiology 18 (3), 343–364.

Peng, R., Lin, G., Li, J., 2016. Potential pitfalls of CRISPR/Cas9-mediated genome editing. FEBS J. 283 (7), 1218–1231.

Ran, F.A., Hsu, P.D., Lin, C.Y., Gootenberg, J.S., Konermann, S., Trevino, A.E., … Zhang, F., 2013. Double nicking by RNA-guided CRISPR Cas9 for enhanced genome editing specificity. Cell 154 (6), 1380–1389.

Roy, S., 2016. Function of MYB domain transcription factors in abiotic stress and epigenetic control of stress response in plant genome. Plant Signal. Behav. 11 (1), e1117723.

Rulten, S.L., Fisher, A.E., Robert, I., Zuma, M.C., Rouleau, M., Ju, L., ... Caldecott, K.W., 2011. PARP-3 and APLF function together to accelerate nonhomologous end-joining. Mol. Cell 41 (1), 33–45.
Sah, S.K., Reddy, K.R., Li, J., 2016. Abscisic acid and abiotic stress tolerance in crop plants. Front. Plant Sci. 7, 571.
Senthil-Kumar, M., Ajith, A., Uppalapati, S.R., Mysore, K.S., 2008. Virus-induced gene silencing and its applications. CAB Rev. Perspect. Agric. Vet. Sci. Nutr. Nat. Res. 3, 1–18.
Shavrukov, Y., Kurishbayev, A., Jatayev, S., Shvidchenko, V., Zotova, L., Koekemoer, F., ... Langridge, P., 2017. Early flowering as a drought escape mechanism in plants: how can it aid wheat production? Front. Plant Sci. 8, 1950.
Shen, B., Zhang, W., Zhang, J., Zhou, J., Wang, J., Chen, L., ... Skarnes, W.C., 2014. Efficient genome modification by CRISPR-Cas9 nickase with minimal off-target effects. Nat. Methods 11 (4), 399.
Shen, J., Lv, B., Luo, L., He, J., Mao, C., Xi, D., Ming, F., 2017. The NAC-type transcription factor OsNAC2 regulates ABA-dependent genes and abiotic stress tolerance in rice. Sci. Rep. 7, 40641.
Shu, Y., Liu, Y., Zhang, J., Song, L., Guo, C., 2016. Genome-wide analysis of the AP2/ERF superfamily genes and their responses to abiotic stress in *Medicago truncatula*. Front. Plant Sci. 6, 1247.
Singh, A.K., Kumar, M., Choudhary, D., Rane, J., Singh, N.P., 2018. Virus induced gene silencing approach: a potential functional genomics tool for rapid validation of function of genes associated with abiotic stress tolerance in crop plants. In: Biotechnologies of Crop Improvement. vol. 2. Springer, Cham, pp. 25–39.
Smith, J.M., 1978. Optimization theory in evolution. Annu. Rev. Ecol. Syst. 9 (1), 31–56.
Sternberg, S.H., Redding, S., Jinek, M., Greene, E.C., Doudna, J.A., 2014. DNA interrogation by the CRISPR RNA-guided endonuclease Cas9. Nature 507 (7490), 62.
Sternberg, S.H., LaFrance, B., Kaplan, M., Doudna, J.A., 2015. Conformational control of DNA target cleavage by CRISPR–Cas9. Nature 527 (7576), 110.
Szczepek, M., Brondani, V., Büchel, J., Serrano, L., Segal, D.J., Cathomen, T., 2007. Structure-based redesign of the dimerization interface reduces the toxicity of zinc-finger nucleases. Nat. Biotechnol. 25 (7), 786.
Tuteja, N., 2007. Mechanisms of high salinity tolerance in plants. Methods Enzymol. 428, 419–438.
Urnov, F.D., Rebar, E.J., Holmes, M.C., Zhang, H.S., Gregory, P.D., 2010. Genome editing with engineered zinc finger nucleases. Nat. Rev. Genet. 11 (9), 636.
Van Attikum, H., Gasser, S.M., 2009. Crosstalk between histone modifications during the DNA damage response. Trends Cell Biol. 19 (5), 207–217.
Vanamee, É.S., Santagata, S., Aggarwal, A.K., 2001. FokI requires two specific DNA sites for cleavage. J. Mol. Biol. 309 (1), 69–78.
Wang, Z., Zhang, N., Zhou, X., Fan, Q., Si, H., Wang, D., 2015. Isolation and characterization of StERF transcription factor genes from potato (*Solanum tuberosum* L.). C. R. Biol. 338 (4), 219–226.
Wang, H., La Russa, M., Qi, L.S., 2016a. CRISPR/Cas9 in genome editing and beyond. Annu. Rev. Biochem. 85, 227–264.
Wang, L., Chen, L., Li, R., Zhao, R., Yang, M., Sheng, J., Shen, L., 2017. Reduced drought tolerance by CRISPR/Cas9-mediated SlMAPK3 mutagenesis in tomato plants. J. Agric. Food Chem. 65 (39), 8674–8682.
Wei, Q., Luo, Q., Wang, R., Zhang, F., He, Y., Zhang, Y., ... He, G., 2017. A wheat R2R3-type MYB transcription factor TaODORANT1 positively regulates drought and salt stress responses in transgenic tobacco plants. Front. Plant Sci. 8, 1374.

Wolfe, S.A., Nekludova, L., Pabo, C.O., 2000. DNA recognition by Cys2His2 zinc finger proteins. Annu. Rev. Biophys. Biomol. Struct. 29 (1), 183–212.
Yin, Q., Jacobsen, S.B., Yamashita, K., Blichert-Toft, J., Télouk, P., Albarede, F., 2002. A short timescale for terrestrial planet formation from Hf–W chronometry of meteorites. Nature 418 (6901), 949.
Yin, M., Wang, Y., Zhang, L., Li, J., Quan, W., Yang, L., … Chan, Z., 2017. The Arabidopsis Cys2/His2 zinc finger transcription factor ZAT18 is a positive regulator of plant tolerance to drought stress. J. Exp. Bot. 68 (11), 2991–3005.
You, J., Chan, Z., 2015. ROS regulation during abiotic stress responses in crop plants. Front. Plant Sci. 6, 1092.
Zalatan, J.G., Lee, M.E., Almeida, R., Gilbert, L.A., Whitehead, E.H., La Russa, M., … Lim, W.A., 2015. Engineering complex synthetic transcriptional programs with CRISPR RNA scaffolds. Cell 160 (1–2), 339–350.
Zandalinas, S.I., Mittler, R., Balfagón, D., Arbona, V., Gómez-Cadenas, A., 2018. Plant adaptations to the combination of drought and high temperatures. Physiol. Plant. 162 (1), 2–12.
Zhao, C., Zhu, J.K., 2016. The broad roles of CBF genes: from development to abiotic stress. Plant Signal. Behav. 11 (8), e1215794.
Zhao, C., Zhang, Z., Xie, S., Si, T., Li, Y., Zhu, J.K., 2016. Mutational evidence for the critical role of CBF transcription factors in cold acclimation in Arabidopsis. Plant Physiol. 171 (4), 2744–2759.
Zhao, T., Hu, J., Gao, Y., Wang, Z., Bao, Y., Zhang, X., … Li, J., 2018. Silencing of the SL-ZH13 transcription factor gene decreases the salt stress tolerance of tomato. J. Am. Soc. Hortic. Sci. 143 (5), 391–396.
Zhou, Z., Dang, Y., Zhou, M., Li, L., Yu, C.H., Fu, J., … Liu, Y., 2016. Codon usage is an important determinant of gene expression levels largely through its effects on transcription. Proc. Natl. Acad. Sci. 113 (41), E6117–E6125.

CHAPTER 6

Characterization of transcription factors for drought tolerance in maize

Muhammad Qudrat Ullah Farooqi[a], Muhammad Afzal[b], Amna Parveen[c], and Ma Shi Jun[d]

[a]Department of Soil Science, School of Agriculture and Environment, Faculty of Science, The University of Western Australia, Perth, WA, Australia
[b]Plant Production Department, College of Food and Agricultural Sciences, King Saud University, Riyadh, Saudi Arabia
[c]College of Pharmacy, Gachon University, Incheon, Republic of Korea
[d]Department of Applied Plant Sciences, College of Agriculture and Life Sciences, Kangwon National University, Chuncheon, Republic of Korea

1. Introduction

Maize is an important staple food in the world. It also exhibits the higher grain production for biofuel and animal feeds (Haley, 2011). There are various abiotic stresses that affect the production of maize, but the major stress types are drought, salinity, cold and heat stress, etc. Abiotic stress is a major problem to reduce crop production worldwide (Mittler, 2006). Around 20%–25% of world maize production is affected every year due to drought stress (Heisey and Edmeades, 1999). At the flowering stage, the maize crop is vulnerable to stress when pollination, kernel set, and silk growth occur. In particular, drought stress severely affects the grain filling of maize if the time period of drought extends throughout grain filling (Edmeades et al., 2000). The Maize plant has adapted various defensive mechanisms (molecular, cellular, metabolic, and physiological) to survive in a stressful environment (Bohnert et al., 1995).

The basic important step to overcome abiotic stresses in the plant is to develop good phenotyping frameworks. It is essential for farmers and breeders to control initial damage due to stress by implementing appropriate phenotyping methodologies in plants (Salekdeh et al., 2009). Liu et al. (2013) and Karaba et al. (2007) revealed that the environment is a basic factor to increase the chance of stress mechanism in the crop plant. The environmental variation can be reduced by selecting experimental sites under controlled conditions such as pots, hydroponic culture, and soil filled pipes for

Transcription Factors for Abiotic Stress Tolerance in Plants
https://doi.org/10.1016/B978-0-12-819334-1.00006-X

water drainage. However, it is difficult to obtain high yield under controlled conditions (Hu et al., 2006). For this, identification of abiotic stresses like drought phenotyping sites is very important. Molecular breeding and geographical information system (GIS) are important tools to improve the groundwork for phenotyping in the maize plant (Okono et al., 2014).

The mechanism for physiological adjustment and adaptation can help the crop plant to tolerate or avoid stress. Plants may have the capability to adapt different mechanisms to ensure their survival under stress (Mickelbart et al., 2015). The stress tolerance mechanism is caused by cellular and biochemical modification by which the plant can survive against stress conditions. In contrast, the stress avoidance mechanism may happen due to variation in physiological and morphological characters in the entire plant; due to this, the plant is unable to survive, and suppress its growth and development (Subbaiah and Sachs, 2003). Genetic diversity is an essential investigation for environmental adaptation and acclimation of model plant species. The maize plant responds to stress against the tropical environment (Vinocur and Altman, 2005). It was found that the high stress tolerance in the maize plant is greatly associated with lower plant-to-plant variation. Thus, the genetic improvement of maize occurred with low plant-to-plant variation (Tollenaar and Wu, 1999).

The identification of quantitative trait loci (QTLs) is important for the detection of genomic region associated with traits linked with stress tolerance. Quantitative trait loci are important to configure the physiological mechanism and to control heritable variability under drought stress. QTLs can be detected in specific environmental conditions or strongly expressed with various levels of environmental factors (Collins et al., 2008). Among abiotic stresses, drought is the most devastating stress in the world, and it is associated with plant root architecture (Merewitz et al., 2012). Thus, QTLs for specific traits are linked with temperature, salinity, drought, nutrient, and chemical pollutant. Crop performance and yield improvement can be analyzed in a better way with the findings of specific loci linked with drought stress under harsh environmental conditions (Rodriguez et al., 2014). Therefore, the genetic dissection of stress linked quantitative loci is an important factor to improve sustainability of crop yield under severe environmental conditions (Welcker et al., 2007).

In this book chapter, we describe different aspects of drought stress and genetic response in the maize plant. The primary objective of this chapter is to report recent advancement in drought stress response to the maize plant. Great emphasis is being given to genetic engineering and transcriptional factor activity to control drought stress in the maize plant.

2. Maize response to drought stress

The abiotic stress response to the crop plant that is associated with many biochemical and molecular mechanisms is schematically represented in Fig. 1. The plant shows various metabolic responses under abiotic stress. In severe stress, membrane permeability, photosynthetic rate, and osmotic potential have been deeply affected. The clarification of metabolic activities in response to abiotic stress is essential for the optimization of maize cultivation and breeding to obtain high yield. Various metabolites are involved in controlling maize plant growth under drought stress. The metabolic profile of drought-stressed maize plants had resembled to salt- or heat-stressed type plants. The maize plant response to different growth conditions can be considered as plasticity that allows acclimating metabolic ranges in changing conditions (Sun et al., 2016).

Maize is isohydric and reacts rapidly to develop water stress. It was suggested from previous studies that the combined heat and drought stress effect was greater than the effect of each stress individually (Araus et al., 2008). Factors affecting the stress mechanism in plants are important to investigate within field sites homogeneity. For example, in drought and water stress experiments, the soil depth and water holding capacity difference can affect the stress imposition across the field (Farooqi et al., 2012; Masuka et al., 2012). Sandy soil has a much lower water holding capacity as compared to clayey soil. The amount of sand, silt, and clay is important to determine the water holding capacity and water penetration characteristics of soils. The increase in the resistance of soil penetration is determined by soil texture during water stress. Thus, genotypes with high drought resistance have greater access to uptake of soil water to maintain stomatal conductance in leaves (Mullan and Reynolds, 2010). Drought stress induced large developmental splicing changes in ear and leaf but few in the tassels of the maize plant.

Under water deficit conditions, abscisic acid (ABA) coming through chemical signals plays a central role in activating the defensive mechanism. ABA regulates gene expression including late embryogenesis abundant protein (LEA) in response to drought tolerance in the plant. Excess of water, such as waterlogging or flooding to the rhizosphere, also causes available oxygen reduction (Hypoxia, Anoxia) (Kizis and Pages, 2002; Scanlon et al., 1997). Under waterlogging conditions, ATP has to be produced by fermentation, resulting in acidification and the accumulation of toxic products (Van Dongen et al., 2009). It was demonstrated that ABA is dependent on transcriptional regulation and responsible for the activation of branched

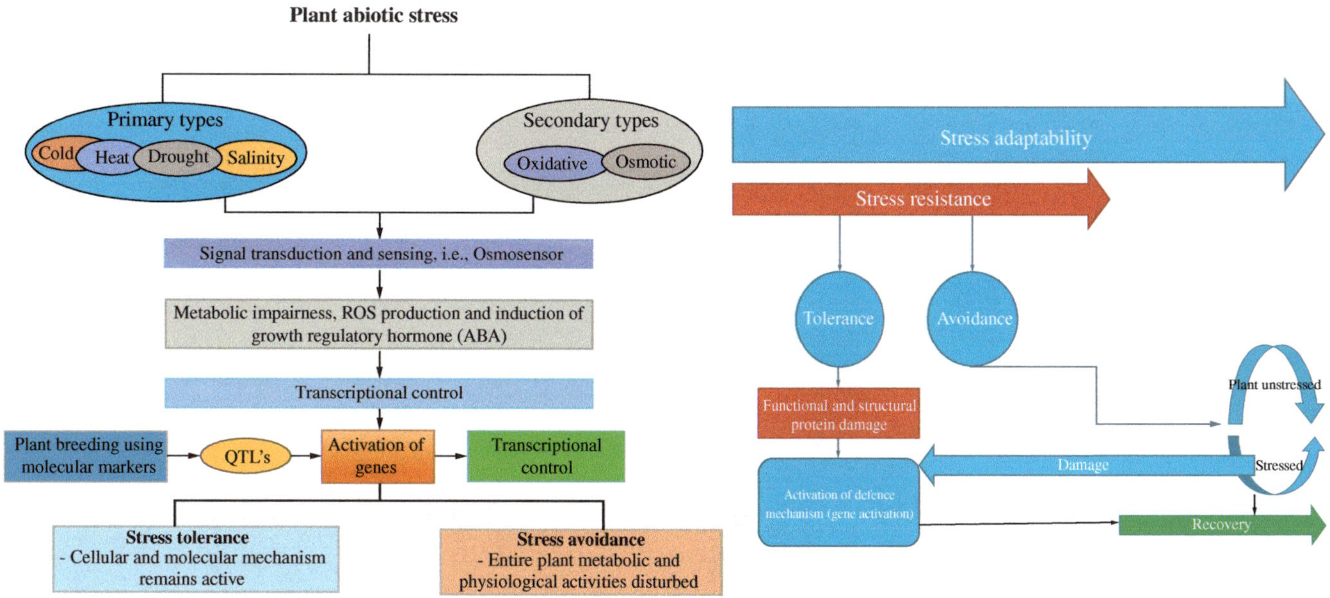

Fig. 1 Schematic diagrams of abiotic stress response in crop plant.

chain amino acids, polyamine, and proline biosynthesis. Salinity stress causes abscisic acid transportation to guard cells to close stomata that results further in photo-inhibition and oxidative stress. In the maize plant, tissue-specific difference was observed in roots and shoots due to salt stress (Krasensky and Jonak, 2012; Verslues et al., 2006).

3. Engineering of drought stress by gene overexpression

Genetic engineering is a faster way to insert useful genes responsive to abiotic stress tolerance than molecular breeding. Several stress-induced genes are introduced by genetic engineering in the maize plant to increase tolerance to cold, drought, and salt stresses. For example, ZmNF-YB2 is stress-induced genes in maize that respond due to physiological changes in the photosynthesis system and polysaccharide metabolism. The ZmNF-YB2 genes were expressed and showed improved drought tolerance in elite maize inbred lines as compared to wild-type plants (Nelson et al., 2007). In another study, Zhang et al. (2011) revealed that the ABP9 gene in maize that encodes the transcription factor bZIP in the Arabidopsis plant can enhance salt and drought tolerance, including the high level of oxidative stress associated with physical exertion of training in a cold environment (Thomashow, 1999). Trehalose treatments are an important genetic engineering approach to reduce adverse salinity effect in the maize seedling. Thus, the maize plant has transformed with a high level of trehalose with the trehalose phosphate synthase gene to improve drought and osmotic stress tolerance (Zeid, 2009).

The maize transgenic lines development is important to enhance water use efficiency and photosynthetic capacity for higher yield under abiotic stress. In this regard, the first ever transgenic plant with abiotic stress tolerance was maize (*Zea mays* MON87460) in 2013 from the United States (Liang et al., 2014). The genetic engineering tools for the development of transgenes like agrobacterium transformation and genetic modification (GM) are important to overcome drought stress in maize plant. The agrobacterium transformation of climate sensitive maize is state-of-the-art in transgenic research. It was reported in a few maize genotypes, especially in tropical areas that prefer to synthesize in a strong Agrobacterium system. In this way, the posttranscriptional gene regulation is important to improve stress tolerance (Frame et al., 2002). However, the genetically modified (GM) maize was observed to be more sensitive as compared to non-GM varieties under abiotic stress (Gullì et al., 2015).

There are a number of gene(s) which are responsible for abiotic stress tolerance in maize when transferred through genetic engineering (Table 1). The overexpression of the mitogen-activated protein kinase (MAPKK) gene in maize is responsible for drought and salinity tolerance (Shou et al., 2004). Liu et al. (1998) had studied 18 ZmDREB genes in

Table 1 Important genes and QTLs detected under drought stress conditions in maize plant.

Gene name/ID/QTLs	Definitions/function	Chr #	Source
Hsp26	Heat shock protein	1	Hao et al. (2010)
GRMZM2G005435	Ethylene-responsive element binding factor	1	Zhang et al. (2016)
SOD4	Superoxide dismutase	1	Hao et al. (2010)
kin4	Kinesin-related protein	1	Hao et al. (2010)
ppr53	Pentatricopeptide repeat 53	1	Zaidi et al. (2015)
BADH1	Betaine aldehyde dehydrogenase	1	Hao et al. (2010)
DBF1	DRE-binding factor	1	Hao et al. (2010)
cpn2	Chaperon	1	Hao et al. (2010)
vp14	Viviparous	1	Hao et al. (2010)
Cytochrome P450	Biosynthesis of endogenous compound	1	Zaidi et al. (2015)
mQTL_GY_1	Grain yield	1	Almeida et al. (2013)
grf1	General regulatory factor	2	Hao et al. (2010)
CBF3	C-repeat-binding factor	2	Hao et al. (2010)
emp2	Empty pericarp	2	Scanlon et al. (1997)
dbf3	DRE-binding protein	2	Kizis and Pages (2002)
aba1	Abscisic stress protein homolog	3	Hao et al. (2010)
hsp18f	Heat shock protein 18f	3	Hao et al. (2010)
su1	Debranching enzyme	4	Hao et al. (2010)
cld1	Cold regulated protein homolog1	4	Hao et al. (2010)
mQTL_GY_4	Grain yield	4	Almeida et al. (2013)
bip1	Binding protein homolog	5	Hao et al. (2010)
ivr2	Invertase	5	Hao et al. (2010)
bt1	ADP-glucose pyrophosphorylase	5	Hao et al. (2010)

Table 1 Important genes and QTLs detected under drought stress conditions in maize plant—cont'd

Gene name/ID/QTLs	Definitions/function	Chr #	Source
mip1	Major intrinsic membrane protein	5	Hao et al. (2010)
MYB3	Drought-responsive cis-acting element	5	Hao et al. (2010)
pcna1	Proliferating cell nuclear antigen	5	Hao et al. (2010)
mQTL_GY_5	Grain yield	5	Almeida et al. (2013)
dhn1	Dehydrin	6	Hao et al. (2010)
SOD3	Superoxide dismutase	6	Hao et al. (2010)
esp1	Embryo specific protein	6	Hao et al. (2010)
Hsp101	Heat shock protein	6	Hao et al. (2010)
mlg3	LEA protein group	6	Hao et al. (2010)
psk1	Phytosulfokine peptide precursor	7	Hao et al. (2010)
kpp1	Kinase-associated protein	7	Hao et al. (2010)
cyp6	Cytochrome P450	7	Hao et al. (2010)
psy3	Phytoene synthase	7	Hao et al. (2010)
sip1	Stress-induced protein	7	Zhang et al. (2016)
SOD2	Superoxide dismutase	7	Hao et al. (2010)
mQTL_GY_7	Grain yield	7	Almeida et al. (2013)
Hsp18c	Heat shock protein	8	Hao et al. (2010)
SOD3c	Superoxide dismutase	8	Hao et al. (2010)
p5cs	1-Pyrroline-5-carboxylate synthetase	8	Hao et al. (2010)
spp1	Sucrose phosphatase	8	Hao et al. (2010)
vgt1	Vegetative-to-generative transition	8	Hao et al. (2010)
sps1	Sucrose phosphate synthase	8	Hao et al. (2010)
bz1	Bronze	9	Hao et al. (2010)
baf1	Barren stalk fastigiate	9	Hao et al. (2010)
hyp1	Hybrid proline rich protein	9	Hao et al. (2010)
wx1	Granule bound starch synthase	9	Hao et al. (2010)
DREB1	DREB-like protein	9	Hao et al. (2010)
mQTL_GY_10	Grain yield	10	Almeida et al. (2013)

the B73 maize genome that was systematically analyzed to determine the phylogenetic relationship in a drought-responsive gene expression pattern. The transcription factor DREB plays a central role in drought tolerance where the phenotypic variability among traits is largely unknown. In total, 368 maize genotypes from tropical and temperate regions were used to assess nucleic acid variability of each ZmDREB gene with drought tolerance. A significant association of genetic variation and drought tolerance was identified in the ZmDREB2.7 promoter region at the maize seedling stage. The overexpression of ZmDREB2.7 in transgenic Arabidopsis displayed tolerance to drought stress. Furthermore, a favorable allele of ZmDREB2.7 was identified in maize tolerant varieties. Thatcher et al. (2016) used 94 RNA sequencing libraries from tassel, ear, and leaves of maize inbred line (B73) to assess drought stress effect at four developmental stages under drought stress and control conditions.

4. Drought-related transcriptional factors activity in maize

Drought stress influences the expression of genes in a very large number that may induce other genes through the transcriptional network. Stress due to environmental changes can initiate the transcriptional factor (TFs) mediated expression of a variety of genes including bZIP, NAC, and WRKY (Mittal et al., 2018). In this way, Wang et al. (2018) had conducted research to find out how the early development of maize tassels is initiated by transcriptional regulation under drought stress. Nineteen genes encoding putative transcription factors were noted in drought stress of young maize tassels. Maize plant expressed most of its metabolic activities at the RNA level, especially the regulation of exon-junction complexes. Several genes have been identified, which can help to elucidate the mechanism of transcriptional regulation related to drought stress.

Transcriptional factors (TFs) like MYB, MYC, NAC, and WRKYs are the key regulators that may act as activator or repressor in altering stress response in plants. WRKY transcriptional factor may play an important role in plant development under drought stress like in pathogen attack. Zhang et al. (2017) has assessed the WRKY genes activity in response to drought stress in maize. The drought-responsive expression profiles of ZmWRKYs were evaluated through protein-protein interaction analysis. Overall, 58 ZmWRKY genes were induced in drought stress through RNA-Seq profiling. Furthermore, ZmWRKY genes were expressed through cis-element analysis in which the putative promoter of each gene includes drought-responsive MYB-binding sites (MBSs).

Plant hormones like Abscisic acid (ABA) are the major contributor in plant development, and in signal transduction of many abiotic stresses. Kizis and Pages (2002) investigated the expression of abscisic acid-responsive gene rab17 during late embryogenesis. TFs like ABRE and DRE cis elements were involved in gene regulation induced by ABA and drought. Mittal et al. (2018) revealed that TFs play a major role in drought tolerance regulation of crop plants by encoding 1,36 genes to characterize the major drought-related TF families in maize. Furthermore, TFs were evaluated based on their structure, function, and evolution of drought-related TF families with their utility in drought breeding of maize. Drought-responsive cis-regulatory elements such as DRE1, ABREB, and ABREA are important to support photosynthetic activity and stomatal regulation in maize under drought conditions. Transcriptional factors like bZIP and GRMZM2G140355 are the potential contributor in maize leaf growth and in the stability of the maize plant under drought.

5. Transcriptional factors activity manipulation in maize

Crop plants respond to a variety of biochemical, physiological, and biochemical changes due to abiotic stress. Crop plants face severe stress due to the overproduction of reactive oxygen species. Enzymes such as ascorbate peroxidase and superoxide dismutase are associated with antioxidant capacity (Sanghera et al., 2011). Some important stress-induced protein types are late embryogenesis abundant (LEA) and heat shock protein (HSK) associated with stress linked genes in maize. LEA proteins are activated under abiotic stress. LEA proteins are important to response drought, cold, and salinity stress. Genetic engineering of molecular chaperons (i.e., Heat shock protein), genes encoding transcription factors, and proteins required for ion homeostasis are important tools to minimize abiotic stress resistance in plants (Wang et al., 2003).

Many genes required cis-, trans-regulatory features to change their expression levels during stress conditions. The pattern of gene expression and protein products can be changed due to the stressful environment. The cis-regulatory variation was observed to be more common to cold and heat stress by evaluating allele specific transcript abundance in F1 hybrids (produced by crossing of B73 with other inbred lines used in that study). Waters et al. (2017) performed RNA sequencing to assess transcriptome variability in a 14-day-old seedling of various maize inbred lines under control, cold, and heat stress conditions. Genes responded differentially between parental inbred lines due to abiotic stress. Makarevitch et al. (2015) used transposable elements in the maize genome to assess the regulatory variation of gene expression. The novel insertion of transposable elements provided a potential mechanism for obtaining a cis-regulatory effect on the genetic variation in response to abiotic stress. Certain insertions of transposable elements are often polymorphic in plant species.

Maize in a plant with many families of transposable elements is located in between genes. The transposable elements have harmful effects through the insertion of protein encoded genes to contribute regulatory variation in gene expression. In another experiment, Liu et al. (1998) used RT-qPCR (reverse transcription quantitative polymerase chain reaction) to measure gene expression in maize due to abiotic stress. Ten candidate reference genes containing two potential candidate genes with eight internal control genes and microarray data sets were used in expression level in maize due to various abiotic stresses (cold, heat, drought, and salinity). To evaluate the stability of reference gene expression, three analytical software packages (geNorm, NormFinder, and Bestkeeper) were used. The results indicated that the appropriate reference gene was used for the normalization of gene expression in maize under certain abiotic stress conditions.

6. Conclusion

Transcriptional factors play an important role in the drought tolerance regulation of crop plants. An effort has been made to characterize the drought-related TF families in the maize plant. The overexpression of genes like WRKY, DRE1, MYB, and NAC may enhance tolerance to drought stress in maize. Thus, the comparative studies of transcriptional factors can help to elucidate the drought-responsive molecular mechanism to improve stress tolerance in maize.

References

Almeida, G.D., Makumbi, D., Magorokosho, C., Nair, S., Borem, A., Ribaut, J.M., Banziger, M., Prasanna, B.M., Crossa, J., Babu, R., 2013. QTL mapping in three tropical maize populations reveals a set of constitutive adaptive genomic regions for drought tolerance. Theor. Appl. Genet. 126, 583–600.

Araus, J.L., Slafer, G.A., Royo, C., Serret, M.D., 2008. Breeding for yield potential and stress adaptation in cereals. Crit. Rev. Plant Sci. 27, 1–36.

Bohnert, H.J., Nelson, D.E., Jensen, R.G., 1995. Adaptations to environmental stresses. Plant Cell 7, 1099–1111.

Collins, N.C., Tardieu, F., Tuberosa, R., 2008. Quantitative trait loci and crop performance under abiotic stress: where do we stand? Plant Physiol. 147, 469–486.

Edmeades, G.O., Bolanos, J., Elings, A., Ribaut, J.M., Banziger, M., Westgate, M.E., 2000. The role and regulation of the anthesis silking interval in maize. In: Westgate, M.E., Boote, K.J. (Eds.), Physiology and Modeling Kernel Set in Maize. CSSA Special Publication No. 29, CSSA, Madison, WI, pp. 43–73.

Farooqi, M.Q.U., Ahmad, R., Waraich, E.A., Arfan, M., 2012. Effect of supplemental foliar applied potassium on yield and grain quality of autumn planted maize under water stress. J. Food Agric. Vet. Sci. 2, 8–12.

Frame, B.R., Shou, H., Chikwamba, R.K., Zhang, Z., Xiang, C., Fonger, T.M., Pegg, S.E.K., Li, B., Nettleton, D.S., Pei, D., Wang, K., 2002. Agrobacterium tumefaciens-mediated transformation of maize embryos using a standard binary vector system. Plant Physiol. 129, 13–22.

Gullì, M., Salvatori, E., Fusaro, L., Pellacani, C., Manes, F., Marmiroli, N., 2015. Comparison of drought stress response and gene expression between a GM maize variety and a near-isogenic non-GM variety. PLoS ONE 10, e0117073.

Haley, C., 2011. A cornucopia of maize genes. Nat. Genet. 43, 87–88.

Hao, Z., Li, X., Liu, X., Xie, C., Li, M., Zhang, D., Zhnag, S., 2010. Meta-analysis of constitutive and adaptive QTL for drought tolerance in maize. Euphytica 174, 165–177.

Heisey, P.W., Edmeades, G.O., 1999. CIMMYT 1997/98 World Maize Facts and Trends; Maize Production in Drought-Stressed Environments: Technical Options and Research Resource allocation. CIMMYT, Mexico D.F.

Hu, H., Dai, M., Yao, J., Xiao, B., Li, X., Zhang, Q., Xiong, L., 2006. Overexpression a NAM, ATAF, and CUC(NAC) transcription factor enhances drought resistance and salt tolerance in rice. Proc. Natl. Acad. Sci. U. S. A. 103, 12987–12992.

Karaba, A., Dixit, S., Greco, R., Aharonia, A., Trijamtmiko, K.R., Marsch-Martinez, N., Krishnan, A., Nataraja, K.N., Udayakumar, M., Pereira, A., 2007. Improvement of water use efficiency in rice expression of HARDY, an Arabidopsis drought and salt tolerance gene. Proc. Natl. Acad. Sci. U. S. A. 104, 15270–15275.

Kizis, D., Pages, M., 2002. Maize DRE binding proteins DBF1 and DBF2 are involved in rab17 regulation through the drought responsive element in an ABA dependent pathway. Plant J. 30, 679–689.

Krasensky, J., Jonak, C., 2012. Drought, salt, and temperature stress-induced metabolic rearrangements and regulatory networks. J. Exp. Bot. 63, 1593–1608.

Liang, C., Prins, T.W., van de Wiel, C.C.M., Kok, E.J., 2014. Safety aspects of genetically modified crops with abiotic stress tolerance. Trends Food Sci. Technol. 40, 115–122.

Liu, Q., Kasuga, M., Sakuma, Y., Abe, H., Miura, S., Yamaguchi-Shinozaki, K., Shinozaki, K., 1998. Two transcription factors, DREB1 and DREB2, with an EREBP/AP2DNA binding domain separate two cellular signal transduction pathways in drought and low temperature responsive gene expression, respectively, in Arabidopsis. Plant Cell 10, 1391–1406.

Liu, S., Wang, X., Wang, H., Xin, H., Yang, X., Yan, J., Li, J., 2013. Genome wide analysis of ZmDREB genes and their association with natural variation in drought tolerance at seedling stage of *Zea mays* L. PLoS Genet. 9, e1003790.

Makarevitch, I., Waters, A.J., West, P.T., Stitzer, M., Hirsch, C.N., Ibarra, J.R., Springer, N.M., 2015. Transposable elements contribute to activation of maize genes in response to abiotic stress. PLoS Genet. 11, e1004915.

Masuka, B., Araus, J.L., Das, B., Sonder, K., Cairns, J.E., 2012. Phenotyping for abiotic stress tolerance in maize. J. Integr. Plant Biol. 54, 238–249.

Merewitz, E.B., Belanger, F.C., Warnke, S.E., Huang, B., 2012. Identification of quantitative trait loci linked to drought tolerance in a colonial X creeping bentgrass hybrid population. Crop Sci. 52, 1891–1901.

Mickelbart, M.V., Hasegawa, P.M., Serres, J.B., 2015. Genetic mechanism of abiotic stress tolerance that translate to crop yield stability. Nat. Rev. Genet. 16, 237–251.

Mittal, S., Banduni, P., Mallikarjuna, M.G., Rao, A.R., Jain, P.A., Dash, P.K., Thirunavukkarasu, N., 2018. Structural, functional, and evolutionary characterization of major drought transcription factors families in maize. Front. Chem. 6, 177.

Mittler, R., 2006. Abiotic stress, the field environment and stress combination. Trends Plant Sci. 11, 15–19.

Mullan, D.J., Reynolds, M.P., 2010. Quantifying genetic effects of ground cover on soil water evaporation using digital imaging. Funct. Plant Biol. 37, 703–712.

Nelson, D.E., Repetti, P.P., Adams, T.R., Creelman, R.A., Wu, J., Warner, D.C., Anstrom, D.C., Bensen, R.J., Castiglioni, P.P., Donnarummo, M.G., Hinchey, B.S., Kumimoto, R.W., Maszle, D.R., Canales, R.D., Krolikowski, K.A., Dotson, S.B., Gutterson, N., Ratcliffe, O.J., Heard, J., 2007. Plant nuclear factor Y (NF-Y) B subunits confer drought tolerance and lead to improved corn yield on water-limited acres. Proc. Natl. Acad. Sci. U. S. A. 104, 16450–16455.

Okono, A., Monneveux, P., Ribaut, J.M., 2014. Facing the challenges of global agricultural today: what can we do about drought? Front. Physiol. 4, 6–7.

Rodriguez, V.M., Butron, A., Rady, M.O.A., Soengas, P., Revilla, P., 2014. Identification of quantitative trait loci involved in the response to cold stress in maize (*Zea mays* L.). Mol. Breed. 33, 363–371.

Salekdeh, G.H., Reynolds, M., Bennett, J., Boyer, J., 2009. Conceptual framework for drought phenotyping during molecular breeding. Trends Plant Sci. 14, 488–496.

Sanghera, G.S., Wani, S.H., Hussain, W., Singh, N.B., 2011. Engineering cold stress tolerance in crop plants. Curr. Genomics 12, 30–43.

Scanlon, M.J., Myers, A.M., Schneerberger, R.G., Freeling, M., 1997. The maize gene empty pericarp-2 is required for progression beyond early stages of embryogenesis. Plant J. 12, 901–909.

Shou, H., Bordallo, P., Wang, K., 2004. Expression of the Nicotiana protein kinase (NPK1) enhanced drought tolerance in transgenic maize. J. Exp. Bot. 55, 1013–1019.

Subbaiah, C.C., Sachs, M.M., 2003. Molecular and cellular adaptations of maize to flooding stress. Ann. Bot. 91, 119–127.

Sun, C.X., Li, M.Q., Gao, X.X., Liu, L.N., Wu, X.F., Zhou, J.H., 2016. Metabolic response of maize plants to multi-factorial abiotic stresses. Plant Biol. (Stuttg.) 18, 120–129.

Thatcher, S.R., Danilevskaya, O.N., Meng, X., Beatty, M., Hayes, G.Z., Harris, C., Allen, B.V., Habben, J., Li, B., 2016. Genome-wide analysis of alternative splicing during development and drought stress in maize. Plant Physiol. 170, 586–599.

Thomashow, M.F., 1999. Plant cold acclimation: freezing tolerance genes and regulatory mechanisms. Annu. Rev. Plant Physiol. Plant Mol. Biol. 50, 571–599.

Tollenaar, M., Wu, J., 1999. Yield improvement in temperature is attributable to greater stress tolerance. Crop Sci. 39, 1597–1604.

Van Dongen, J.T., Frohlich, A., Ramirez-Aguilar, S.J., Schauer, N., Fernie, A.R., Erban, A., Kopka, J., Clark, J., Langer, A., Geigenberger, P., 2009. Transcript and metabolite profiling of the adaptive response to mild decrease profiling of adaptive response to mild decreases in oxygen concentration in the roots of Arabidopsis plants. Ann. Bot. 103, 269–280.

Verslues, P.E., Agarwal, M., Katiyar-Agarwal, S., Zhu, J., Zhu, J.K., 2006. Methods and concepts in quantifying resistance to drought, salt and freezing, abiotic stresses that affect plant water status. Plant J. 45, 523–539.

Vinocur, B., Altman, A., 2005. Recent advances in engineering plant tolerance to abiotic stress: achievements and limitations. Curr. Opin. Biotechnol. 16, 123–132.
Wang, W., Vinocur, B., Altman, A., 2003. Plant responses to drought, salinity and extreme temperatures: towards genetic engineering for stress tolerance. Planta 218, 1–14.
Wang, N., Li, L., Gao, W.W., Wu, Y.B., Yong, H.J., Weng, J.F., Li, M.S., Zhang, D.G., HAO, Z.F., Li, X.H., 2018. Transcriptomes of early developing tassels under drought stress reveal differential expression of genes related to drought tolerance in maize. J. Integr. Agric. 17 (6), 1276–1288.
Waters, A.K., Makarevitch, I., Noshay, J., Burghardt, L.T., Hirsch, C.N., Hirsch, C.D., Springer, N.M., 2017. Natural variation for gene expression responses to abiotic stress in maize. Plant J. 89, 706–717.
Welcker, C., Boussuge, B., Benciveni, C., Ribaut, J.M., Tardieu, F., 2007. Are source and sink strengths genetically linked in maize plants subjected to water deficit? A QTL study of the responses of leaf growth and anthesis-silking interval to water deficit. J. Exp. Bot. 58, 339–349.
Zaidi, P.H., Rashid, Z., Vinayan, M.T., Almeida, G.D., Phagna, R.K., Babu, R., 2015. QTL mapping of agronomic waterlogging tolerance using recombinant inbred lines derived from tropical maize (Zea mays L.) germplasm. PLoS ONE 10, e0124350.
Zeid, I.M., 2009. Trehalose as osmoprotectant for maize under salinity induced stress. Res. J. Agric. Biol. Sci. 5, 613–622.
Zhang, X., Wang, L., Meng, H., Wen, H., Fan, Y., Zhao, J., 2011. Maize ABP9 enhances tolerance to multiple stresses in transgenic Arabidopsis by modulating ABA signaling and cellular levels of reactive oxygen species. Plant Mol. Biol. 75, 365–378.
Zhang, X., Warburton, M.L., Setter, T., Liu, H., Xue, Y., Yang, N., Yan, J., Xiao, Y., 2016. Genome-wide association studies of drought-related metabolic changes in maize using an enlarged SNP panel. Theor. Appl. Genet. 129, 1449–1463.
Zhang, T., Tan, D., Zhang, L., Zhnag, X., Han, Z., 2017. Phylogenetic analysis and drought-responsive expression profiles of the WRKY transcription factor family in maize. Agri Gene 3, 99–108.

CHAPTER 7

Major transcription factor families involved in salinity stress tolerance in plants

V.C. Dilukshi Fernando
Department of Biological Sciences, University of Manitoba, Winnipeg, MB, Canada

1. Introduction

Plants are exposed to environmental stresses throughout its life cycle. Salt stress severely impacts the growth and photosynthetic efficiency of plants and eventually causes death. Salinity has been a serious threat to about 75 countries including Australia, China, Egypt, India, and the United States and severely impacted food production all over the world (Szabolcs, 1994). Thus, salt tolerance mechanisms play a major role in mitigating the harmful effects of salinity with regard to crop production. These salt tolerance mechanisms highly depend on abiotic stress signaling pathways regulated by transcription factors (tfs). Tfs bind to regulatory regions upstream of the gene and upregulate expression of stress-responsive genes, thereby inducing cellular responses leading to salt tolerance. Different tfs bind to different, very specific regions on the promoter of stress-responsive genes. Tfs recognize consensus sequences on the genome and the specificity of binding onto these *cis* elements on the target genes is important for proper cellular response (Todeschini et al., 2014).

In order to address the need of successfully engineering crops to tolerate abiotic stresses, genes encoding tfs have often been selected as targets of genetic engineering (Wang et al., 2016). Most tfs are involved in multiple stress signaling pathways and therefore are potentially good candidates to confer stress tolerance to multiple stresses.

Tfs bind to specific upstream promoter elements of the gene and make transcription more efficient. Tfs only bind to specific sequences in the promoter region of the gene so that they regulate only specific genes. The role of tfs in abiotic stress tolerance is to regulate the transcription of stress-responsive genes and to restore cellular ion homeostasis. Plant hormones

Transcription Factors for Abiotic Stress Tolerance in Plants
https://doi.org/10.1016/B978-0-12-819334-1.00007-1

act as key integrators linking plant growth, development, and other stress signaling pathways (Gonzalez, 2016; Golldack et al., 2014). Hormones bind to extracellular receptors and induce a cascade of signaling events that lead to activation of transcription factors that increase expression of stress-responsive genes. The products of these genes are involved in various cellular responses including modifications in lipid signaling to maintain the structure and function of plasma membranes, detoxification of reactive oxygen species (ROS), thickening of the extracellular matrix (ECM), biosynthesis of cell walls, and regulation of guard cells and thereby maintaining the stomatal aperture. Some gene products such as dehydrin also act as dehydration tolerance proteins (Golldack et al., 2014; Fernando and Schroeder, 2016).

Abscisic acid (ABA) is a plant hormone involved in regulation of multiple stresses. ABA plays a pivotal role in maintaining plant water balance under salt and drought stress. ABA-responsive gene expression under abiotic stress conditions involves various transcription factors, protein kinases, secondary messengers, etc. (Fujita et al., 2011). However, plant stress responses are not always ABA-dependent and ABA-independent pathways have also been identified in regulation of certain stress conditions. In addition, there is lot of cross talk between stress signaling pathways and regulatory networks involved in stress signaling (reviewed in Fernando and Schroeder, 2016). In this review, we will focus only on different transcription factor families characterized in salt tolerance pathways.

2. Effects of salt stress on plants and salt stress signaling pathways

Exposure to salt stress mainly induces two types of responses in plants, namely ionic imbalance and osmotic stress due to water deprivation. Therefore, salt stress conditions induce not only salt stress tolerance but also drought tolerance. Due to these common effects of drought stress and salt stress, tfs that are involved in drought stress signaling overlap with salt stress signaling (Xiong et al., 2002; Zhu, 2002). Salt stress usually increases osmolarity in plant cells, and therefore salt tolerance mechanisms decrease sensitivity to osmolarity (Kang et al., 2002).

The effect of Na^+ has a more dramatic effect on the plant leaves and affects photosynthesis and other metabolic processes (Munns and Tester, 2008). Various mechanisms have been proposed for salt tolerance in plants such as exclusion of salt from the plant and soils, upregulation of stress tolerant genes, and accumulation of proline, which is a protective compound that maintains plasma membrane integrity and scavenges free radicals. In addition, phytohormones and antioxidants are produced in plants under stress (Wang et al., 2016; Fernando and Schroeder, 2016).

3. Major transcription factor families in stress response

Key tf families that modulate adaptive responses in stress signaling pathways can be broadly classified into major families based on their DNA-binding domains. They range from basic leucine zipper (bZIP), Apetala (AP2), basic helix-loop-helix (bHLH), and Zn finger tf families to myeloblastosis (MYB), WRKY, and ABA-binding factor (ABF) families of tfs. Specificity of tfs depends on homo and heterodimerization as well as posttranscriptional and posttranslational modifications driven by small RNA (miRNA). DNA methylation and histone modifications play a key role in regulation of gene expression in stress conditions. Reactive oxygen species signaling also affects the specificity of tfs during environmental stresses. Tfs regulate gene expression by binding to *cis*-acting elements in the promoter regions of specific genes. Therefore, tfs are regulated by multiple mechanisms in terms of its efficiency in environmental stresses (Golldack et al., 2014). This review will discuss the role of different tf families in salt stress and tolerance.

ABA-mediated stress response pathways are mainly regulated by the ABA-responsive element binding factor (AREB/ABF) family of tfs while ABA-independent pathways are regulated by the dehydration responsive element binding (DREB) factor tf family (Choi et al., 2000; Yoshida et al., 2014). In addition to these MYB, NAC and WRKY are two major tf families involved in salt stress responses as described below in detail (Golldack et al., 2014; Wang et al., 2016).

3.1 ABF transcription factors

ABFs are a subfamily of basic leucine zipper tfs (bZIP tfs described in detail in Section 3.3) that bind to conserved *cis*-acting elements called ABA-responsive elements (ABRE) (PyACGTGG/TC) in the promoter region of stress-responsive genes. Activation of downstream ABA-responsive genes therefore depends on ABF transcription factors (Sakuma et al., 2002). The ABF/AREB subfamily of tfs has highly conserved bZIP domains in the C terminal and N terminal. ABFs usually from dimers for effective binding with DNA when regulating the function of downstream gene expression.

Transcriptional activation of ABFs is mediated by ABA-dependent phosphorylation via kinases (Kim, 2006). ABFs are highly homologous to the abscisic acid binding factor 5 (ABI5) tfs. ABFs are expressed in response to ABA, salt, and water stress and are usually expressed in vegetative tissues of the plant (Busk and Pagès, 1998; Kang et al., 2002). ABF1,3,4 have somewhat overlapping functions in Arabidopsis and *ABF3* seems to be the gene that is highly upregulated in salt stress in vegetative tissues as well as in seeds (Kilian et al., 2007; Fernando et al., 2018). Studies on function of ABFs in *Thellungiella salsuginea*—a close relative of Arabidopsis with tolerance to several stress conditions—showed that ABFs were highly upregulated in salt stress and the effect was more dramatic in the roots than in the shoots. ABFs in this extremophile plant were highly homologous to *At*ABFs and ABF2,3,4 were identified as master regulators in salt stress (Vysotskii et al., 2013). *Thellungiella* was identified as a model plant to study the expression and effects of *ABF* genes.

3.2 DREB transcription factors

The core *cis*-acting sequence on to which DREB tfs bind is dehydration responsive element (DRE) found in stress-responsive genes (Sakuma et al., 2002). DREB tfs belong to the APETLA2/ethylene responsive element binding protein (AP2/EREBP) family. AP2 tfs contain a conserved APETLA2/ERF binding domain (Kizis et al., 2001; Mizoi et al., 2012). The DREB1/C-repeat binding factor (CBF) family of tfs is mostly implicated in cold stress rather than salt stress (Fernando, 2018). DREB2 induces the expression of salt and osmotic stress tolerance genes. It has been identified to play an important role in salt, drought, and heat shock response. Homologs have been identified in soybean GmDREB2 with improved stress tolerance (Chen et al., 2016). DREB2 in mung bean (*Vigna radiata* L.) was found to be highly induced by salt and drought stress. In addition, homologs of *DREB2* genes have been identified in many species in the Grass family including rice, wheat, and maize. In particular, *DREB2A* has been identified to play a role in salt stress inducing stress responses. Downstream gene products of DREB2A were found to be molecular chaperones and catabolic enzymes of toxins with potential roles in detoxification. About 373 genes were upregulated in Arabidopsis plants overexpressing *DREB2A* and dehydrin. LEA proteins were also identified among the gene products (Singh and Laxmi, 2015; Maruyama et al., 2009; Nakashima et al., 2009).

3.3 bZIP transcription factors

The basic leucine Zipper transcription factor family of tfs is a large family of tfs in plants mainly involved in the ABA signaling pathway. bZIP tfs have been classified into 11 groups and have been widely studied in Arabidopsis and rice. Group A bZIP tfs bind to ABREs and are therefore described as ABFs (Nijhawan et al., 2008; Liu et al., 2014).

bZIP tfs bind to DNA as homo and heterodimers and this heterodimerization plays a critical role in activation of tfs and their functions. In Arabidopsis, heterodimerization of bZIP tfs has been shown to be involved in combinatorial gene expression of target genes providing flexibility in transcriptional systems (Weltmeier et al., 2006).

In Arabidopsis, bZIP24 regulates growth and development in stress conditions by regulating a wide array of salinity-induced genes (Yang et al., 2009). bZIP71 in rice has been shown to enhance tolerance to multiple stresses (Liu et al., 2014). In soybean, *bZIP110* is highly induced by salt and has been implicated in salt tolerance. bZIP110 reduces uptake of Na^+ ions in soybean by increasing the uptake of other harmless ions such as Ca^{2+}. bZIP110 is found to be involved in ABA-dependent stress signaling and has some role in ion transport. Thus, it has been shown to confer salt tolerance in soybean (Xu et al., 2016).

3.4 MYB transcription factor family

MYB tfs are one of the largest families of tfs identified in plants. MYB transcription factors are characterized by a conserved MYB DNA-binding domain and have diverse functions in eukaryotes. Some of these functions include defense from biotic factors and abiotic stresses. It has been shown that transgenic plants expressing MYB tfs confer abiotic stress tolerance. MYB tfs in wheat, rice, soybean, and Arabidopsis have been identified to have a role in salt and drought tolerance by regulating stress response genes (Li et al., 2015). Wheat salinity-induced R2R3-MYB transcription factor (TaSIM) enhances salt stress tolerance. Overexpression of TaSIM in Arabidopsis has been studied in this respect to identify TaSIM as a potential candidate to confer salt stress tolerance in wheat (Yu et al., 2017).

In rice, the high-affinity potassium transporter (HKT) family of transport proteins is regulated by MyBc tfs and is shown to reduce Na^+ accumulation in leaves and thereby contribute to salt tolerance in rice (Wang et al., 2015). OsMyBc tfs are upregulated by salt stress and bind directly to OsHKT1 during salt stress. MYB genes have been studied widely in drought stress response, and in Arabidopsis, about 51% MYB genes were upregulated while 41% were downregulated in response to drought stress (reviewed in Singh and Laxmi, 2015).

3.5 WRKY transcription factor family

WRKY tfs belong to one of the largest tf families and are characterized by the WRKY amino acid sequence in the N-terminus of the highly conserved DNA-binding domain (Rushton et al., 2012). They bind to the (T)TGAC(C/T) (W box) regulatory sequence of target genes and are known to induce salt and drought responsive genes. WRKY tfs have multiple roles including plant immune response to biotic stress, protection from herbivory, enhanced tolerance to salt and drought stress. They are therefore expressed by ABA and ABA-related stress conditions. Overexpression of *WRKY45* has been shown to induce salt tolerance in rice and Arabidopsis by enhancing salt stress signaling pathways (Qiu and Yu, 2009). *WRKY93* has been shown to be induced by both ABA and salt and confers resistance to salt in wheat. Their functions probably relate to changes in osmolarity, membrane stability, and upregulating stress-responsive gene expression (Qin et al., 2015). Another study by Qin et al. (2013) shows that overexpression of wheat WRKY79 tfs in Arabidopsis produces salt tolerant plants. These transgenic plants had more lateral roots, longer primary roots, increased expression of salt stress-responsive genes, and produced more proline as an adaptation to osmotic stress induced by salinity. Wheat *WRKY79* was identified as an early salt responsive gene (Qin et al., 2013). In soybean, three *WRKY* genes (WRKY21, WRKY54, WRKY13) were identified with altered stress responses. Overexpression of *GmWRKY54* in Arabidopsis confers tolerance to salt and drought stress while *GmWRKY21* induces cold stress. Interestingly, overexpression of *GmWRKY54* and *GmWRKY21* in Arabidopsis did not show any phenotypic differences under normal conditions, although the two transcription factors seem to be involved in two difference stress response pathways (Qiu and Yu, 2009; Qin et al., 2013).

Another study showed that rice WRKY tfs are induced by salt stress and drought stress. *OsWRKY45* overexpression in Arabidopsis increases expression of stress-responsive genes and decreases sensitivity to ABA (Qiu and Yu, 2009). Moreover, *WRKY63* was shown to be a positive regulator of *AREB1/ABF2*, implicating its role in ABA-mediated stress response pathways (reviewed in Singh and Laxmi, 2015).

3.6 NAC transcription factors

NAM, ATAF1,2, CUC2 (NAC) tfs are a group of tfs limited to plants and have multiple biological and developmental functions. They are characterized by the conserved N-terminus NAC domain. NAC tfs are a very large family of tfs and some of the *NAC* gene functions are yet to be characterized (Olsen et al., 2005). In Arabidopsis, there are about 33 *NAC* genes that have been identified to be involved in stress tolerance (Jiang and Deyholos, 2006), while six NAC genes have been identified in rice. *NAC022* tf is upregulated in salt and drought stress and has been shown to have a role in stress tolerance, particularly salt tolerance in rice via the ABA-dependent pathway. Homolog of *OsNAC022* was earlier identified in Arabidopsis as AtNAC036 (Kato et al., 2010). Transgenic rice overexpressing *NAC022* showed high sensitivity to ABA, accumulated less Na^+ in roots and shoots, and also accumulated more protective compounds like proline. These studies demonstrate that *NAC* acts as a stress-responsive gene and also as a transcriptional activator in an ABA-dependent manner (Hong et al., 2016).

NAC29 tf was identified in wheat (*Triticum aestivum*) and it is upregulated upon exposure of wheat plants to salt, drought, and ABA. Overexpression of *TaNAC29* in Arabidopsis shows an ABA hypersensitive phenotype with enhanced salt and drought tolerance via activating antioxidant enzymes (Huang et al., 2015). Various other NAC tfs have been identified in other crops as a potential candidate of regulating salt stress tolerance as described in Table 1.

Table 1 Major transcription factor families in salt stress signaling (references are indicated in the text below).

Transcription factor	Crop plant tf was identified in	Induced by	Confers tolerance to
Tf families in ABA-dependent pathways			
WRKY79	Wheat	Salt, ABA	salinity and ionic tolerance
WRKY93	Wheat		
WRKY45	Rice	Salt, drought	
NAC022	Rice	Salt, drought, ABA	salt and drought tolerance
NAC29	Wheat	Salt, drought, ABA	salt and drought tolerance

Continued

Table 1 Major transcription factor families in salt stress signaling (references are indicated in the text below)—cont'd

Transcription factor	Crop plant tf was identified in	Induced by	Confers tolerance to
NAC4	Chick pea	Salt and drought	salt and drought tolerance (reviewed in Shao et al., 2015)
NAC20	Soybean	Salt and freezing tolerance	salt and drought tolerance
NAC4,5,6,10	Rice	High salinity, drought	salt and drought tolerance
bZIP24	Arabidopsis	Salt	
bZIP71	Rice	Multiple stresses	
bZIP110	Soybean	Salt	
Tf families in ABA-independent pathways			
DREB2	Arabidopsis	Salt, drought	
DREB2	Soybean	Salt, drought	

4. Conclusions and perspectives

Therefore, tfs are the master regulators of stress signaling pathways that play a pivotal role in plant salinity stress tolerance. Tfs mostly act via the ABA-dependent pathway but also can function independent of ABA, depending on the stress condition. In adverse conditions such as salinity in the soil, plants can be engineered to tolerate high saline soils by manipulating tfs. Most of the tfs that have been implicated in abiotic stress tolerance are well characterized so far. The downside of using tfs in engineering crops for stress tolerance is that sometimes it may not be coupled with increasing the grain yield in adverse conditions in food crops. For instance, overexpression of *NAC022* in rice makes the rice plants tolerant to abiotic stress but does not have any effect on restoring the grain quantity and quality of rice as in plants grown in normal conditions (Hong et al., 2016). Moreover, over-expression more often leads to phenotypic changes like dwarf plants or even more severe growth impairments (Nakashima et al., 2009). Interaction and cross talk between multiple stress signaling pathways is a hot topic in the present world, and thus tfs that have overlapping roles in multiple stresses are identified to play a major role in this regard.

References

Busk, P.K., Pagès, M., 1998. Regulation of abscisic acid-induced transcription. Plant Mol. Biol. 37 (3), 425–435.

Chen, H., Liu, L., Wang, L., Wang, S., Cheng, X., 2016. VrDREB2A, a DREB-binding transcription factor from *Vigna radiata*, increased drought and high-salt tolerance in transgenic *Arabidopsis thaliana*. J. Plant Res. 129 (2), 263–273.

Choi, H.I., Hong, J.H., Ha, J.O., Kang, J.Y., Kim, S.Y., 2000. ABFs, a family of ABA-responsive element binding factors. J. Biol. Chem. 275 (3), 1723–1730.

Fernando, V.C.D., 2018. Cross talk between cold stress response signaling pathway and other stress response pathways. In: Cold Tolerance in Plants. Springer, Cham, pp. 103–109.

Fernando, V.C.D., Schroeder, D.F., 2016. Role of ABA in Arabidopsis salt, drought, and desiccation tolerance. In: Abiotic and Biotic Stress in Plants—Recent Advances and Future Perspectives. IntechOpen.

Fernando, V.C.D., Al Khateeb, W., Belmonte, M.F., Schroeder, D.F., 2018. Role of Arabidopsis ABF1/3/4 during det1 germination in salt and osmotic stress conditions. Plant Mol. Biol. 97 (1–2), 149–163.

Fujita, Y., Fujita, M., Shinozaki, K., Yamaguchi-Shinozaki, K., 2011. ABA-mediated transcriptional regulation in response to osmotic stress in plants. J. Plant Res. 124 (4), 509–525.

Golldack, D., Li, C., Mohan, H., Probst, N., 2014. Tolerance to drought and salt stress in plants: unraveling the signaling networks. Front. Plant Sci. 5, 151.

Gonzalez, D.H., 2016. Plant Transcription Factors: Evolutionary, Structural and Functional Aspects. Academic Press, an imprint of Elsevier, London Wall.

Hong, Y., Zhang, H., Huang, L., Li, D., Song, F., 2016. Overexpression of a stress-responsive NAC transcription factor gene ONAC022 improves drought and salt tolerance in rice. Front. Plant Sci. 7, 4.

Huang, Q., Wang, Y., Li, B., Chang, J., Chen, M., Li, K., et al., 2015. TaNAC29, a NAC transcription factor from wheat, enhances salt and drought tolerance in transgenic Arabidopsis. BMC Plant Biol. 15 (1), 268.

Jiang, Y., Deyholos, M.K., 2006. Comprehensive transcriptional profiling of NaCl-stressed Arabidopsis roots reveals novel classes of responsive genes. BMC Plant Biol. 6 (1), 25.

Kang, J.Y., Choi, H.I., Im, M.Y., Kim, S.Y., 2002. Arabidopsis basic leucine zipper proteins that mediate stress-responsive abscisic acid signaling. Plant Cell 14 (2), 343–357.

Kato, H., Motomura, T., Komeda, Y., Saito, T., Kato, A., 2010. Overexpression of the NAC transcription factor family gene ANAC036 results in a dwarf phenotype in *Arabidopsis thaliana*. J. Plant Physiol. 167 (7), 571–577.

Kilian, J., Whitehead, D., Horak, J., Wanke, D., Weinl, S., Batistic, O., et al., 2007. The AtGenExpress global stress expression data set: protocols, evaluation and model data analysis of UV-B light, drought and cold stress responses. Plant J. 50 (2), 347–363.

Kim, S.Y., 2006. The role of ABF family bZIP class transcription factors in stress response. Physiol. Plant. 126, 519–527.

Kizis, D., Lumbreras, V., Pagès, M., 2001. Role of AP2/EREBP transcription factors in gene regulation during abiotic stress. FEBS Lett. 498 (2–3), 187–189.

Li, C., Ng, C.K.Y., Fan, L.M., 2015. MYB transcription factors, active players in abiotic stress signaling. Environ. Exp. Bot. 114, 80–91.

Liu, C., Mao, B., Ou, S., Wang, W., Liu, L., Wu, Y., et al., 2014. OsbZIP71, a bZIP transcription factor, confers salinity and drought tolerance in rice. Plant Mol. Biol. 84 (1–2), 19–36.

Maruyama, K., Takeda, M., Kidokoro, S., Yamada, K., Sakuma, Y., Urano, K., et al., 2009. Metabolic pathways involved in cold acclimation identified by integrated analysis of metabolites and transcripts regulated by DREB1A and DREB2A. Plant Physiol. 150, 1972–1980.

Mizoi, J., Shinozaki, K., Yamaguchi-Shinozaki, K., 2012. AP2/ERF family transcription factors in plant abiotic stress responses. Biochim. Biophys. Acta 1819 (2), 86–96.

Munns, R., Tester, M., 2008. Mechanisms of salinity tolerance. Annu. Rev. Plant Biol. 59, 651–681.

Nakashima, K., Ito, Y., Yamaguchi-Shinozaki, K., 2009. Transcriptional regulatory networks in response to abiotic stresses in Arabidopsis and grasses. Plant Physiol. 149, 88–95.

Nijhawan, A., Jain, M., Tyagi, A.K., Khurana, J.P., 2008. Genomic survey and gene expression analysis of the basic leucine zipper transcription factor family in rice. Plant Physiol. 146 (2), 333–350.

Olsen, A.N., Ernst, H.A., Leggio, L.L., Skriver, K., 2005. NAC transcription factors: structurally distinct, functionally diverse. Trends Plant Sci. 10 (2), 79–87.

Qin, Y., Tian, Y., Han, L., Yang, X., 2013. Constitutive expression of a salinity-induced wheat WRKY transcription factor enhances salinity and ionic stress tolerance in transgenic *Arabidopsis thaliana*. Biochem. Biophys. Res. Commun. 441 (2), 476–481.

Qin, Y., Tian, Y., Liu, X., 2015. A wheat salinity-induced WRKY transcription factor TaWRKY93 confers multiple abiotic stress tolerance in Arabidopsis thaliana. Biochem. Biophys. Res. Commun. 464 (2), 428–433.

Qiu, Y., Yu, D., 2009. Over-expression of the stress-induced OsWRKY45 enhances disease resistance and drought tolerance in Arabidopsis. Environ. Exp. Bot. 65 (1), 35–47.

Rushton, D.L., Tripathi, P., Rabara, R.C., Lin, J., Ringler, P., Boken, A.K., et al., 2012. WRKY transcription factors: key components in abscisic acid signalling. Plant Biotechnol. J. 10 (1), 2–11.

Sakuma, Y., Liu, Q., Dubouzet, J.G., Abe, H., Shinozaki, K., Yamaguchi-Shinozaki, K., 2002. DNA-binding specificity of the ERF/AP2 domain of Arabidopsis DREBs, transcription factors involved in dehydration-and cold-inducible gene expression. Biochem. Biophys. Res. Commun. 290 (3), 998–1009.

Shao, H., Wang, H., Tang, X., 2015. NAC transcription factors in plant multiple abiotic stress responses: progress and prospects. Front. Plant Sci. 6, 902.

Singh, D., Laxmi, A., 2015. Transcriptional regulation of drought response: a tortuous network of transcriptional factors. Front. Plant Sci. 6, 895.

Szabolcs, I., 1994. The concept of soil resilience. In: Greenland, D.J., Szabolcs, I. (Eds.), Soil Resilience and Sustainable Land Use. CAB International and Willingford, pp. 33–39.

Todeschini, A.L., Georges, A., Veitia, R.A., 2014. Transcription factors: specific DNA binding and specific gene regulation. Trends Genet. 30 (6), 211–219.

Vysotskii, D.A., de Vries-van Leeuwen, I.J., Souer, E., Babakov, A.V., de Boer, A.H., 2013. ABF transcription factors of *Thellungiella salsuginea*: structure, expression profiles and interaction with 14-3-3 regulatory proteins. Plant Signal. Behav. 8(1), e22672. https://doi.org/10.4161/psb.22672.

Wang, R., Jing, W., Xiao, L., Jin, Y., Shen, L., Zhang, W., 2015. The rice high-affinity potassium transporter1; 1 is involved in salt tolerance and regulated by an MYB-type transcription factor. Plant Physiol. 168 (3), 1076–1090.

Wang, H., Wang, H., Shao, H., Tang, X., 2016. Recent advances in utilizing transcription factors to improve plant abiotic stress tolerance by transgenic technology. Front. Plant Sci. 7, 67.

Weltmeier, F., Ehlert, A., Mayer, C.S., Dietrich, K., Wang, X., Schütze, K., et al., 2006. Combinatorial control of Arabidopsis proline dehydrogenase transcription by specific heterodimerisation of bZIP transcription factors. EMBO J. 25 (13), 3133–3143.

Xiong, L., Schumaker, K.S., Zhu, J.K., 2002. Cell signaling during cold, drought, and salt stress. Plant Cell 14 (suppl. 1), S165–S183.

Xu, Z., Ali, Z., Xu, L., He, X., Huang, Y., Yi, J., et al., 2016. The nuclear protein GmbZIP110 has transcription activation activity and plays important roles in the response to salinity stress in soybean. Sci. Rep. 6, 20366.

Yang, O., Popova, O.V., Süthoff, U., Lüking, I., Dietz, K.J., Golldack, D., 2009. The Arabidopsis basic leucine zipper transcription factor AtbZIP24 regulates complex transcriptional networks involved in abiotic stress resistance. Gene 436 (1–2), 45–55.

Yoshida, T., Mogami, J., Yamaguchi-Shinozaki, K., 2014. ABA-dependent and ABA-independent signaling in response to osmotic stress in plants. Curr. Opin. Plant Biol. 21, 133–139.

Yu, Y., Ni, Z., Chen, Q., Qu, Y., 2017. The wheat salinity-induced R2R3-MYB transcription factor TaSIM confers salt stress tolerance in Arabidopsis thaliana. Biochem. Biophys. Res. Commun. 491 (3), 642–648.

Zhu, J.K., 2002. Salt and drought stress signal transduction in plants. Annu. Rev. Plant Biol. 53 (1), 247–273.

CHAPTER 8

Regulation of inducible promoters during salinity stress in plants

Aditya Banerjee and Aryadeep Roychoudhury
Department of Biotechnology, St. Xavier's College (Autonomous), Kolkata, West Bengal, India

1. Introduction

Plants are passive organisms which have to endure a multitude of abiotic stresses like salinity, drought, extreme temperature, heavy metal toxicity, etc. by remaining attached to the substratum (Banerjee et al., 2018, 2019; Banerjee and Roychoudhury, 2018a,b,c,d,e). Among all types of abiotic stress, salinity is globally the most prevalent suboptimal condition. Such high salt content in the soil drastically reduces crop productivity and also destroys the physiological balance in susceptible plant species (Banerjee and Roychoudhury, 2018f,g). These sensitive plant species or varieties cannot maintain the ionic equilibrium during salinity due to excess influx of Na^+ and uncontrolled xylem loading (Tester and Davenport, 2003). The tolerant varieties, on the contrary, regulate salt-induced toxicity by upregulating an array of osmotic stress-responsive genes involved in transport and also antioxidant synthesis (Banerjee and Roychoudhury, 2018h,i). Such specialized genes are upregulated due to the presence of sophisticated stress-inducible promoter elements and efficient recruitment of the RNA polymerase II to these promoters during stress (Agarwal et al., 2018; Banerjee and Roychoudhury, 2019). The existence of different cis-acting elements determines the stochastic and temporal distribution of transcription factors (TFs) across a promoter. Potenza et al. (2004) discussed the existence of three types of promoters in plants: (i) constitutive (active in all tissues and in all situations), (ii) spatiotemporal (having tissue-dependent or developmental stage-responsive induction), and (iii) inducible (activated in the presence of an external chemical or physical cue). In this chapter, we have concisely discussed the recent major developments in the search for salt-inducible promoters across representative plant species.

Transcription Factors for Abiotic Stress Tolerance in Plants
https://doi.org/10.1016/B978-0-12-819334-1.00008-3

2. Halophytes as a resource for sieving out salt-responsive genes: Analysis of promoter in these genes

Halophytes exhibit salt resistance or salt tolerance with remarkable physiological flexibility (Grigore et al., 2014). The search for new stress-inducible genes revealed that the model plant, *Arabidopsis thaliana*, or any model glycophyte cannot be always used to screen novel molecular candidates. Though species like *Suaeda* and *Atriplex* have been investigated to some extent, modern salinity studies prefer *Thellungiella halophila* as the model halophyte for the abovementioned purpose (Singh et al., 2016). *Cakile maritima* is also often considered for transcript profiling experiments, whereas *Salicornia brachiata* is treated as a new resource for identifying novel salt-responsive genes (Udawat et al., 2016, 2017).

Exhaustive transcriptomic analyses revealed the presence of a more efficient transcriptional machinery in *T. halophila* compared to *Arabidopsis* because several stress-associated genes were expressed as if in a constitutive fashion in the halophyte compared to *Arabidopsis* (Tiwari et al., 2014, 2016). Promoter characterization of the *betaine aldehyde dehydrogenase* (*BADH*) gene from the halophyte *Atriplex centralasiatica* revealed the existence of two salt-inducible enhancers within −1150 to −890 and −462 to −230. A silencer region was also found to exist between −890 and −641 (Yin et al., 2002). The upstream 300-bp promoter fragment of the *BADH* gene in *Suaeda liaotungensis* exhibited a close to sixfold induction in response to 400 mM NaCl stress as compared to the control plants (Zhang et al., 2008). Sun et al. (2010) identified a 130 bp cis-acting element in the *viviparous 1* (*VP1*) gene of *T. halophila*. Tagging the promoter with the reporter *β-glucuronidase* (*GUS*) gene showed high *GUS* expression in transgenic *Arabidopsis* exposed to salinization. An almost 18-fold salt-induced increase in *GUS* expression was observed in *Arabidopsis* plants transformed with the 897 bp promoter region from *S. liaotungensis* fused with *GUS* (Li et al., 2016). Thus, partial promoters containing the necessary cis-acting elements can autonomously trigger the fused gene under a stress situation. Difference in the cis elements was observed in case of the *stress-associated protein* (*SAP*) gene from *Aeluropus littoralis* (halophyte) and *Oryza sativa* (glycophyte) (Saad et al., 2011). The regulatory regions of the promoters of the same *SAP* gene from two different species had varying specificity of regulation during salt stress (Mishra and Tanna, 2017). Tiwari et al. (2014) determined that the *ascorbate peroxidase* (*APX*) gene from *S. brachiata* possessed enhancers, repressors, and cis-regulatory element binding sites in the upstream promoter region. The presence of novel cis-acting motifs was reported in the *S. brachiata tau class glutathione-S-transferase* promoter. These cis elements regulated high expression of *GSTU* under salt stress (Tiwari et al., 2016).

3. Evidences of inducible promoter regulating ion transport in glycophytes during salt stress

The Na^+/H^+ antiporter (NHX1) is involved in compartmentalization of excess Na^+ ions into vacuoles during high salt stress. Hence, the gene encoding this transporter is induced during salinization. The reporter study with *GUS* fusion showed *NHX1* expression in all tissues of *Arabidopsis* except root tips. High expression in root hair cells was probably due to the occurrence of large vacuoles which act as reservoirs of excess Na^+ ions (Agarwal et al., 2018). This gene was upregulated in the presence of salts like NaCl, KCl, and also by the universal stress hormone, abscisic acid (ABA) (Shi and Zhu, 2002). Adler et al. (2010) showed that *NHX1* expression was concomitantly high in almost all the tissues of salt-tolerant species *Beta vulgaris*. Long-term exposure to mild stress rather than short-term exposure to severe salt shock better activated the *NHX1* gene in *B. vulgaris*. However, unlike in *Arabidopsis*, the *NHX1* in *B. vulgaris* was not significantly responsive to ABA treatments (Adler et al., 2010). Five *NHX* genes have been identified in rice, of which the *NHX1* promoter was expressed in guard cells and trichomes, whereas the *NHX5* promoter was induced in root tips and pollens (Fukuda et al., 2011). Detailed promoter analysis of *NHX* homologs in *Arabidopsis* and rice revealed significant differences in the presence of stress-inducible cis-acting elements. A complex signaling process ushers *NHX* expression in both the species. ABA-responsive element (ABRE) with the core ACGT motif was present in *OsNHX1-3* illustrating direct ABA-mediated regulation. However, an MBC/MYB TF binding site was found instead of a signature ABRE in *AtNHX1*, *AtNHX2*, and *OsNHX5*, highlighting an indirect ABA-mediated upregulation of these genes during salt stress (Fukuda et al., 2011).

Salt overly sensitive 1 (SOS1) is a known molecular target for salinity stress due to its efficient role in Na^+ transport. SOS1 is activated via the SOS2/SOS3 pathway. Tobacco plants overexpressing *SOS1* from *Salicornia brachiata* exhibited enhanced xylem loading of Na^+ ions (Yadav et al., 2012). *SOS1-GUS* fusion construct designed using the *SOS1* promoter from *S. brachiata* showed high *GUS* transcript accumulation in response to salt, but not by ABA or cold (Goyal et al., 2013). Oh et al. (2010) identified conserved motifs in *T. parvula* and *T. salsuginea SOS1* promoters, but could not find homology between the *SOS1* promoter from *Arabidopsis* and that from either of the halophytes. Feki et al. (2014) reported a higher role of the *SOS1* promoter in Na^+ exclusion in roots compared to the *NHX1* promoter. Kaur et al. (2015) reported that the *SOS2* promoter isolated from *Brassica juncea* revealed high GUS activity during ABA, salinity, and desiccation treatments in the shoots and roots. SOS promoters are known to contain ABREs as well as MYB and MYC TF-binding elements. Thus, these genes usually act in ABA-dependent fashion in most plant species (Kaur et al., 2015). Furthermore, the *SOS* promoter is regulated by other phytohormones apart from ABA as well. Gibberellic acid-responsive elements (GARE) and auxin-responsive element (Aux RR-core) sequences have also been identified in the upstream promoter stretches of *SOS* genes (Agarwal et al., 2018).

The six c-subunits belonging to the V_o domain of proton transporters, V-ATPase and V-PPase, situated in the vacuoles have proven roles during salt stress (Gao et al., 2011). Exposure to salt stress promoted *V-ATPase c subunit* expression in shoots and roots of *Pennisetum glaucum* (Tyagi et al., 2005). Tagging the promoter with *GUS* showed high reporter activity in trichomes involved in salt secretion in *P. glaucum* (Tyagi et al., 2005). The ABA-mediated expression of V-ATPase could be deciphered from the presence of ABREs and dehydration-responsive elements (DREs) (Gao et al., 2010). V-PPases have been grouped as type I AVP1 and type II AVP2 in *Arabidopsis*. *GUS* fusion experiments using *AVP1* and *T. halophila VP1* promoters showed higher *GUS* expression in case of the halophyte promoter-reporter fusion construct (Sun et al., 2010). The importance of promoters in this case is extremely important, since 75% of the sequences of *AVP1* and *ThVP1* were similar. The *VP1* promoter from the halophyte contained light-responsive elements (LREs), ABRE, heat shock-responsive element (HSE), anaerobic induction responsive element (ARE), and HD-Zip1 (linked to palisade-mesophyll cells) (Sun et al., 2010). A type II V-PPase gene was isolated from *Zea mays* which exhibited strong induction to salinity, desiccation, and cold stresses (Hou et al., 2016).

High affinity potassium transporters (*HKTs*) encode important transporters regulating Na^+ passage or even Na^+-K^+ symport. Two repeats (R1 and R2) located at a distal position in the *Arabidopsis HKT1* promoter showed affinity to a repressor or an enhancer in a condition-specific manner (Baek et al., 2011). This promoter region is also targeted by small RNAs and is differentially methylated in the roots and shoots (Baek et al., 2011). The absence of the TATA box in a near isogenic line of *Arabidopsis* resulted in a lower salt-induced expression of HKT1 compared to that in the near isogenic line, containing a TATA box in the promoter (Asins et al., 2013). The promoter sequence of *HKT1* in rice was reported to contain a MYB TF-binding domain at −76 to −68 positions. This enabled the isolation of the MYBc protein from the yeast-one-hybrid assay (Wang et al., 2015). The presence of ABRE also endorses the ABA-dependent expression of *HKT1* across the majority of glycophtes. Babgohari et al. (2014) reported the presence of a jasmonic acid (JA)-responsive element in the *HKT1* promoter of *Triticum monococcum*. The presence of a higher number of TATA and CCAAT core sequences in the promoter enabled higher *HKT1* expression in *T. monococcum* compared to *T. aestivum*. It was also determined that the presence of TATA boxes at position −30 of the *HKT1* promoter dictates higher salt responsiveness compared to the presence of TATA boxes in the rest of the promoter regions (Babgohari et al., 2014).

4. Evidences of genes harboring inducible promoters, associated with signaling and osmotic protection during salt stress

The information regarding positional occurrence of cis-acting regulatory elements has been deciphered using online softwares like PLACE and PlantCARE tools (Basu and Roychoudhury, 2014). A large number of stress-inducible motifs have been identified and tabulated for genes like *NAC1* [*no apical meristem, Petunia (**N**AM), Arabidopsis thaliana activating factor (ATAF1-2), and cup-shaped cotyledon, Arabidopsis (**C**UC2)*], *Oryza sativa basic leucine zipper 8* (*OSBZ8*), *DRE binding protein 2* (*DREBP2*), *C-repeat binding/DRE binding protein* (*CRT/DRE BP*), WRKY24, WRKY51, and WRKY71 (encoding TFs), *S-adenosylmethionine decarboxylase* (*SAMDC*: encoding the rate-limiting enzyme for polyamine biosynthesis) and *responsive to ABA 16A* [*Rab16A*: encoding a dehydrin, i.e., a type II late embryogenesis abundant protein (LEA)]. The salt-inducible elements found in these genes were ABRE-, DRE-, W-box-, MYB-binding element, HSE, and long-terminal repeats (LTR) (Basu and Roychoudhury, 2014a; Banerjee and Roychoudhury, 2015). Salt stress strongly induced the expression of these genes except *CRT/DREBP* in the leaves of salt-tolerant rice cultivar Nonabokra compared to the susceptible variety IR-29 (Basu and Roychoudhury, 2014b). Basu and Roychoudhury (2014c) created a synthetic promoter 4X DRE, containing four tandem repeats of DRE from *Arabidopsis responsive to dehydration 29A* (*Rd29A*) fused to the basal elements from the cauliflower mosaic virus 35S (CaMV35S) promoter. This synthetic promoter fused with *GUS* was introgressed into tobacco plants. Salt stress imposition induced the *GUS* gene driven by tandem repeats of DRE along with minimal promoter sequences ranging between −70 and +5 bps of CaMV35S in the transgenic tobacco lines (Basu and Roychoudhury, 2014c). In another study, 4X ABRE (consisting of four tandem repeats of ABRE from the *Em* gene) and 2X ABRC (ABA-responsive complex) (consisting of two tandem repeats of ABRE plus two repeats of the GC-rich coupling element, CE1 from the barley *HVA22* gene) was fused to the reporter *GUS* gene and introgressed in rice by particle bombardment (Ganguly et al., 2011). The transgenic plants exhibited high salt-inducible *GUS* expression for both the promoters. However, the highest *GUS* transcript accumulation was in the embryo and aleurone layer of endosperm in the 2X ABRC containing transgenics in response to salt stress and exogenous ABA treatment (Ganguly et al., 2011).

Roychoudhury and Sengupta (2009) reported the binding of a cognate trans-acting factor from the tobacco nuclear extract to the ABREs of the *Oryza sativa early embryonic* (*Osem*) promoter. *Osem* encodes a dehydrin which conserves water molecules via their positively charged amino acids and maintains cellular osmotic homeostasis during desiccation. Dehydrins also exhibit moonlighting and confer protection to the system during salt stress. A 26 kDa polypeptide was found to interact with the upstream ABRE sequences of *Osem* (Roychoudhury and Sengupta, 2009). It is already reported that salt stress upregulated the expression of the master trans-acting regulatory factor *OSBZ8* (Basu and Roychoudhury, 2014a). Roychoudhury et al. (2008) verified the binding of OSBZ8 (in its phosphorylated state) to Motif I/Motif II of *Rab16A* and Motif A-1/Motif B of *Osem*, both of which encode crucial salt-responsive proteins. Apart from OSBZ8, another unidentified 29 kDa protein was found to bind to Motif IIa (found to act as a coupling element) of *Rab16A* (Roychoudhury et al., 2008). Overexpression of *Rab16A* cDNA from salt-tolerant rice Pokkali, driven by an endogenous promoter, led to the salt-inducible expression of the gene, thereby generating salt-tolerant transgenic tobacco and rice (Roychoudhury et al., 2007; Ganguly et al., 2012). Zhang et al. (2018) analyzed the upstream 974 bp promoter sequence of the dehydrin gene *DHN* from *Ipomoea pes-caprae*. In addition, the MYB-binding site (MBS), ABRE, Skn-1 motif, and TC-rich repeats were identified. These cis-acting elements were suggested to upregulate the *IpDHN* expression in response to salt stress (Zhang et al., 2018). Analysis of the promoter region of *DHN4* from *Cynodon dactylon* revealed the presence of cis-acting elements like the TATA box,

ABRE, MBS, LTRE, and GT1 motif, which triggered its high expression during salinity in an ABA-dependent pathway (Lv et al., 2017). Histochemical studies showed that promoters of *DHN1*, *DHN2,* and *DHN3* from *Triticum aestivum* × *Thinopyrum ponticum* amphiploid fused with the *GUS* gene exhibited high reporter transcript accumulation on exposure to salinity, cold, and ABA treatments (Qin and Qin, 2016). In another study, segmental or tandem duplication was observed in 33% of all *LEA* genes in *Citrus sinensis* (Pedrosa et al., 2015). Analysis of the upstream promoter region revealed the presence of crucial cis-acting elements like ABRE, DRE/CRT, MBS, and LTRE. RNA-seq analysis illustrated high induction of these *LEA* genes in response to salinity and drought stress (Pedrosa et al., 2015). The dehydrin encoding the gene *cold binding factor 7* (*CBF7*) in *Deschampsia antarctica* was found to contain a CRT/DRE core sequence which enabled its high induction during salinity, cold, and drought stresses (Byun et al., 2015). Fusion construct of the *DHN1* promoter from the banana and reporter *GUS* gene verified high accumulation of the reporter transcript during multiple abiotic stresses like salt, drought, cold, and heavy metal toxicity (Shekhawat et al., 2011).

Polyamines (PAs) are a group of crucial compatible solutes which chiefly maintain osmotic pressure and membrane stability during salt stress in plants. *SAMDC* encoding the rate-limiting enzyme is largely regulated by a number of inducible motifs (Basu et al., 2014a). The cis-acting elements identified in the promoter of *SAMDC* were TATA-, CAAT-, ABRE-, MYB-binding element, W-box, light-responsive element (LTRE), GATA, GT1, and I-box (Basu et al., 2014b). The presence of such regulatory elements promoted high induction of *SAMDC* in rice cultivars exposed to salt stress (Basu et al., 2014b).

5. Conclusion

Toxic salt concentration in the soil has largely affected the growth and development of crops and plants. Accumulation of high amounts of salt within the tissue biomass largely disturbs the K^+/Na^+ ratio and triggers severe osmotic imbalance. As a result, a large subset of *osmotic stress-responsive* (*OR*) genes are upregulated due to the existence of salt-inducible promoters. It is clear that promoters of the same gene differ in their sequence among glycophytes and halophytes. As a result, the promoters of the *OR* genes contain a greater number of novel cis-acting elements, which allow the plants to maintain a constitutive or higher expression of the corresponding genes compared to glycophytes. Several GUS-associated studies have been highlighted to show the temporal, condition-specific, and tissue-specific activity of *OR* genes due to differential promoter activity. The genes involved in ion transport are crucial as they encode transporters involved in sequestering excess Na^+ ions into vacuoles or in their exclusion from the cells. Variable promoter sequences allow spatial expression of these genes. It was also illustrated that most of these promoters are activated in an ABA-dependent pathway during salt stress.

6. Future perspectives

The field of promoter research has made a tremendous leap in the recent years. Future perspectives involve the generation of transgenic plants carrying the particular *OR* gene tagged to a strong salt-inducible promoter possibly from a halophyte species. Another way could be to increase the number of TF-binding elements and create a synthetic promoter which would be more strongly responsive during salt stress. This would enable the system to better tackle salt stress since a strong salt-responsive promoter would help in higher transcript accumulation of the corresponding gene, which in turn would encode more amount of the protein. Such a strategy can be used to confer salt tolerance in a more regulated and efficient fashion.

Acknowledgments

Financial assistance from the Science and Engineering Research Board, Government of India through the grant [EMR/2016/004799] and Department of Higher Education, Science and Technology and Biotechnology, Government of West Bengal, through the grant [264(Sanc.)/ST/P/S&T/1G-80/2017] to Dr. Aryadeep Roychoudhury is gratefully acknowledged. The authors are thankful to the University Grants Commission (UGC), Government of India, for providing a Senior Research Fellowship to Aditya Banerjee.

References

Adler, G., Blumwald, E., Bar-Zvi, D., 2010. The sugar beet gene encoding the *sodium/proton exchanger 1* (*BvNHX1*) is regulated by a MYB transcription factor. Planta 232, 187–195.

Agarwal, P.K., Dave, A., Agarwal, P., 2018. Transcriptional regulation of salinity stress: role and spatio-temporal expressions of ion-transporter gene promoters. Biol. Plant. 62, 641–646.

Asins, M.J., Villalta, I., Aly, M.M., Olias, R., Alvarez de Morales, P., Huertas, R., Li, J., Jaime-Pérez, N., Haro, R., Raga, V., Carbonell, E.A., Belver, A., 2013. Two closely linked tomato HKT coding genes are positional candidates for the major tomato QTL involved in Na^+/K^+ homeostasis. Plant Cell Environ. 36, 1171–1191.

Babgohari, M., Ebrahimie, E., Niazi, A., 2014. *In silico* analysis of high affinity potassium transporter (HKT) isoforms in different plants. Aquat. Biosyst. 10, 9.

Baek, D., Jiang, J., Chung, J.S., Wang, B., Chen, J., Xin, Z., Shi, H., 2011. Regulated *AtHKT1* gene expression by a distal enhancer element and DNA methylation in the promoter plays an important role in salt tolerance. Plant Cell Physiol. 52, 149–161.

Banerjee, A., Roychoudhury, A., 2015. WRKY proteins: signaling and regulation of expression during abiotic stress responses. Sci. World J. 2015, 807560.

Banerjee, A., Roychoudhury, A., 2018a. Interactions of brassinosteroids with major phytohormones: antagonistic effects. J. Plant Growth Regul. 37, 1025–1032.

Banerjee, A., Roychoudhury, A., 2018b. Strigolactones: multi-level regulation of biosynthesis and diverse responses in plant abiotic stresses. Acta Physiol. Plant. 40, 86.

Banerjee, A., Roychoudhury, A., 2018c. The gymnastics of epigenomics in rice. Plant Cell Rep. 37, 25–49.

Banerjee, A., Roychoudhury, A., 2018d. Seed priming technology in the amelioration of salinity stress in plants. In: Rakshit, A., Singh, H.B. (Eds.), Advances in Seed Priming. Springer Nature, Singapore, pp. 81–93.

Banerjee, A., Roychoudhury, A., 2018e. Role of beneficial trace elements in salt stress tolerance of plants. In: Hasanuzzaman, M., Fujita, M., Oku, H., Nahar, K., Hawrylak-Nowak, B. (Eds.), Plant Nutrients and Abiotic Stress Tolerance. Springer Nature, Singapore, pp. 377–390.

Banerjee, A., Roychoudhury, A., 2018f. Small heat shock proteins: structural assembly and functional responses against heat stress in plants. In: Ahmad, P., Ahanger, M.A., Singh, V.P., Tripathi, D.K., Alam, P., Alyemeni, M.N. (Eds.), Plant Metabolites and Regulation Under Abiotic Stress. Academic Press, Elsevier, United Kingdom and USA, pp. 367–374.

Banerjee, A., Roychoudhury, A., 2018g. Effect of salinity stress on growth and physiology of medicinal plants. In: Ghorbanpour, M. et al., (Eds.), Medicinal Plants and Environmental Challenges. Springer International Publishing AG, Cham, Switzerland, pp. 177–188.

Banerjee, A., Roychoudhury, A., 2018h. Regulation of photosynthesis under salinity and drought stress. In: Singh, V.P., Singh, S., Singh, R., Prasad, S.M. (Eds.), Environment and Photosynthesis: A Future Prospect. Studium Press (India) Pvt. Ltd, pp. 134–144.

Banerjee, A., Roychoudhury, A., 2018i. Abiotic stress, generation of reactive oxygen species, and their consequences: an overview. In: Singh, V.P., Singh, S., Tripathi, D., Mohan Prasad, S., Chauhan, D.K. (Eds.), Revisiting the Role of Reactive Oxygen Species (ROS) in Plants: ROS Boon or Bane for Plants? John Wiley & Sons, Inc., USA, pp. 23–50.

Banerjee, A., Roychoudhury, A., 2019. Structural introspection of a putative fluoride transporter in plants. 3 Biotech 9, 103.

Banerjee, A., Tripathi, D.K., Roychoudhury, A., 2018. Hydrogen sulphide trapeze: environmental stress amelioration and phytohormone crosstalk. Plant Physiol. Biochem. 132, 46–53.

Banerjee, A., Tripathi, D.K., Roychoudhury, A., 2019. The karrikin 'callisthenics': can compounds derived from smoke help in stress tolerance? Physiol. Plant. 165, 290–302.

Basu, S., Roychoudhury, A., 2014a. Computational analysis of abiotic stress inducible genes and proteins from rice (*Oryza sativa* L.). Int. J. Pharm. Bio. Sci. 5, 718–735.

Basu, S., Roychoudhury, A., 2014b. Expression profiling of abiotic stress-inducible genes in response to multiple stresses in rice (*Oryza sativa* L.) varieties with contrasting level of stress tolerance. Biomed. Res. Int. 2014, 706890.

Basu, S., Roychoudhury, A., 2014c. Inducibility of dehydration responsive element (DRE)-based promoter through *gusA* expression in transgenic tobacco. Indian J. Biotechnol. 13, 172–177.

Basu, S., Roychoudhury, A., Sengupta, D.N., 2014a. Deciphering the role of various cis-acting regulatory elements in controlling *SamDC* gene expression in rice. Plant Signal. Behav. 9, e28391.

Basu, S., Roychoudhury, A., Sengupta, D.N., 2014b. Identification of trans-acting factors regulating *SamDC* expression in *Oryza sativa*. Biochem. Biophys. Res. Commun. 445, 398–403.

Byun, M.Y., Lee, J., Cui, L.H., Kang, Y., Oh, T.K., Park, H., Lee, H., Kim, W.T., 2015. Constitutive expression of *DaCBF7*, an Antarctic vascular plant *Deschampsia antarctica* CBF homolog, resulted in improved cold tolerance in transgenic rice plants. Plant Sci. 236, 61–74.

Feki, K., Brini, F., Ben Amar, S., Saibi, W., Masmoudi, K., 2014. Comparative functional analysis of two wheat Na^+/H^+ antiporter *SOS1* promoters in *Arabidopsis thaliana* under various stress conditions. J. Appl. Genet. 56, 15–26.

Fukuda, A., Nakamura, A., Hara, N., Toki, S., Tanaka, Y., 2011. Molecular and functional analysis of rice NHX-type Na^+/H^+ antiporter genes. Planta 233, 17–188.

Ganguly, M., Roychoudhury, A., Sarkar, S.N., Sengupta, D.N., Datta, S.K., Datta, K., 2011. Inducibility of three salinity/abscisic acid-regulated promoters in transgenic rice with *gusA* reporter gene. Plant Cell Rep. 30, 1617–1625.

Ganguly, M., Datta, K., Roychoudhury, A., Gayen, D., Sengupta, D.N., Datta, S.K., 2012. Overexpression of *Rab16A* gene in indica rice variety for generating enhanced salt tolerance. Plant Signal. Behav. 7, 502–509.

Gao, C., Wang, Y., Liu, G., Wang, C., Jiang, J., Yang, C., 2010. Cloning of ten peroxidase (*POD*) genes from *Tamarix hispida* and characterization of their responses to abiotic stress. Plant Mol. Biol. Report. 28, 77.

Gao, C., Wang, Y., Jiang, B., Liu, G., Yu, L., Wei, Z., Yang, C., 2011. A novel vacuolar membrane H^+-ATPase c subunit gene (*ThVHAc1*) from *Tamarix hispida* confers tolerance to several abiotic stresses in *Saccharomyces cerevisiae*. Mol. Biol. Rep. 38, 957–963.

Goyal, E., Singh, R.S., Kanika, K., 2013. Isolation and functional characterization of salt overly sensitive 1 (*SOS1*) gene promoter from *Salicornia brachiata*. Biol. Plant. 57, 465–473.

Grigore, M.N., Ivanescu, L., Toma, C., 2014. Halophytes: An Integrative Anatomical Study. Springer, New York.

Hou, J., Jiang, P., Qi, S., Zhang, K., He, Q., Xu, C., Ding, Z., Zhang, K., Li, K., 2016. Isolation and functional validation of salinity and osmotic stress inducible promoter from the maize type-II H^+-pyrophosphatase gene by deletion analysis in transgenic tobacco plants. PLoS ONE. 11, e0154041.

Kaur, C., Kumar, G., Kaur, S., Ansari, M.W., Pareek, A., Sopory, S.K., Singla-Pareek, S.L., 2015. Molecular cloning and characterization of salt overly sensitive gene promoter from *Brassica juncea* (*BjSOS2*). Mol. Biol. Rep. 42, 1139–1148.

Li, Q.-L., Xie, J.-H., Ma, X.-Q., Li, D., 2016. Molecular cloning of *Phosphoethanolamine N-methyltransferase* (*PEAMT*) gene and its promoter from the halophyte *Suaeda liaotungensis* and their response to salt stress. Acta Physiol. Plant. 38, 39.

Lv, A., Fan, N., Xie, J., Yuan, S., An, Y., Zhou, P., 2017. Expression of *CdDHN4*, a novel YSK_2-type dehydrin gene from bermudagrass, responses to drought stress through the ABA-dependent signal pathway. Front. Plant Sci. 8, 748.

Mishra, A., Tanna, B., 2017. Halophytes: potential resources for salt stress tolerance genes and promoters. Front. Plant Sci. 8, 829.

Oh, D., Dassanayake, M., Haas, J.S., Kropornika, A., Wright, C., et al., 2010. Genome structures and halophyte-specific gene expression of the extremophile *Thellungiella parvula* in comparison with *Thellungiella salsuginea* (*Thellungiella halophila*) and *Arabidopsis*. Plant Physiol. 154, 1040–1052.

Pedrosa, A.M., Martins, C.P., Goncalves, L.P., Costa, M.G., 2015. Late embryogenesis abundant (LEA) constitutes a large and diverse family of proteins involved in development and abiotic stress responses in sweet orange (*Citrus sinensis* L. Osb.). PLoS ONE. 10, e0145785.

Potenza, C., Aleman, L., Sengupta-Gopalan, C., 2004. Targeting transgene expression in research, agricultural and environmental applications. Promoters used in plant transformation. In Vitro Cell Dev. Biol. Plant 40, 1–22.

Qin, Y.X., Qin, F., 2016. Dehydrins from wheat x *Thinopyrum ponticum* amphiploid increase salinity and drought tolerance under their own inducible promoters without growth retardation. Plant Physiol. Biochem. 99, 142–149.

Roychoudhury, A., Sengupta, D.N., 2009. The promoter-elements of some abiotic stress-inducible genes from cereals interact with a nuclear protein from tobacco. Biol. Plant. 53, 583–587.

Roychoudhury, A., Roy, C., Sengupta, D.N., 2007. Transgenic tobacco plants overexpressing the heterologous *lea* gene *Rab16A* from rice during high salt and water deficit display enhanced tolerance to salinity stress. Plant Cell Rep. 26, 1839–1859.

Roychoudhury, A., Gupta, B., Sengupta, D.N., 2008. Trans-acting factor designated OSBZ8 interacts with both typical abscisic acid responsive elements as well as abscisic acid responsive element-like sequences in the vegetative tissues of indica rice cultivars. Plant Cell Rep. 27, 779–794.

Saad, R.B., Romdhan, W.B., Zouari, N., Azaza, J., Mieulet, D., Verdeil, J.L., et al., 2011. Promoter of the *AlSAP* gene from the halophyte grass *Aeluropus littoralis* directs developmental-regulated, stress-inducible, and organ-specific gene expression in transgenic tobacco. Transgenic Res. 20, 1003–1018.

Shekhawat, U.K., Srinivas, L., Ganapathi, T.R., 2011. *MusaDHN-1*, a novel multiple stress-inducible SK(3)-type dehydrin gene, contributes affirmatively to drought- and salt-stress tolerance in banana. Planta 234, 915–932.

Shi, H., Zhu, J.K., 2002. Regulation of expression of the vacuolar Na^+/H^+ antiporter gene *AtNHX1* by salt stress and abscisic acid. Plant Mol. Biol. 50, 543–550.

Singh, V.K., Mishra, A., Haque, I., Jha, B., 2016. A novel transcription factor-like gene *SbSDR1* acts as a molecular switch and confers salt and osmotic endurance to transgenic tobacco. Sci. Rep. 6, 31686.

Sun, Q., Gao, F., Zhao, L., Li, K., Zhang, J., 2010. Identification of a new 130 bp cis-acting element in the TsVP1 promoter involved in the salt stress response from *Thellungiella halophila*. BMC Plant Biol. 10, 90.

Tester, M., Davenport, R., 2003. Na^+ tolerance and Na^+ transport in higher plants. Ann. Bot. 91, 503–527.

Tiwari, V., Chaturvedi, A.K., Mishra, A., Jha, B., 2014. The transcriptional regulatory mechanism of the *peroxisomal ascorbate peroxidase* (*pAPX*) gene cloned from an extreme halophyte *Salicornia brachiata*. Plant Cell Physiol. 55, 201–217.

Tiwari, V., Patel, M.K., Chaturvedi, A.K., Mishra, A., Jha, B., 2016. Functional characterization of the *tau class Glutathione-S-Transferases* gene (*SbGSTU*) promoter of *Salicornia brachiata* under salinity and osmotic stress. PLoS ONE. 11, e0148494.

Tyagi, W., Rajagopal, D., Singla-Pareek, S.L., Reddy, M.K., Sopory, S.K., 2005. Cloning and regulation of a stress-regulated *Pennisetum glaucum* vacuolar ATPase c gene and characterization of its promoter that is expressed in shoot hairs and floral organs. Plant Cell Physiol. 46, 1411–1422.

Udawat, P., Jha, R.K., Sinha, D., Mishra, A., Jha, B., 2016. Overexpression of a cytosolic abiotic stress responsive universal stress protein (*SbUSP*) mitigates salt and osmotic stress in transgenic tobacco plants. Front. Plant Sci. 7, 518.

Udawat, P., Jha, R.K., Mishra, A., Jha, B., 2017. Overexpression of a plasma membrane-localized SbSRP-like protein enhances salinity and osmotic stress tolerance in transgenic tobacco. Front. Plant Sci. 8, 582.
Wang, R., Jing, W., Xiao, L., Jin, Y., Shen, L., Zhang, W., 2015. The rice high-affinity potassium transporter1;1 is involved in salt tolerance and regulated by an MYB-type transcription factor. Plant Physiol. 168, 1076–1090.
Yadav, N.S., Shukla, P.S., Jha, A., Agarwal, P.K., Jha, B., 2012. The *SbSOS1* gene from the extreme halophyte *Salicornia brachiata* enhances Na^+ loading in xylem and confers salt tolerance in transgenic tobacco. BMC Plant Biol. 12, 188.
Yin, X., Zhao, Y., Luo, D., Zhang, H., 2002. Isolating the promoter of a stress-induced gene encoding betaine aldehyde dehydrogenase from the halophyte *Atriplex centralasiatica* Iljin. BBA Gene Struct. Expr. 1577, 452–456.
Zhang, G.H., Su, Q., An, L.J., Wu, S., 2008. Characterization and expression of a vacuolar Na^+/H^+ antiporter gene from the monocot halophyte *Aeluropus littoralis*. Plant Physiol. Biochem. 46, 117–126.
Zhang, H., Zheng, J., Su, H., Xia, K., Jian, S., Zhang, M., 2018. Molecular cloning and functional characterization of the dehydrin (*IpDHN*) gene from *Ipomoea pes-caprae*. Front. Plant Sci. 9, 1454.

CHAPTER 9

The role of key transcription factors for cold tolerance in plants

Pankaj Sharma[a], Mayur Mukut Murlidhar Sharma[b], Anupam Patra[c], Medhavi Vashisth[d], Sahil Mehta[c], Baljinder Singh[e], Manish Tiwari[e], and Vimal Pandey[e]
[a]Department of Microbiology, CCS Haryana Agricultural University, Hisar, India
[b]Department of Agriculture and Life Industry, Kangwon National University, Chuncheon, Republic of Korea
[c]International Centre for Genetic Engineering and Biotechnology, New Delhi, India
[d]Department of Molecular Biology, Biotechnology and Bioinformatics, CCS Haryana Agricultural University, Hisar, India
[e]National Institute of Plant Genome Research, New Delhi, India

1. Introduction

The global human population is growing at a very disturbing speed and is further assumed to surpass 9.7 billion by 2050 (Mehta et al., 2019). This increased human mass will also need more resources to quench its hunger and would undoubtedly rely on the agricultural resources whereas at the same spell the efficiency of agricultural systems is on a steeper decline owing to the diverse ill effects of mounting environmental constraints because of the global climate change initiated by the activities of this increased population. The increased anthropogenic activities and growth in the industrial sector has significantly affected the plant productivity by altering the optimum environmental conditions required by the plant for best growth and maximum productivity. The altered environmental conditions make the plant suffer from several biotic and abiotic stresses (Singh et al., 2018, 2019a,b; Rani et al., 2019; Mehta et al., 2019; Rahman et al., 2019). Plants, being sessile in nature, are unable to move and thus have to face a lot of stresses like abiotic environmental factors, such as water deficiency (drought), excessive salt (salinity), threshold temperatures (from freezing to scorching), decreased availability of essential nutrients (nutrient starvation), heavy metal/metalloid presence, and variable light conditions. Such factors can either affect the plant simultaneously or at different stages of plant growth thereby significantly harming the plant growth coupled with its productivity (Zhu et al., 2007; Lal et al., 2018; Dhakate et al., 2019). Although plants face a lot of abiotic stresses but one among those prevalent and major stresses that

Transcription Factors for Abiotic Stress Tolerance in Plants
https://doi.org/10.1016/B978-0-12-819334-1.00009-5

plants have to face in their natural as well as agricultural environs is cold stress. Although plants demonstrate a broad range of tolerance to temperature fluctuations but the temperature drop below a threshold level significantly deteriorates the plant health as well as productivity. Moreover, different plant species display different levels of cold tolerance which acts as a major factor governing the spatial distribution of diverse plant species thereby limiting the expansion of the growing areas of a particular crop species into a particular area that is otherwise unsuitable for the production of other crops. The low-temperature stress is defined by its two different components named as chilling, frequently demarcated as subordinate than the usual growing temperatures for a particular plant species but greater than 0°C, and freezing, distinct as less than 0°C. A considerable number of different crops of economic importance, for instance, soybean, maize, cotton, rice, brassica, and tomato are sensitive to chilling stress (Raju et al., 2018). The cold stress affects the plant productivity by instigating different changes at the molecular level such as reduction in the enzymatic activity of plants coupled with a reduction in the permeability of the membrane. The reduction in enzymatic activity of plants grievously affects the functioning of the photosynthetic machinery of the plant. The chilling stress does not affect much the light reactions of the photosynthesis, however, the enzymes of the dark reactions face a steeper decline in their activity which further leads to the photoinhibition of photosystem I and sometimes II which in turn leads to the production of reactive oxygen species (Zoldan et al., 2012; Zhang and Scheller, 2004). The ROS are highly treacherous for plant systems as they affect the permeability, fluidity, and flexibility of the cell membrane by lipid peroxidation. The rigid cell membranes are easily damaged, thereby providing a passage for the leakage of electrolytes. The fact that is of prime importance is that cold stress produces a similar damage throughout different plant species. Throughout the globe, the precipitation index displays a lower extent of rainfall, however, the cold injuries induced by day-to-day temperature oscillations amplified exceptionally, demonstrating the impact as well as a need to cope with the cold stress which is significantly damaging the plants. The inexorable strategy of fulfilling the growing nutritious demands via plant protein sources is clear and has persuaded researchers to pay more attention toward the development of new strategies in plants touching the cold stress.

The field of biotechnology is also growing at a very faster pace and offers several ways that can help to cope with the problem of cold stress by developing transgenic crop plants with improved tolerance to cold stress. The engagement of numerous techniques as well as sophisticated machineries in the research has assisted the scientific community to draw a comprehensive portrait of how plants intellect the environmental stress signal, forward the stress message to the nucleus to subsequently regulate gene expression so that they can generate appropriate responses upon stress exposure. However, a clear intrusion into the insights of molecular events taking place along with the identification of the precise tasks of different participants as well as their interactions with other members within the network generated during the stress period has not been possible yet but several different components involved in the network has been targeted and used for genetic manipulation. Several attempts of genetic engineering have produced favorable outcomes where the transgenic plants displayed improved stress tolerance. The genes that are often targeted for genetic engineering experiments are divided into two groups: genes encoding functional proteins and genes encoding regulatory proteins. The members belonging to the previous cluster are very miscellaneous as they include products working in the upkeep of osmotic balance, antioxidation, and ROS elimination along with the protecting macromolecules such as proteins and DNA from degradation. The second group of genes that accounts for the key role of several members in shaping plant activities during the exposure to stress comprises transcription factors (TFs), kinases, phosphatases, micro-RNAs (miRNAs), and two-component systems (TCSs) (Lan Thi Hoang et al., 2017). The key role of these regulatory associates is to convey the stress signal from external environment to the nucleus (like TFs and TCSs) and directly regulate gene expression (like TFs) via interacting with the promoters of genes. This chapter largely focuses on the important roles of different TFs in conferring cold-tolerant mechanisms in plants and their potential for being used as a resource for crop improvement.

2. Cold stress: Types and effects

Winter damage in addition to freeze as well as frost damage is often regarded as identical. Nevertheless, cold sustained damages have got severe concerns owing to the thrilling temperature oscillation, more willingly than persistent low-temperature situations. The impulsive changes in temperature, for instance, swiftly dropping temperature in addition to the stiff freeze are potent enough to cause stress development and damage in the plants that have attained the state of dormancy, but have not yet copiously acclimatized to the cold (Guy, 1990; Mukhopadhyay and Roychoudhury, 2018).

The plant acclimatization to below-freezing environments can be attained successfully but it is governed by the restriction that the fall in temperature must be slow and steady; however, the continuous exposure of the plant to the prolonged durations of mild winter can also result in deacclimatization. The abrupt exposure of the plant to exceedingly low temperatures is highly potent to pose a considerable hazard to plants. Such kinds of deacclimatized plants are susceptible to face several tissue injuries coupled with low-temperature stress (Kalberer et al., 2006). The prolonged low temperatures during the winters can also cause grievous damages to the plant systems and the injuries are more pronounced, primarily when the plant is exposed to a temperature drop lower than a convinced tolerance level of the plant. The plant systems that were previously considered weak as per the physiological point of view, possibly due to prior exposure to stressful conditions or as a result of the lack of resilience as well as adaptableness to the punitive environments of a precise topographical spot are more susceptible to undergo winter injuries (Arora and Rowland, 2011). The expression of cold allied damages is exceedingly flexible, and different plant parts show different ranges of susceptibility yet buds demonstrate concentrated vulnerability. The frost and freeze injuries are closely allied as both produce membrane injury due to osmotic shock, dehydration stress, and ice crystal development (Fig. 1).

The occurrence of frost injury takes place during radiation freeze, principally during cool and strong nights, as plant systems release more heat energy than that can be remunerated by the heat received. Consequently, it marks a temperature overturn, in which the colder air close to the ground is trapped by the warmer air layer above it. When the temperature of air at the plant level is near or below freezing, the temperature of the plant systems is unavoidably lower than the ambient environments. Alternatively, freeze damage occurs because of advective freezes, where the air with below-freezing temperature travels into and inhabits a zone, dislodging the warmer air. This leads to a reduction in plant temperature that is low enough to result in the formation of ice crystals, thereby causing tissue injuries (Wani and Herath, 2018).

2.1 Freezing injury

The damage suffered by plants as a result of temperature drop beyond the freezing point, that is, 0°C or 32°F is often deliberated to be chilling injury.

Fig. 1 A picture depicting the common effects of cold stress on the plants.

The plants dwelling tropical or subtropical geographical zones are known to be more susceptible to the chilling injury. The plant parts that are found to be more affected in different delicate plants are leaves, fruits, and flowers. The most common appearance is in the form of purple or reddish wilting of leaves. The term freezing injury primarily targets the physiological plant dysfunctions produced as a result of the freezing of the water present in plant tissues. The water freezing in plant tissues is unsurprisingly in consort with the ice formation, which can be intracellular or extracellular (Pearce, 2001). The intracellular freezing is considered to be more austere and destructive owing to its attitude of disrupting membranal integrity which advocates its lethality. The formation of intracellular ice in tender plants is common; on the other hand, hardy plants are also affected before getting acclimatized. The formation of intracellular ice may take place on a spontaneous basis from centers of nucleation in the cytoplasm or in cell walls neighboring intercellular spaces. The plasmodesmatal connections also act as channels for transferring ice from cell to cell. However, plasmalemma restricts the entry of ice and, therefore, can partially prevent the loss of water but the cells as well as the organelles have a tendency to shrink and give way to freezing damage to some degree. The thick cuticle also acts as an operative armor for protecting plantlets from extracellular ice. The forfeiture of membranal integrity marks the leakage of metabolites and the distresses in plasmolysis as well as deplasmolysis characterize the tissue injury due to freezing stress (Mukhopadhyay and Roychoudhury, 2018).

The plant systems have also evolved with several mechanisms which can protect them from the sufferings of cold stress. The evolution of some plant tissues with the phenomenon of deep supercooling is a chief mechanism which assists plant systems in avoidance of freezing damages. The ability of water present in plant tissues to maintain its state of being liquid, surprisingly, even below subzero temperature is regarded as supercooling. This lack of the transition from liquid to solid phase prevents the formation of crystals which in turn leads to cold-allied injuries (Wisniewski et al., 2008). The ability of different fruit trees along with some hardwoods to supercool down to −35°C can prevent them from damage; nevertheless, a drop in temperature below −40°C marks the spontaneity of the process of ice crystallization. However, the crystallization resulting in the formation of smaller crystals due to quick freezing is not considered to be much harmful for plants as the crystals smaller in size often melt before the instigation of cold injuries. As a consequence, water has the potential to indeterminately persist in a supercooled form, until the temperature drops below the homogeneous ice nucleation temperature. If the exterior ice accomplishes nucleation, it quickly binges through vascular tissues, and the intensity as well as localization of nucleations are governed by the initial level of supercooling attained by the plant systems. If the level of supercooling attained is higher than a threshold level, numerous ice nucleation positions are accessible to the exterior ice for its intrusion (Wani and Herath, 2018). The occurrence of different solutes is also known to play a key role in lowering the spontaneous nucleation point. In several tissues, for instance, flower buds and xylem parenchymatous tissues, the penetration of ice from neighboring frozen tissues is prevented by a barricade produced by undifferentiated cells between floral primordia in bud and nearby frozen stem tissues. Such kind of hurdles which assist the plant systems in preventing the transmission of ice to healthy tissues may encompass the finer microcapillaries of cell wall other than the antinucleating biochemical present in the protoplasm.

2.2 Chilling injury

The damage suffered by chilling-sensitive plants at the temperatures above the freezing point of tissues but below 15°C is referred to as chilling injury. The plants demonstrating optical appearances of damage at temperatures above 15°C are designated enormously chilling sensitive (Lukatkin et al., 2012). Consequently, plants can be categorized as either chilling sensitive which are harshly impaired at temperatures between 0°C and 15°C or chilling resilient based on their potential to put up with low temperature up to a tolerance threshold and display marks of damage only during the ice formation. The controlling response for any chilling injury is marked by the alteration in the physical state of cell membranes as they suffer a shift from stretchy liquid crystalline to inflexible gel structure at an acute temperature. The temperature drop in plant species sensitive to chilling leads to the solidification of membrane lipids, which generates contraction, triggering cracks and channels and, subsequently, results in an augmented permeability. This distressed permeability results in ionic disproportion as well as ion seepage from the tissues. The enzymatic activity is also hindered due to the unavailability of the optimum conditions for enzyme activity. The temperature-prompted change in the phase of membrane lipids is revocable till deteriorating destruction has been caused to the plant systems (Parkin et al., 1989). The plasma membrane acts as the prime site for detecting chilling-sensitive plants that are facing chilling stress because of the multiple changes occurring in plasma membrane at the time of stress, for instance, diminution in the elasticity of cell membrane, abridged passivity, aversion in the insertion of lipids in membrane configuration, decline in the membrane fluidity, and henceforth elasticity of membrane lipids, and denaturation of the membrane-destined enzymes, comprising H+-ATPase with augmented cross diffusion of sterols, phospholipids, as well as proteins in the cell membrane (Kasamo et al., 1992; Kasamo and Noushi, 1987). The functioning of cellular membrane in chilling stress is largely governed by the lipids comprising the cellular membrane (Routaboul et al., 2000). The plants that are more vulnerable to the attack of chilling stress suffer more damage in the membrane structure owing to the chilling-prompted deterioration of phospholipids as well as galactolipids, which further upshots in an enlarged puddle of free fatty acids. The stressed plant systems also face a distinctive alteration in the level of sterols along with an increased sterols/phospholipids ratio which further triggers a decline in the membrane fluidity during low temperature (Whitaker, 1993). A noticeable upsurge in the level of unsaturated fatty acids, accrual of phospholipids in the plant tissues, and reduction in the level of sterols as well as sterol esters are different physiological expressions of chilling stress (Kojima et al., 1998; Kaniuga et al., 1999). The membrane transport also gets severely altered because of a steeper decline in the membrane permeability concomitant with the amplified viscidness due to low temperatures. Therefore, the affected plant species also suffers a reduction in the water uptake as well as sugar translocation. In addition to it, the membrane proteins also undergo discrete alterations during the plant exposure to chilling stress. Moreover, the proteins suffer a loss in their conformation, and the nonprotein constituents of the enzymes face a release and separation from their proteinous counterpart, which further results in alterations in the allosteric control of activity as well as kinetic constraints. The low temperature-prompted loss of the enzymatic activity is also arbitrated by protein-lipid communications in the membrane. The chilling stress also alters the molecular ordering of membrane lipids. Such variations are convoyed by a lowered level of cellular ATP coupled with the augmented membrane permeability.

3. Cold stress sensation

It is a proven fact that cellular membrane undergoes several changes during the exposure of plant to cold stress. The plant systems are persuasive enough to sense cold stress by the abovementioned low temperature-prompted changes in the cellular, protein as well as nucleic acid conformation along with the concentration of several metabolites that are secreted particularly during cold stress. The inelastic attitude of plasma membrane has been considered to induce the expression of COR (cold responsive) genes which accounts for cold acclimatization in different plants (Örvar et al., 2000; Sangwan et al., 2001). The exposure to cold stress also induces an upsurge in the level of calcium ions in the cytosol which can further be arbitrated through membrane rigidification-triggered mechano-profound or ligand-stimulated calcium ion channels. Consequently, the gradual intensification in the calcium signaling along with the phospholipid signaling may perhaps be engaged in cold-induced stress signaling (Vergnolle et al., 2005; Williams et al., 2005; Chinnusamy et al., 2006). The other tributary signals, for instance, abscisic acid along with the reactive oxygen species are also known to encourage calcium ion signs that further influence the cold-induced cell signaling (Fig. 2) (Llorente et al., 2000). The perpetual accumulation of reactive oxygen species in the cells facing different kinds of abiotic stresses found

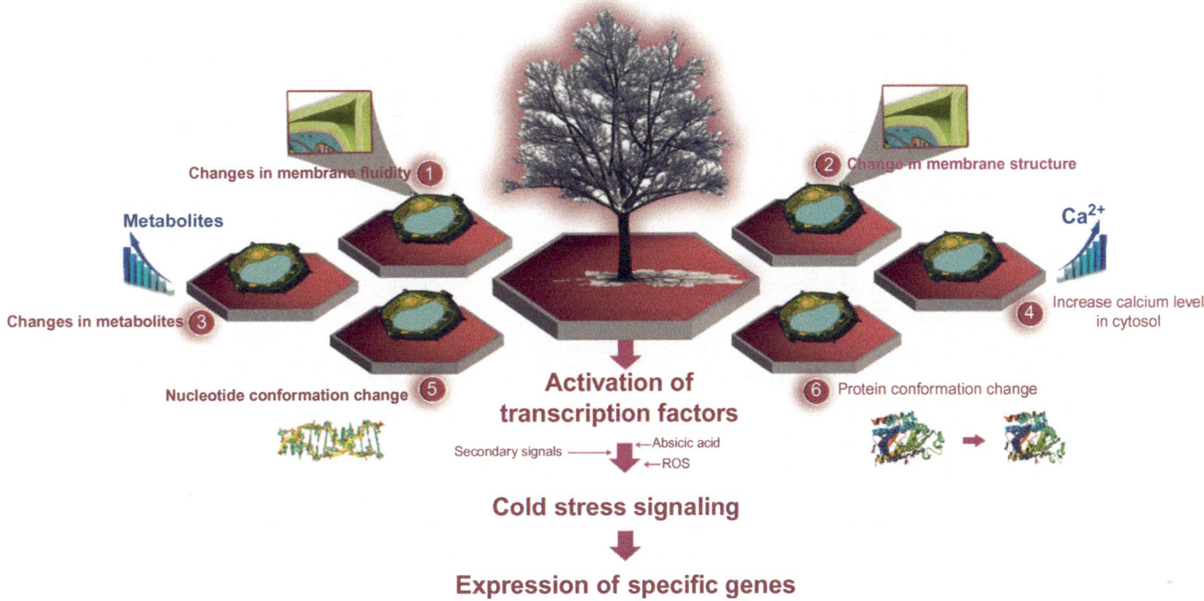

Fig. 2 A portrayal depicting the cascade of events created due to the plant exposure to cold stress.

to have a robust effect on cold-induced regulation of gene expression. For instance, the mutant *Arabidopsis* fro1, which demonstrates accumulative behavior for ROS at upper degrees, displays compromised expression of COR genes along with a higher level of sensitivity to chilling as well as freezing. The FRO1 is known for encoding the Fe–S subunit of complex I of the electron transfer chain in mitochondria, and the interruption of this chain results in the generation of a high-intensity ROS (Lee et al., 2002). Although, ROS has a considerable effect on the calcium signatures they also directly exercise their effects by activating the redox-receptive proteins, for instance, transcription factors and enzymes such as protein kinases.

The temperature drop is also known to affect the uptake of water as well as other essential nutrients, and also affects metabolic reactions drastically. It can either directly harm the metabolic machinery by plummeting the speeds of different biochemical reactions taking place inside the cell or can indirectly affect the cell by reprogramming the gene expression. The metabolic fingerprinting of plants, for instance, *Arabidopsis* has revealed that the process of cold acclimatization upsurges approximately 75% of the 434 metabolites identified in the plant systems (Cook et al., 2004; Kaplan et al., 2004). Besides, their behavior as osmolytes and osmoprotectants, definite metabolites induced throughout the cold acclimatization possibly act as signals meant for the reconfiguration of gene expression during the cold stress. For instance, the accretion of proline, which is a well-regarded osmoprotectant, is induced during the plant exposure to cold stress. Surprisingly, as revealed by microarray as well as RNA gel blot studies, it has been proven that proline has the potential to induce expression of several genes, having proline-responsive element (PRE, ACTCAT) in their promoters.

4. Cold stress tolerance

Plant systems have evolved different mechanisms in the quest to survive under the situations of freezing or to tolerate cold stress. The level of tolerance exhibited by the plant systems toward low temperatures is often regarded as one of the parsimoniously as well as environmentally imperative restrictions directing the topographical dispersal of the plants along with the growing seasons best suited for their growth. The level of tolerance to cold stress fluctuates on a very significant basis throughout the plant kingdom, where different plant species have evolved different resistance mechanisms. The maintenance of the fluid nature of cell membrane in accumulative cold as well as oxidative cellular environs is of prime concern for the plant systems. The response of variable plant species to the cold stress largely comprises deviations in the types as well as the levels of desaturation of different membrane lipids, however, the specific lipids affected by the cold stress tend to differ by the plant species. The regulation of changes taking place in the cellular membrane coupled with other cold tolerance factors is a highly controlled and genetically regulated process which is checked by both transcriptional as well as posttranscriptional mechanisms (Raju et al., 2018).

The mainstream of the plant plasma membrane is made up of the glycerolipids which further comprises a polar head attached via a glycerol molecule to two tails made up of fatty acids. The fatty acid molecules comprising the structure of the membrane may have complete levels of saturation, i.e., have no double bonds along the tail, or may also be have unsaturation at precise locations. The degree of saturation present in the fatty acid affects fitting of the fatty acid in the membrane thereby impacting the membrane fluidity at diverse temperature ranges. Furthermore, the comparative size of a lipid head as compared to its tail also marks the fitting as well as the fluid nature of the membrane (Melser et al., 2011). Consequently, it can be considered that alterations in the head molecule of fatty acids or complete fatty acids lay a significant effect on the overall fluid nature of cellular membrane. Sphingolipids have also been shown to play significant roles in the tolerance of cold stress (Chen et al., 2012; Dutilleul et al., 2012), conversely, they are not much quantified in studies targeting cold stress as opposed to the glycerolipids (Michaelson et al., 2016). The glycerolipids experience an increase in the degree of unsaturation of the fatty acid tails during cold stress (Miquel et al., 1993; Steponkus, 1996; Raju et al., 2018) coupled with transformations amid diverse whole lipid head groups, for instance, aggregation in triacylglycerol, also known as oil, or elimination of lipids like monogalactosyldiacylglycerol (MGDG) that weaken the cellular membranes.

There are several other natural mechanisms for cold stress tolerance which are often employed by the plant systems. One among them is represented by the state of endodormancy which is a distinctive physiological practice sensitive to cold. Although, the state of endodormancy can either occur in nonappearance of cold or due to sporadic cold environments, however continual above-freezing temperature marks the endodormancy release. The cold environments of late autumn and early winter leads to the probable accrual of FT (Flowering Locus T) (Lin et al., 2007) which further lays an effect on the scheduling of endodormancy discharge and bud flush. The Flowering Locus T is a moveable proteinaceous element that moves from leaves to shoot peaks, and therefore, the establishment of callose masses during autumn constrains its translocation. Such masses are often detached during the freezing winter triggered by the instigation of gibberellic acid-reactive genes that control GH17 proteins, which break down the callose, and subsequently movement as well as accumulation of FT succeed (Rinne et al., 2001). Therefore, the state of endodormancy can be considered as one among the different endurance policies of cold stressed plants in the temperate regions, and its key regulator FT has evolved from lineage-specific duplications. Thus, it can be concluded that several machineries of plant systems get affected by the cold stress and plant systems also react to this state by changing the gene expression, altering the membrane configuration, manufacturing of harmonious solutes, cold shock proteins, and antioxidant enzymes, that are considered to play a key role in the defense of plant cells against freezing injury. Predominantly when plant systems face a gradual exposure to the cold stress, these variations at a cellular level can assist the plant systems in tolerating the cold stress, the process is known as "cold acclimation."

5. Cold-induced protein expression

5.1 Cold shock proteins

The exposure to cold stress, undoubtedly, initiates a cascade of events in the plant systems and one among them is the expression of several unique proteinous molecules which are expressed only under chilling temperatures and are often regarded as cold shock proteins. The initial studies regarding cold shock proteins find their base in different microorganisms especially in *Escherichia coli*. A number of studies have established that the exposure to low temperatures in *E. coli* is accompanied by a remarkable accretion of nucleic acid-binding cold shock proteins (Graumann and Marahiel, 1998; Yamanaka et al., 1998; Bae et al., 2000). The most prominent of the nine-member cold shock protein family of *E. coli* is CspA, which accrues up to 10% of total proteins during the exposure to cold stress (Jiang et al., 1997). This protein prevents the formation of secondary structures in the RNA (Jiang et al., 1997), thus assists in improving translation at the reduced temperature ranges (Karlson and Imai, 2003). These proteins are also known to possess glycine-rich regions. *Arabidopsis* has been known to possess two such cold shock proteins, namely, AtGRP2 and AtGRP2b; a single protein NtGRP has been found in the tobacco plant while WCSP1 has been found as important cold shock protein in the wheat plant. The transcription of AtGRP2 upsurges markedly in *Arabidopsis thaliana* during the periods of cold stress. The expression of another cold shock protein GRP2 at 11°C supports the growth of roots in plants. GRP2 has the potential to melt the secondary structures of RNA which marks its possible role in RNA processing during cold conditions (Maleki and Ghorbanpour, 2018).

5.2 Heat-shock proteins

Heat-shock proteins is a class of proteins that, on the basis of their function, are associated with the comprehensive and innumerable families of molecular chaperones that are defined by their proficiency to differentiate and attach substrate proteins present in an unstable or inactive form. These are considered to be molecular chaperons that function in protein folding and assembly, protein intracellular localization as well as discharge, along with the degradation of misfolded or shortened protein molecules (Hu et al., 2009). The HSPs belonging to the class of small heat-shock proteins behold an essential part in plant adaptation to innumerable stressful environments (Timperio et al., 2008), encompassing the cold stress. The oxidative stress induced as a result of cold stress triggers the expression of proteins engaged in defending against the oxidative-stress, for instance, chaperonins and HSPs (Larkindale and Knight, 2002). The low-temperature stress has already been proved to upsurge the levels of both chaperonins and HSPs, particularly HSP90, HSP70, and the smHSPs (Wang et al., 2004). Nevertheless, it is not an established fact that cold stress always enhances the levels of HSPs as a decline has also been reported in the concentrations of HSP90 under cold stress conditions (Vítámvás et al., 2012). However, a few HSPs and mHSPs have the potential to defend the plant systems against the cold stress damages by averting the denaturation of proteins due to freezing by circumventing the accumulation of denatured proteins (Yan et al., 2006). The smHSPs are also known for their role in refolding denatured proteins as a mechanism of molecular chaperoning (Lopez-Matas et al., 2004). Therefore, the smHSPs act not only as cryoprotectants but also as molecular chaperones.

5.3 Antifreeze proteins

These proteins are another kind of evolutionary diverse class of polypeptides fashioned by definite living beings and are considered to be significant constituents of the acclimatization systems to freezing temperatures. These are also referred to as ice binding proteins which adsorb superficially on the surface layer of ice crystals, thereby elevating the regular melting point of water to some extent along with a concomitant decline of thermal hysteresis as well as the inhibition of ice nucleators. The genes encoding for antifreeze proteins are often upregulated throughout the period of freezing as well as other biotic and abiotic stresses. A cohort of proteins possibly allied with pathogenic activities has been recognized during cold stress that exhibits some antifreeze actions. The freeze-forbearing approach shaped by the protein LpAFP has been pronounced in a perennial ryegrass showing similarities to the antifreeze proteins from other freeze-circumventing creatures (Lauersen et al., 2011). The TaIRI proteins of wheat are also a type of AFPs having leucine-rich repeat receptor domain of receptor-like kinases along with additional domain-targeting reticence of ice recrystallization. It has also been anticipated that the plant hormones like ethylene play a key role in the regulation of antifreeze proteinaceous molecules in plants like winter rye throughout the plant exposure to freezing as well as drought stress (Atıcı and Nalbantoğlu, 2003). The view targeting the conduction of antifreeze proteins from other plants to the frost-sensitive plants established lower degrees of thermal hysteresis, inhibition of the process of recrystallization process, or abridged electrolyte seepage. However, a major proportion of such altered plants are not able to reduce the freezing temperatures (Lin et al., 2011).

5.4 Dehydrins

The other kind of plant's systemic response to cold stress marks the accrual of several hydrophilic proteinaceous elements which are supposed to form an amphipathic α-helix structure (Shabala, 2017). A number of genes coding for such kind of proteins, on characterization, were found to be reactive to cold stress as well as other different types of stresses faced by the plants. Such proteins are encompassed by the dehydrin proteins, which further explain group II of late embryogenesis abundant (LEA) proteins (Shabala, 2017). These proteins are also designated to play key roles in assisting plant systems for tolerating freezing stress, maybe by averting the disruption of cell membrane that mainly happens because of the alteration in osmotic contraction allied with freezing. These proteins are amassed in the later phases of seed growth and are also found to be allied with the accomplishment of drought forbearance (Liu et al., 2017). These are further classified into seven diverse clusters (Battaglia et al., 2008). The dehydrins, associated with group II LEA proteins, are considered for their possible involvement in the plants' defensive responses toward dehydration, and therefore are regarded as stress proteins (Liu et al., 2017). Therefore, they can also be designated as hydrophilins.

These group II LEA proteins are also considered to play imperative parts in late embryogenesis and can also be tempted in vegetative parts for rewarding the plant with cold acclimatization. The perpetual increase in the level of dehydrin expression is also motivated by hormone abscisic acid, therefore, they are also referred to as responsive to ABA (RAB) proteins (Hanin et al., 2011). Their regulation by miRNAs has also been established earlier. A vast array of studies has designated the role of these proteins in different molecular functions, such as protection of cellular membrane from toxicities due to peroxidation of lipid molecules, their role as chaperones, their activity as scavengers for ROS, and adsorption of metals as well as lipid vesicles. They are also known to play key roles in facilitating stress signaling paths, for instance, jasmonate signaling (Brini et al., 2011). Their level increases considerably during the plant's exposure to cold stress. Khazaei et al. (2015) reported a significant enhance in *DHN* transcripts in acclimated chickpea plants, as relative CT response. Different plant species are known to secrete different dehydrins during cold stress which play different roles in acclimatizing the plant to cold stress, for instance, the WCOR410 (SK3) protein of wheat plays a key role in membrane binding, P80/DHN5 (K9) of barley and PCA 60 (Y2K9) of peach are known for cryoprotection of the enzyme lactate dehydrogenas, the CuCOR19 (SK3) protein of citrus plants is helpful in cryoprotection of catalase and lactate dehydrogenase. Furthermore, they are also helpful in reducing the levels of lipid peroxidation via scavenging peroxyl as well as hydroxyl radicals. The dehydrin CuCOR15 (SK2) of citrus binds well with different metal ions, such as Ni^{2+}, Fe^{3+}, Cu^{2+}, Co^{2+}, and Zn^{2+} (Kazemi-Shahandashti and Maali-Amiri, 2018).

6. Transcription factors for cold tolerance

It is a proven fact that the plant systems have developed different sophisticated mechanisms for tolerating as well as adapting to the cold stress. These mechanisms come into progression by the alteration of different physiological as well as biochemical processes. A number of different genes have been identified which are often expressed under cold stress, therefore are known as cold-responsive genes. Such genes include phosphatases, protein kinases, transcription factors, which collectively confer the plant the ability of cold tolerance. Generally, such responses find their regulation either by activation or by inhibition of gene-explicit expression. One among these gene-explicit expression paths is the interaction of transcription factors (TFs) with *cis*-acting elements to precisely express genes engaged in the stress to uphold the normal life happenings of plant systems (Sun et al., 2018).

The low-temperature stress promptly lead to the expression of a number of transcription factors, for instance, MYB, C-repeat-binding factor/dehydration-responsive binding protein (CBF/DREB), ABA-responsive element-binding protein/ABA-binding factor, APETALA2 (AP2)/ethylene responsive factor, basic region/leucine zipper motif (bZIP), basic helix-loop-helix (bHLH), and NAM, ATAF1, 2, and CUC2 (NAC) families, which sequentially bind the promoter of various stress-reactive genetic elements, thereby controlling their expression (Table 1) (Zhao et al., 2018). Therefore, the expression of these transcription factors under stressful conditions make them suitable elements for creating genetically altered cold tolerant crops (Fig. 3).

6.1 bHLH TF superfamily

The basic helix-loop-helix protein (bHLH) is the supreme and widespread class of transcription factors in eukaryotes, which has the potential to regulate the gene expression by interacting with the exact motifs in the target genes. This family represents a large class of proteins having different functions in plant as well as animal kingdom (Garrell and Campuzano, 1991; Ledent and Vervoort, 2001). Such proteinaceous elements consist of conserved domain of 60 amino acids (AA) in length, which further encompasses two separate zones assorted on the basis of their function. One of these is represented by a simple region comprehending 13–17 AAs present at the N-terminal and serves as the DNA-binding domain, whereas the other is present at the C-terminus and is referred to as HLH region that subsidizes the establishment of hetero or homodimers (Heim et al., 2003; Toledo-Ortiz et al., 2003). It is an established fact that the bHLHs finding their origin in plant systems significantly regulate the gene transcriptional grids encompassing light as well as hormonal signaling, biosynthesis of flavonoids, timings concerning the flowering of plants, stomatal, trichome as well as the root hair materialization, along with various abiotic stress retorts counting the cold stress (Zhao et al., 2018).

A total of 167 bHLH genes have been recognized in the model plant system *Arabidopsis* whereas the number is 177 in case of rice (Mao et al., 2017). Conversely, the portrayal of the functional behavior of most bHLHs is

Table 1 Role of different transcription factors in enhancing cold tolerance in different crops.

S. no.	TF	Family	Plant	Function	References
1	JcDREB	AP2/EREBP	*Jatropha curcas*	Enhancement in tolerance toward salt and freezing stress	Tang et al. (2011)
2	FTL1/DDF1	CBF/DREB1	*Arabidopsis thaliana*	Increased tolerance to cold, heat, and drought stresses	Kang et al. (2011)
3	AtGT2L	GT	*Arabidopsis thaliana*	Regulate plant response toward cold and salt stress	Xi et al. (2012)
4	SlICE1a	MYC	Tomato	Cold, osmotic and salt tolerance in transgenic tobacco	Feng et al. (2013)
5	GmMYBJ1	MYB	*Glycine max*	Cold and drought tolerance	Su et al. (2014)
6	DaCBF7	CBF	*Deschampsia antarctica*	Improved cold tolerance in transgenic rice plants	Byun et al. (2015)
7	MfNAC3	NAC	*Medicago truncatula*	Cold tolerance	Qu et al. (2016)
8	CsWRKY46	WRKY	Cucumber	Cold tolerance via ABA-dependent pathway	Zhang et al. (2016)
9	SlNAM1	NAC	Tomato	Improvement in chilling stress tolerance of transgenic tobacco	Li et al. (2016)
10	MeMYB2	MYB	Cassava	Tolerance to freezing and drought stress	Ruan et al. (2017)
11	MdHY5	bZIP	Apple calli and *Arabidopsis*	Cold tolerance via CBF-dependent and CBF-independent pathway	An et al. (2017)
12	BdCBF3	CBF	*Brachypodium distachyon*	Cold acclimatization	Hao et al. (2017)

Continued

Table 1 Role of different transcription factors in enhancing cold tolerance in different crops.—cont'd

S. no.	TF	Family	Plant	Function	References
13	ZmmICE1	MYC	*Zea mays* ssp. *mexicana* L	Freezing tolerance in transgenic *Arabidopsis thaliana*	Lu et al. (2017)
14	ZjICE1	MYC	*Zoysia japonica*	Positive plant regulator during cold stress	Zuo et al. (2019)
15	BpERF13	AP2/ERF	*Betula platyphylla*	Regulation of physiological processes essential for cold tolerance	Lv et al. (2020)
16	MdMYB108L	MYB	Apple	Light and cold tolerance	(Wang, Mao, et al., 2019)
17	MeTCP4	TCP	*Manihot esculenta*	Cold stress tolerance	Cheng et al. (2019)
18	CaNAC064	NAC	*Capsicum annuum*	Positive modulation of the plant cold-tolerance	Hou et al. (2020)
19	STCH4	CBF	*Arabidopsis thaliana*	Chilling and freezing tolerance	Yu et al. (2020)
20	AtCRAP2	AP2	*Arabidopsis thaliana*	Cold tolerance along with the induction of flowering under low temperatures	Luo et al. (2020)

Fig. 3 A figure depicting the importance of transcription factors for creating genetically modified plants with increased capability of tolerating cold stress.

comparatively restricted. The functional characterization is limited to only 64 in *Arabidopsis* out of the total of 167 whereas the number is mere 19 in rice (Ji et al., 2016). The transcription factors like *AtbHLH104*, *MUTE* are known to enhance the tolerance for iron scarcity in *Arabidopsis*, and regulate the stomatal differentiation, whereas the other TFs, for instance, *PIF1/PIL5*, *PIF3*, *PIF4*, *PIF5/PIL6*, and *OsbHLH102* are engaged in light as well as gibberellin signaling in *Arabidopsis* and rice (Leivar and Monte, 2014; Zhao et al., 2018). However, the involvement of numerous bHLH genes in different abiotic stress responses including cold tolerance has also been reported in a number of studies. For instance, the TF *OsbHLH1* has been reported for its involvement in the cold-induced signal transduction pathways (Wang et al., 2003) whereas *PtrbHLH* has also been acknowledged for conferring cold tolerance in the plant *Poncirus trifoliate* (Huang et al., 2013). The expression of TF *OrbHLH001* above the normal levels is also known to confer freezing as well as salinity tolerance. Recently, the role of TF *NtbHLH123* has been positively confirmed in the tobacco plant for its imperative assistance in attaining cold resistance (Zhao et al., 2018). This TF is a kind of transcriptional activator which functions by binding to the G-box/*E*-box motif in the NtCBF gene promoter, thereby controlling the expression of ROS clearance-related and stress receptive genes, thus increases the cold tolerance. The overexpression of TFs *VabHLH1* and *VvbHLH1* transcription in genetically modified *Arabidopsis* is also known to augment its tolerance to chilling stress by acting as positive regulators for the expression level of COR genes (Xu et al., 2014). A distinctive MYC-like basic helix-loop-helix (bHLH) TF encoded by ICE1 (inducer of CBF expression 1) has also been proved as a critical constituent of the cold signaling pathways in *Arabidopsis* plant. ICE1 protein is potent enough to bind the MYC *cis*-element (CANNTG) precisely at the promoter region of the CBF1–3 gene and further supports the gene expression during cold stress (Chinnusamy et al., 2003; Ding et al., 2015; Yang et al., 2019a,b).

6.2 Dehydration-responsive element-binding

The dehydration-responsive element-binding (DREB) proteins are in the possession of AP2/ERF (APETALA2/ethylene-responsive element-binding factor) gene family, which consist of one of the major collection of plant transcription factors that are engaged in different plant stress responses. The genes for DREB were initially retrieved from the plant *A. thaliana*, but now they have been extensively acknowledged in numerous plants, for instance, rice, soybean, and maize (Li et al., 2019). The TFs belonging to the group of DREBs are mainly allocated into two major subgroups, namely, DREB1 and DREB2. These TFs possess only a single AP2/ERF domain. DREB TFs bind explicitly to the DNA sequence motif G/ACCGAC (known as the dehydration responsive element (DRE)/C-repeat; DRE/CRT) found in various cold- and drought-responsive genes. The DREB1/CBF TFs work predominantly during the low-temperature stress, whereas the DREB2 TFs become functional under drought as well as salt stress. The DREB transcription factors that are rapidly induced as a result of low-temperature stress are AtDREB1A/CBF3, AtDREB1B/CBF1, and AtDREB1C/CBF2 where the constitutive expression of AtDREB1A/CBF1, CBF3, as well as CBF4 imparts the plant with improved tolerance toward the freezing stress in case of *Arabidopsis*. The DREB subfamily of AP2/ERF comprises 57 genes in rice genome (Moon et al., 2019). OsDREB1G, a member of DREB, is known for conferring cold tolerance to rice plant and its overexpression in the plant confers the plant with the ability of improved cold tolerance (Moon et al., 2019). The constitutive ectopic expression of DREB genes in different species of plants has resulted in an upsurged degree of cold tolerance. For instance, the constitutive overexpression of wheat *TaDREB2* and *TaDREB3* in barley ensue a noticeable enhancement in the cold tolerance as well as in the expression of genes coding for LEA/COR/DHN proteins, which are further recognized for protecting the plant cells from injury as well as death due to cold stress (Morran et al., 2011). The overexpression of peach DREB, *PpCBF1*, in apple accounted for a 4–6°C surge in the cold robustness in both the acclimatized as well as nonacclimatized conditions as parallel with the untransformed plants (Wisniewski et al., 2011). The expression of DREB TF *RdreB1BI* in the cold-sensitive strawberry improves the plant efficiency to tolerate cold stress by acting as the dominant regulator of low-temperature stress responses, thereby advocating the superiority as well as cold hardiness of *RdreB1BI* transgenic strawberry (Gu et al., 2015). The gene for DREB1-type transcription factor, *ZjDREB1.4*, when inserted into *Arabidopsis* showed improved tolerance toward cold stress. The expression of *ZjDREB1.4* follows a significant increment on exposure to chilling stress. Furthermore, the engineered *Arabidopsis* plant exhibited an increased level of tolerance to the freezing temperatures along with an augmented degree of proline as well as soluble sugar content (Feng et al., 2019). TF SlDREB3 is known to induce cold resistance in transgenic tomato plants. The plants which express SlDREB3 in higher levels have been reported to exhibit high fresh mass, chlorophyll content, and reduced injury in cellular membrane along with a lower level of ROS. TF SlDREB3 is considered to improve the cold tolerance level in transgenic tomato plants by upregulating the expression of SlLEAs expression as high expression levels of late embryogenesis-abundant genes SlLEA9 and SlLEA26 were detected in transgenic plants in response to cold stress (Wang, Xu, et al., 2019).

6.3 Basic leucine zipper

The basic leucine zipper (bZIP) family of transcription factors is one of the largest as well as assorted classes of transcription factors. The transcription factors belonging to the class of bZIP are also found to be involved in numerous biochemical progressions counting expansion of flowers and seeds, seed sprouting, light, hormone as well as sugar signaling. Furthermore, bZIPs also find engrossment in responses to different biotic as well as abiotic hassles (Sornaraj et al., 2016). The exploration of the structural profile of bZIP revealed that these transcription factors comprehend an extremely preserved bZIP domain which is poised of a basic region at the N-terminal and a leucine zipper region at the C-terminal (Pan et al., 2019). The basic region is almost 16 amino acid residues long and is held accountable for binding to DNA whereas the other region, leucine zipper is not so preserved and contains a heptad repeat of leucine amino acid or other hydrophobic amino acids and finds its role in explicit recognition and dimerization (Yang et al., 2019a,b). The investigation of plant genomes recommends that there are numerous integers of bZIPs in plant genomes. For instance, there are 75 bZIP genes in the genome of *Arabidopsis* plant, 92 in rice, 125 in maize, and 187 in wheat. The phylogenetic assays suggests that bZIP homologues present in angiosperms can be divided into 13 subcategories (Cai et al., 2018). A number of transcriptions factors belonging to the bZIP groups are known for their role in cold tolerance mechanisms of the plant systems. For instance, the cold-induced expression of the transcript *TabZIP1* suggests its role in cold tolerance (Zhang et al., 2008). The ectopically expressed transcripts *TaAREB3* and *TaABL1* in the plant *Arabidopsis* enhanced the freezing tolerance of transgenic *Arabidopsis* seedlings (Wang et al., 2016). In *Brassica rapa* also, the levels of the *BrbZIP* genes were found to be boosted during the cold stress (Hwang et al., 2014). In case of soybean, the overexpression of TF *GmbZIP1* possibly improves the plant tolerance to low-temperature stress in transgenic plants (Gao et al., 2011). TF *TabZIP60* from wheat when engineered into the *Arabidopsis* showed enhanced tolerance to freezing stress upon its overexpression (Zhang et al., 2015). The CsbZIP genes of tea plants show a tissue-specific expression and several genes have been reported to code for the TFs involved in the cold tolerance (Cao et al., 2015). The rice TF bZIP73^{Jap} is also known to play an imperative part for conferring the cold adaptation at the seedling stage. It has also been proved that the rice TF bZIP73^{Jap} also provides the plant with the ability to tolerate low-temperature stress at the propagative phase. TF bZIP73^{Jap} upregulates during exposure to cold stress and primarily expresses in the panicles during the initial binucleate and flowering stages. The dimerization of bZIP73^{Jap} and bZIP71, as well as the co-expression of bZIP73^{Jap} along with bZIP71 has a remarkable effect on the seed-setting rate coupled with the grain yield under the conditions of low-temperature stress. The co-expression of TFs bZIP73^{Jap} and bZIP71 works not only by the repression of abscisic acid in the anthers, but also by augmenting the transport of soluble sugars from anthers to pollens which furthers accounts for an improved pollen grain fertility (Liu et al., 2019).

7. Conclusion

The nature has not awarded the plants with the ability to move at their own will, therefore, they have to face several biotic as well as abiotic stresses.

These stresses are extremely important to manage in case of agricultural crops (cereals, pulses, oil yielders, fiber yielders, masticatory plants). The low-temperature stress is a major concern for the scientific community as it restricts the plant growth, damages the cell membrane, and halts the plant photosynthetic machinery, thereby reducing the overall yield of the plant systems. Although, plants try their best to cope with the low-temperature stress by introducing different mechanism/changes such as modification of the membrane lipids, triggering "supercooling" phenomenon, and producing several compatible solutes. Like other stresses, the cold stress sensing in the plant systems is initiated by a cascade of events which trigger the reaction and response accordingly. The plants try to resist the cold stress by the phenomenon of cold "acclimatization," however, this result in huge economic losses to the human beings directly. A little exploration of the plant biology realm under cold stress has revealed that plant alters its gene expression under cold stress by overexpressing different kinds of genes (regulated by different transcription factors). These genetic elements act as a boon for genetic engineering experiments as their overexpression can largely solve the problem of cold stress. Although, a lot of TFs have been identified, the complexity of plant biology along with the higher ploidy levels of plant genomes restricts the scientific community to draft a clear picture of molecular events taking place during the cold stress. A more lucid understanding is required for producing frost-tolerant genetically engineered plant systems for better yields. Therefore, more emphasis should be given to this particular thrust area as population is increasing at a very alarming rate and a major part of global agriculture is still facing the decline in crop yields due to cold stress.

References

An, J.P., Yao, J.F., Wang, X.N., You, C.X., Wang, X.F., Hao, Y.J., 2017. MdHY5 positively regulates cold tolerance via CBF-dependent and CBF-independent pathways in apple. J. Plant Physiol. 218, 275–281.

Arora, R., Rowland, L.J., 2011. Physiological research on winter-hardiness: deacclimation resistance, reacclimation ability, photoprotection strategies, and a cold acclimation protocol design. In: Hort Science.46, pp. 1070–1078. https://doi.org/10.17660/ActaHortic.2013.990.6.

Atıcı, Ö., Nalbantoğlu, B., 2003. Antifreeze proteins in higher plants. Phytochemistry 64 (7), 1187–1196.

Bae, W., Xia, B., Inouye, M., Severinov, K., 2000. Escherichia coli CspA-family RNA chaperones are transcription antiterminators. Proc. Natl. Acad. Sci. U. S. A. 97 (14), 7784–7789.

Battaglia, M., Olvera-Carrillo, Y., Garciarrubio, A., Campos, F., Covarrubias, A.A., 2008. The enigmatic LEA proteins and other hydrophilins. Plant Physiol. 148 (1), 6–24.

Brini, F., Yamamoto, A., Jlaiel, L., Takeda, S., Hobo, T., Dinh, H.Q., Hattori, T., Masmoudi, K., Hanin, M., 2011. Pleiotropic effects of the wheat dehydrin DHN-5 on stress responses in Arabidopsis. Plant Cell Physiol. 52 (4), 676–688.

Byun, M.Y., Lee, J., Cui, L.H., Kang, Y., Oh, T.K., Park, H., Lee, H., Kim, W.T., 2015. Constitutive expression of DaCBF7, an Antarctic vascular plant *Deschampsia antarctica* CBF homolog, resulted in improved cold tolerance in transgenic rice plants. Plant Sci. 236, 61–74.

Cai, W., Yang, Y., Wang, W., Guo, G., Liu, W., Bi, C., 2018. Overexpression of a wheat (*Triticum aestivum* L.) bZIP transcription factor gene, TabZIP6, decreased the freezing tolerance of transgenic Arabidopsis seedlings by down-regulating the expression of CBFs. Plant Physiol. Biochem. 124, 100–111.

Cao, H., Wang, L., Yue, C., Hao, X., Wang, X., Yang, Y., 2015. Isolation and expression analysis of 18 CsbZIP genes implicated in abiotic stress responses in the tea plant (*Camellia sinensis*). Plant Physiol. Biochem. 97, 432–442.

Chen, M., Markham, J.E., Cahoon, E.B., 2012. Sphingolipid Δ8 unsaturation is important for glucosylceramide biosynthesis and low-temperature performance in Arabidopsis. Plant J. 69 (5), 769–781.

Cheng, Z., Lei, N., Li, S., Liao, W., Shen, J., Peng, M., 2019. The regulatory effects of MeTCP4 on cold stress tolerance in Arabidopsis thaliana: a transcriptome analysis. Plant Physiol. Biochem. 138, 9–16.

Chinnusamy, V., Ohta, M., Kanrar, S., Lee, B.H., Hong, X., Agarwal, M., Zhu, J.K., 2003. ICE1: a regulator of cold-induced transcriptome and freezing tolerance in Arabidopsis. Genes Dev. 17 (8), 1043–1054.

Chinnusamy, V., Zhu, J., Zhu, J.K., 2006. Gene regulation during cold acclimation in plants. Physiol. Plant. 126 (1), 52–61.

Cook, D., Fowler, S., Fiehn, O., Thomashow, M.F., 2004. A prominent role for the CBF cold response pathway in configuring the low-temperature metabolome of Arabidopsis. Proc. Natl. Acad. Sci. 101 (42), 15243–15248.

Dhakate, P., Sharma, P., Mehta, S., Akter, J., Bhatt, V., Chandanshive, S., Chakravarty, D., Rahman, M., Islam, M.A., 2019. Beneficial role of metalloids in plants: molecular understanding and applicability. In: Wani, S.H. (Ed.), Recent Approaches in Omics for Plant Resilience to Climate Change. Springer, Cham, pp. 357–387.

Ding, Y., Li, H., Zhang, X., Xie, Q., Gong, Z., Yang, S., 2015. OST1 kinase modulates freezing tolerance by enhancing ICE1 stability in Arabidopsis. Dev. Cell 32 (3), 278–289.

Dutilleul, C., Benhassaine-Kesri, G., Demandre, C., Rézé, N., Launay, A., Pelletier, S., Renou, J.P., Zachowski, A., Baudouin, E., Guillas, I., 2012. Phytosphingosine-phosphate is a signal for AtMPK6 activation and Arabidopsis response to chilling. New Phytol. 194 (1), 181–191.

Feng, H.L., Ma, N.N., Meng, X., Zhang, S., Wang, J.R., Chai, S., Meng, Q.W., 2013. A novel tomato MYC-type ICE1-like transcription factor, SlICE1a, confers cold, osmotic and salt tolerance in transgenic tobacco. Plant Physiol. Biochem. 73, 309–320.

Feng, W., Li, J., Long, S., Wei, S., 2019. A DREB1 gene from zoysiagrass enhances Arabidopsis tolerance to temperature stresses without growth inhibition. Plant Sci. 278, 20–31.

Gao, S.Q., Chen, M., Xu, Z.S., Zhao, C.P., Li, L., Xu, H.J., Tang, Y.M., Zhao, X., Ma, Y.Z., 2011. The soybean GmbZIP1 transcription factor enhances multiple abiotic stress tolerances in transgenic plants. Plant Mol. Biol. 75 (6), 537–553.

Garrell, J., Campuzano, S., 1991. The helix-loop-helix domain: a common motif for bristles, muscles and sex. BioEssays 13 (10), 493–498.

Graumann, P.L., Marahiel, M.A., 1998. A superfamily of proteins that contain the cold-shock domain. Trends Biochem. Sci. 23 (8), 286–290.

Gu, X., Chen, Y., Gao, Z., Qiao, Y., Wang, X., 2015. Transcription factors and anthocyanin genes related to low-temperature tolerance in rd29A: RdreB1BI transgenic strawberry. Plant Physiol. Biochem. 89, 31–43.

Guy, C.L., 1990. Cold acclimation and freezing tolerance: role of protein metabolism. Annu. Rev. Plant Physiol. Plant Mol. Biol. 41, 187–223.

Hanin, M., Brini, F., Ebel, C., Toda, Y., Takeda, S., Masmoudi, K., 2011. Plant dehydrins and stress tolerance: versatile proteins for complex mechanisms. Plant Signal. Behav. 6 (10), 1503–1509.

Hao, J., Yang, J., Dong, J., Fei, S.Z., 2017. Characterization of BdCBF genes and genome-wide transcriptome profiling of BdCBF3-dependent and-independent cold stress responses in Brachypodium distachyon. Plant Sci. 262, 52–61.

Heim, M.A., Jakoby, M., Werber, M., Martin, C., Weisshaar, B., Bailey, P.C., 2003. The basic helix–loop–helix transcription factor family in plants: a genome-wide study of protein structure and functional diversity. Mol. Biol. Evol. 20 (5), 735–747.

Hou, X.M., Zhang, H.F., Liu, S.Y., Wang, X.K., Zhang, Y.M., Meng, Y.C., Luo, D., Chen, R.G., 2020. The NAC transcription factor CaNAC064 is a regulator of cold stress tolerance in peppers. Plant Sci. 291, 110346.

Hu, W., Hu, G., Han, B., 2009. Genome-wide survey and expression profiling of heat shock proteins and heat shock factors revealed overlapped and stress specific response under abiotic stresses in rice. Plant Sci. 176, 583–590.

Huang, X.S., Wang, W., Zhang, Q., Liu, J.H., 2013. A basic helix-loop-helix transcription factor, PtrbHLH, of *Poncirus trifoliata* confers cold tolerance and modulates peroxidase-mediated scavenging of hydrogen peroxide. Plant Physiol. 162 (2), 1178–1194.

Hwang, I., Jung, H.J., Park, J.I., Yang, T.J., Nou, I.S., 2014. Transcriptome analysis of newly classified bZIP transcription factors of Brassica rapa in cold stress response. Genomics 104 (3), 194–202.

Ji, X., Nie, X., Liu, Y., Zheng, L., Zhao, H., Zhang, B., Huo, L., Wang, Y., 2016. A bHLH gene from Tamarix hispida improves abiotic stress tolerance by enhancing osmotic potential and decreasing reactive oxygen species accumulation. Tree Physiol. 36 (2), 193–207.

Jiang, W., Hou, Y., Inouye, M., 1997. CspA, the major cold-shock protein of *Escherichia coli*, is an RNA chaperone. J. Biol. Chem. 272 (1), 196–202.

Kalberer, S.R., Wisniewski, M., Arora, R., 2006. Deacclimation and reacclimation of cold-hardy plants: current understandings and emerging concepts. Plant Sci. 171, 3–16.

Kang, H.G., Kim, J., Kim, B., Jeong, H., Choi, S.H., Kim, E.K., Lee, H.Y., Lim, P.O., 2011. Overexpression of FTL1/DDF1, an AP2 transcription factor, enhances tolerance to cold, drought, and heat stresses in Arabidopsis thaliana. Plant Sci. 180 (4), 634–641.

Kaniuga, Z., Saczynska, V., Miskiewicz, E., Garstka, M., 1999. The fatty acid composition of phosphatidylglycerol and sulfoquinovosyldiacylglycerol of Zea mays genotypes differing in chilling susceptibility. J. Plant Physiol. 154, 256–263.

Kaplan, F., Kopka, J., Haskell, D.W., Zhao, W., Schiller, K.C., Gatzke, N., Sung, D.Y., Guy, C.L., 2004. Exploring the temperature-stress metabolome of Arabidopsis. Plant Physiol. 136 (4), 4159–4168.

Karlson, D., Imai, R., 2003. Conservation of the cold shock domain protein family in plants. Plant Physiol. 131 (1), 12–15.

Kasamo, K., Noushi, I., 1987. The role of phospholipids in plasma membrane ATPase activity in *Vigna radiata* L. (mung bean) roots and hypocotyls. Plant Physiol. 83, 323–328.

Kasamo, K., Kagita, F., Yamanishi, H., Sasaki, T., 1992. Low temperature-induced changes in the thermotropic properties and fatty acid composition of the plasma membrane and tonoplast of cultured rice (*Oryza sativa* L) cells. Plant Cell Physiol. 33, 609–616.

Kazemi-Shahandashti, S.S., Maali-Amiri, R., 2018. Global insights of protein responses to cold stress in plants: signaling, defence, and degradation. J. Plant Physiol. 226, 123–135.

Khazaei, M., Maali-Amiri, R., Talei, A.R., Ramezanpour, S., 2015. Differential transcript accumulation of dhydrin and beta-glucosidase genes to cold-induced oxidative stress in chickpea. J. Agric. Sci. Technol. 17 (3), 725–734.
Kojima, M., Suzuki, H., Ohnishi, M., Ito, S., 1998. Effects of growth temperature on lipids of adzuki bean cells. Phytochemistry 47, 1483–1487.
Lal, S.K., Kumar, S., Sheri, V., Mehta, S., Varakumar, P., Ram, B., Borphukan, B., James, D., Fartyal, D., Reddy, M.K., 2018. Seed priming: an emerging technology to impart abiotic stress tolerance in crop plants. In: Advances in Seed Priming. Springer, Singapore, pp. 41–50.
Lan Thi Hoang, X., Du Nhi, N.H., Binh Anh Thu, N., Phuong Thao, N., Phan Tran, L.S., 2017. Transcription factors and their roles in signal transduction in plants under abiotic stresses. Curr. Genomics 18 (6), 483–497.
Larkindale, J., Knight, M.R., 2002. Protection against heat stress-induced oxidative damage in Arabidopsis involves calcium, abscisic acid, ethylene, and salicylic acid. Plant Physiol. 128, 682–695.
Lauersen, K.J., Brown, A., Middleton, A., Davies, P.L., Walker, V.K., 2011. Expression and characterization of an antifreeze protein from the perennial rye grass, Lolium perenne. Cryobiology 62, 194–201.
Ledent, V., Vervoort, M., 2001. The basic helix-loop-helix protein family: comparative genomics and phylogenetic analysis. Genome Res. 11 (5), 754–770.
Lee, B.H., Lee, H., Xiong, L., Zhu, J.K., 2002. A mitochondrial complex I defect impairs cold-regulated nuclear gene expression. Plant Cell 14 (6), 1235–1251.
Leivar, P., Monte, E., 2014. PIFs: systems integrators in plant development. Plant Cell 26 (1), 56–78.
Li, X.D., Zhuang, K.Y., Liu, Z.M., Yang, D.Y., Ma, N.N., Meng, Q.W., 2016. Overexpression of a novel NAC-type tomato transcription factor, SlNAM1, enhances the chilling stress tolerance of transgenic tobacco. J. Plant Physiol. 204, 54–65.
Li, X., Liang, Y., Gao, B., Mijiti, M., Bozorov, T.A., Yang, H., Zhang, D., Wood, A.J., 2019. ScDREB10, an A-5c type of DREB gene of the desert moss Syntrichia caninervis, confers osmotic and salt tolerances to Arabidopsis. Genes 10 (2), 146.
Lin, M.K., Belanger, H., Lee, Y.J., Varkonyi-Gasic, E., Takoa, K., Miura, E., Xoconostle-Cazares, B., Gendler, K., Jorgensen, R.A., Phinney, B., Lough, T.J., Lucas, W.J., 2007. FLOWERING LOCUS T protein may act as the long-distance florigenic signal in the cucurbits. Plant Cell 19, 1488–1506.
Lin, X., Wisniewski, M.E., Duman, J.G., 2011. Expression of two self-enhancing antifreeze proteins from the beetle *Dendroides canadensis* in Arabidopsis thaliana. Plant Mol. Biol. Report. 29 (4), 802–813.
Liu, Y., Song, Q., Li, D., Yang, X., Li, D., 2017. Multifunctional roles of plant dehydrins in response to environmental stresses. Front. Plant Sci. 8, 1018.
Liu, C., Schläppi, M.R., Mao, B., Wang, W., Wang, A., Chu, C., 2019. The bZIP73 transcription factor controls rice cold tolerance at the reproductive stage. In: Plant Biotechnol. J.17 (9), pp. 1834–1849. https://doi.org/10.1111/pbi.13104.
Llorente, F., Oliveros, J.C., Martínez-Zapater, J.M., Salinas, J., 2000. A freezing-sensitive mutant of Arabidopsis, frs1, is a new aba3 allele. Planta 211 (5), 648–655.
Lopez-Matas, M.A., Nunez, P., Soto, A., Allona, I., Casado, R., Collada, C., Guevara, M.A., Aragoncillo, C., Gomez, L., 2004. Protein cryoprotective activity of a cytosolic small heat shock protein that accumulates constitutively in chestnut stems and is up-regulated by low and high temperatures. Plant Physiol. 134 (4), 1708–1717.
Lu, X., Yang, L., Yu, M., Lai, J., Wang, C., McNeil, D., Zhou, M., Yang, C., 2017. A novel Zea mays ssp. mexicana L. MYC-type ICE-like transcription factor gene ZmmICE1, enhances freezing tolerance in transgenic Arabidopsis thaliana. Plant Physiol. Biochem. 113, 78–88.

Lukatkin, A.S., Brazaityte, A., Bobinas, C., Duchovskis, P., 2012. Chilling injury in chilling-sensitive plants: a review. Zemdirbyste-Agriculture 99, 111–124.
Luo, C., Liu, H., Ren, J., Chen, D., Cheng, X., Sun, W., Hong, B., Huang, C., 2020. Cold-inducible expression of an Arabidopsis thaliana AP2 transcription factor gene, AtCRAP2, promotes flowering under unsuitable low-temperatures in chrysanthemum. Plant Physiol. Biochem. 146, 220–230.
Lv, K., Li, J., Zhao, K., et al., 2020. Overexpression of an AP2/ERF family gene, BpERF13, in birch enhances cold tolerance through upregulating CBF genes and mitigating reactive oxygen species. In: Plant Sci.292, p. 110375. https://doi.org/10.1016/j.plantsci.2019.110375.
Maleki, M., Ghorbanpour, M., 2018. Cold tolerance in plants: molecular machinery deciphered. In: Biochemical, Physiological and Molecular Avenues for Combating Abiotic Stress Tolerance in Plants Academic Press.pp. 57–71. https://doi.org/10.1016/b978-0-12-813066-7.00004-8.
Mao, K., Dong, Q., Li, C., Liu, C., Ma, F., 2017. Genome wide identification and characterization of apple bHLH transcription factors and expression analysis in response to drought and salt stress. Front. Plant Sci. 8, 480.
Mehta, S., Singh, B., Dhakate, P., Rahman, M., Islam, M.A., 2019. Rice, marker-assisted breeding, and disease resistance. In: Disease Resistance in Crop Plants. Springer, Cham, pp. 83–111.
Melser, S., Molino, D., Batailler, B., Peypelut, M., Laloi, M., Wattelet-Boyer, V., Bellec, Y., Faure, J.D., Moreau, P., 2011. Links between lipid homeostasis, organelle morphodynamics and protein trafficking in eukaryotic and plant secretory pathways. Plant Cell Rep. 30 (2), 177–193.
Michaelson, L.V., Napier, J.A., Molino, D., Faure, J.D., 2016. Plant sphingolipids: their importance in cellular organization and adaption. Biochim. Biophys. Acta Mol. Cell Biol. Lipids 1861 (9), 1329–1335.
Miquel, M., James, D., Dooner, H., 1993. Arabidopsis requires polyunsaturated lipids for low-temperature survival. Proc. Natl. Acad. Sci. 90 (13), 6208–6212.
Moon, S.J., Min, M.K., Kim, J., Kim, D.Y., Yoon, I.S., Kwon, T.R., Byun, M.O., Kim, B.G., 2019. Ectopic expression of OsDREB1G, a member of the OsDREB1 subfamily, confers cold stress tolerance in rice. Front. Plant Sci. 10, 297.
Morran, S., Eini, O., Pyvovarenko, T., Parent, B., Singh, R., Ismagul, A., Eliby, S., Shirley, N., Langridge, P., Lopato, S., 2011. Improvement of stress tolerance of wheat and barley by modulation of expression of DREB/CBF factors. Plant Biotechnol. J. 9 (2), 230–249.
Mukhopadhyay, J., Roychoudhury, A., 2018. Cold-induced injuries and signaling responses in plants. In: Cold Tolerance in Plants. Springer, Cham, pp. 1–35.
Örvar, B.L., Sangwan, V., Omann, F., Dhindsa, R.S., 2000. Early steps in cold sensing by plant cells: the role of actin cytoskeleton and membrane fluidity. Plant J. 23 (6), 785–794.
Pan, F., Wu, M., Hu, W., Liu, R., Yan, H., Xiang, Y., 2019. Genome-wide identification and expression analyses of the bZIP transcription factor genes in moso bamboo (*Phyllostachys edulis*). Int. J. Mol. Sci. 20 (9), 2203.
Parkin, K.L., Marangoni, A., Jackman, R.L., Yada, R.Y., Stanley, D.W., 1989. Chilling injury: a review of possible mechanism. J. Food Biochem. 13, 1745–4514.
Pearce, R.S., 2001. Plant freezing and damage. Ann. Bot. 87, 417–424.
Qu, Y., Duan, M., Zhang, Z., Dong, J., Wang, T., 2016. Overexpression of the *Medicago falcata* NAC transcription factor MfNAC3 enhances cold tolerance in Medicago truncatula. Environ. Exp. Bot. 129, 67–76.
Rahman, M., Sultana, S., Nath, D., Kalita, S., Chakravarty, D., Mehta, S., Wani, S.H., Islam, M.A., 2019. Molecular breeding approaches for disease resistance in sugarcane. In: Disease Resistance in Crop Plants. Springer, Cham, pp. 131–155.

Raju, S.K.K., et al., 2018. Low-temperature tolerance in land plants: are transcript and membrane responses conserved. In: Plant Sci.276, pp. 73–86.
Rani, K., Sharma, P., Kumar, S., Wati, L., Kumar, R., Gurjar, D.S., Kumar, D., 2019. Legumes for sustainable soil and crop management. In: Sustainable Management of Soil and Environment. Springer, Singapore, pp. 193–215.
Rinne, P.L., Kaikuranta, P.M., van der Schoot, C., 2001. The shoot apical meristem restores its symplasmic organization during chilling-induced release from dormancy. Plant J. 26, 249–264.
Routaboul, J.M., Fischer, S., Browse, J., 2000. Trienoic fatty acids are required for photosynthesis at low temperatures. Plant Physiol. 124, 1697–1705.
Ruan, M.B., Guo, X., Wang, B., Yang, Y.L., Li, W.Q., Yu, X.L., Zhang, P., Peng, M., 2017. Genome-wide characterization and expression analysis enables identification of abiotic stress-responsive MYB transcription factors in cassava (Manihot esculenta). J. Exp. Bot. 68 (13), 3657–3672.
Sangwan, V., Foulds, I., Singh, J., Dhindsa, R.S., 2001. Cold-activation of *Brassica napus* BN115 promoter is mediated by structural changes in membranes and cytoskeleton, and requires Ca^{2+} influx. Plant J. 27 (1), 1–2.
Shabala, S. (Ed.), 2017. Plant Stress Physiology. Cabi.
Singh, B., Mehta, S., Tiwari, M., Bhatia, S., 2018. Legume breeding for fungal resistance: a lesson to learn. In: Molecular Approaches for Plant Improvement. Kalpaz Publication, New Delhi, pp. 159–180.
Singh, A., Sharma, P., Kumari, A., Kumar, R., Pathak, D.V., 2019a. Management of root-knot nematode in different crops using microorganisms. In: Plant Biotic Interactions. Springer, Cham, pp. 85–99.
Singh, B., Mehta, S., Aggarwal, S.K., Tiwari, M., Bhuyan, S.I., Bhatia, S., Islam, M.A., 2019b. Barley, disease resistance, and molecular breeding approaches. In: Disease Resistance in Crop Plants. Springer, Cham, pp. 261–299.
Sornaraj, P., Luang, S., Lopato, S., Hrmova, M., 2016. Basic leucine zipper (bZIP) transcription factors involved in abiotic stresses: a molecular model of a wheat bZIP factor and implications of its structure in function. Biochim. Biophys. Acta Gen. Subj. 1860 (1), 46–56.
Steponkus, P.L., 1996. Advances in Low-Temperature Biology. Elsevier.
Su, L.T., Li, J.W., Liu, D.Q., Zhai, Y., Zhang, H.J., Li, X.W., Zhang, Q.L., Wang, Y., Wang, Q.Y., 2014. A novel MYB transcription factor, GmMYBJ1, from soybean confers drought and cold tolerance in Arabidopsis thaliana. Gene 538 (1), 46–55.
Sun, X., Wang, Y., Sui, N., 2018. Transcriptional regulation of bHLH during plant response to stress. Biochem. Biophys. Res. Commun. 503 (2), 397–401.
Tang, M., Liu, X., Deng, H., Shen, S., 2011. Over-expression of JcDREB, a putative AP2/EREBP domain-containing transcription factor gene in woody biodiesel plant Jatropha curcas, enhances salt and freezing tolerance in transgenic Arabidopsis thaliana. Plant Sci. 181 (6), 623–631.
Timperio, A.M., Egidi, M.G., Zolla, L., 2008. Proteomics applied on plant abiotic stresses: role of heat shock proteins (HSP). J. Proteome 71, 391–411.
Toledo-Ortiz, G., Huq, E., Quail, P.H., 2003. The Arabidopsis basic/helix-loop-helix transcription factor family. Plant Cell 15 (8), 1749–1770.
Vergnolle, C., Vaultier, M.N., Taconnat, L., Renou, J.P., Kader, J.C., Zachowski, A., Ruelland, E., 2005. The cold-induced early activation of phospholipase C and D pathways determines the response of two distinct clusters of genes in Arabidopsis cell suspensions. Plant Physiol. 139 (3), 1217–1233.
Vítámvás, P., Prášil, I.T., Kosová, K., Planchon, S., Renaut, J., 2012. Analysis of proteome and frost tolerance in chromosome 5A and 5B reciprocal substitution lines between two winter wheats during long-term cold acclimation. Proteomics 12, 68–85.

Wang, Y.G., Xing, Q.H., Deng, Q.Y., Liang, F.S., Yuan, L.P., Weng, M.L., Wang, B., 2003. Fine mapping of the rice thermo-sensitive genic male-sterile gene tms5. Theor. Appl. Genet. 107 (5), 917–921.
Wang, W., Vinocur, B., Shoseyov, O., Altman, A., 2004. Role of plant heat-shock proteins and molecular chaperones in the abiotic stress response. Trends Plant Sci. 9, 244–252.
Wang, J., Li, Q., Mao, X., Li, A., Jing, R., 2016. Wheat transcription factor TaAREB3 participates in drought and freezing tolerances in Arabidopsis. Int. J. Biol. Sci. 12 (2), 257.
Wang, Y., Mao, Z., Jiang, H., Zhang, Z., Chen, X., 2019. A feedback loop involving MdMYB108L and MdHY5 controls apple cold tolerance. Biochem. Biophys. Res. Commun. 512 (2), 381–386.
Wang, G., Xu, X., Wang, H., Liu, Q., Yang, X., Liao, L., Cai, G., 2019. A tomato transcription factor, SlDREB3 enhances the tolerance to chilling in transgenic tomato. Plant Physiol. Biochem. 142, 254–262.
Wani, S.H., Herath, V., 2018. Cold Tolerance in Plants: Physiological, Molecular and Genetic Perspectives. Springer.
Whitaker, B.D., 1993. Lipid changes in bell pepper fruit during chilling and rewarming. HortScience 28, 519.
Williams, M.E., Torabinejad, J., Cohick, E., Parker, K., Drake, E.J., Thompson, J.E., Hortter, M., DeWald, D.B., 2005. Mutations in the Arabidopsis phosphoinositide phosphatase gene SAC9 lead to overaccumulation of PtdIns (4, 5) P2 and constitutive expression of the stress-response pathway. Plant Physiol. 138 (2), 686–700.
Wisniewski, M., Fuller, M., Palta, J., Arora, R., 2008. Ice nucleation, propagation and deep supercooling in woody plants. J. Crop Improv. 10 (1–2), 2004.
Wisniewski, M., Norelli, J., Bassett, C., Artlip, T., Macarisin, D., 2011. Ectopic expression of a novel peach (*Prunus persica*) CBF transcription factor in apple (Malus × domestica) results in short-day induced dormancy and increased cold hardiness. Planta 233 (5), 971–983.
Xi, J., Qiu, Y., Du, L., Poovaiah, B.W., 2012. Plant-specific trihelix transcription factor AtGT2L interacts with calcium/calmodulin and responds to cold and salt stresses. Plant Sci. 185, 274–280.
Xu, W., Zhang, N., Jiao, Y., Li, R., Xiao, D., Wang, Z., 2014. The grapevine basic helix-loop-helix (bHLH) transcription factor positively modulates CBF-pathway and confers tolerance to cold-stress in Arabidopsis. Mol. Biol. Rep. 41 (8), 5329–5342.
Yamanaka, K., Fang, L., Inouye, M., 1998. The CspA family in Escherichia coli: multiple gene duplication for stress adaptation. Mol. Microbiol. 27 (2), 247–255.
Yan, S.P., Zhang, Q.Y., Tang, Z.C., Su, W.A., Sun, W.N., 2006. Comparative proteomic analysis provides new insights into chilling stress responses in rice. Mol. Cell. Proteomics 5 (3), 484–496.
Yang, Z., Sun, J., Chen, Y., Zhu, P., Zhang, L., Wu, S., Ma, D., Cao, Q., Li, Z., Xu, T., 2019a. Genome-wide identification, structural and gene expression analysis of the bZIP transcription factor family in sweet potato wild relative Ipomoea trifida. BMC Genet. 20 (1), 41.
Yang, X., Wang, R., Hu, Q., Li, S., Mao, X., Jing, H., Zhao, J., Hu, G., Fu, J., Liu, C., 2019b. DlICE1, a stress-responsive gene from Dimocarpus longan, enhances cold tolerance in transgenic Arabidopsis. Plant Physiol. Biochem. 142, 490–499.
Yu, H., Kong, X., Huang, H., Wu, W., Park, J., Yun, D.J., Lee, B.H., Shi, H., Zhu, J.K., 2020. STCH4/REIL2 confers cold stress tolerance in Arabidopsis by promoting rRNA processing and CBF protein translation. Cell Rep. 30 (1), 229–242.
Zhang, S., Scheller, H.V., 2004. Photoinhibition of photosystem I at chilling temperature and subsequent recovery in Arabidopsis thaliana. Plant Cell Physiol. 45 (11), 1595–1602.
Zhang, Y., Zhang, G., Xia, N., Wang, X.J., Huang, L.L., Kang, Z.S., 2008. Cloning and characterization of a bZIP transcription factor gene in wheat and its expression in

response to stripe rust pathogen infection and abiotic stresses. Physiol. Mol. Plant Pathol. 73 (4–5), 88–94.

Zhang, L., Zhang, L., Xia, C., Zhao, G., Liu, J., Jia, J., Kong, X., 2015. A novel wheat bZIP transcription factor, TabZIP60, confers multiple abiotic stress tolerances in transgenic Arabidopsis. Physiol. Plant. 153 (4), 538–554.

Zhang, Y., Yu, H., Yang, X., Li, Q., Ling, J., Wang, H., Gu, X., Huang, S., Jiang, W., 2016. CsWRKY46, a WRKY transcription factor from cucumber, confers cold resistance in transgenic-plant by regulating a set of cold-stress responsive genes in an ABA-dependent manner. Plant Physiol. Biochem. 108, 478–487.

Zhao, Q., Xiang, X., Liu, D., Yang, A., Wang, Y., 2018. Tobacco transcription factor NtbHLH123 confers tolerance to cold stress by regulating the NtCBF pathway and reactive oxygen species homeostasis. Front. Plant Sci. 9, 381.

Zhu, J., Dong, C.H., Zhu, J.K., 2007. Interplay between cold-responsive gene regulation, metabolism and RNA processing during plant cold acclimation. Curr. Opin. Plant Biol. 10 (3), 290–295.

Zoldan, D., Band, R.S., Guy, C.L., Porat, R., 2012. Understanding chilling tolerance traits using arabidopsis chilling-sensitive mutants. In: Environmental Adaptations and Stress Tolerance of Plants in the Era of Climate Change. Springer, New York, NY, pp. 159–173.

Zuo, Z.F., Kang, H.G., Park, M.Y., Jeong, H., Sun, H.J., Song, P.S., Lee, H.Y., 2019. Zoysia japonica MYC type transcription factor ZjICE1 regulates cold tolerance in transgenic Arabidopsis. Plant Sci. 289, 110254.

Further reading

Chinnusamy, V., Zhu, J., Zhu, J.K., 2007. Cold stress regulation of gene expression in plants. Trends Plant Sci. 12 (10), 444–451.

Zhu, J.K., 2019. Cell signaling under salt, water and cold stresses. Curr. Opin. Plant Biol. 4 (5), 401–406.

CHAPTER 10

Recent advances in plant heat stress transcription factors

Giridara-Kumar Surabhi and Bijayalaxmi Badajena
Plant Molecular Biology and "OMICS" Laboratory, Regional Plant Resource Centre, Bhubaneswar, Odisha, India

Abbreviations

AP2/ERF APETALA2/ethylene transcription factor
BAK1 BRI1 associate kinase 1
bHLH basic/helix-loop helix
BnTR1 *Brassica napus* thermas resistance gene 1
BRI1 brassinosteroid intensive 1
BRI2 brassinosteroid intensive 2
bZIP basic leucine zipper
BZR1 brassinazole resistant
Cas9 CRISPR-associated protein 9
CRISPR clustered regularly interspaced palindromic repeat
CTADs C-terminal activation domains
DBD DNA-binding domain
DOF DNA binding with one finger
DRE/CRT dehydration responsive element/C-repeat
DREB dehydration responsive element binding
FER feronia
HD-ZIP homeodomain-leucine zipper
HSE heat shock element
Hsfs heat shock factors
HSFs heat stress transcription factors
HSPs heat shock proteins
IPCC Intergovernmental Panel on Climate Change
MYB myeloblastosis oncogene
NAC no apical meristem, *Arabidopsis thaliana* transcription activation factor and cup-shaped cotyledon protein
NES nuclear export signal
NLS nuclear localization signal
ODs oligomerization domains
PIF4 phytochrome interacting factor 4
RBOH1 respiratory burst oxidase homologue 1
RINGv really interesting new gene variance
RNAi RNA interference
RUBISCO ribulose-1,5-biphosphate carboxylase
WMO World Meteorological Organization

Transcription Factors for Abiotic Stress Tolerance in Plants
https://doi.org/10.1016/B978-0-12-819334-1.00010-1

1. Introduction

The current [CO_2] level of approximately 380 μmol mol^{-1} is estimated to reach between 730 and 1020 μmol mol^{-1} by 2100. The changes projected in [CO_2] and other greenhouse gases are expected to increase global air temperature by 2.5–4.5°C during the same period, and a more frequent occurrence of severe heat waves with subsequent disastrous effects on crop productivity and harvest yield (IPCC, 2012). Heat stress temperature always crosses the threshold level for a period of time that is sufficient to cause multifarious alteration in plant growth, development, physiological processes, and crop yield (Driedonks et al., 2015). Under heat stress, crop plants experience a lot of cellular damage or cell death through denaturing of proteins, RNA, cytoskeleton, and alterations in the efficiency of enzymatic reaction in the cell, which may result into catastrophic collapse of cellular organization and metabolic imbalance (Hasanuzzaman et al., 2013; Bita and Gerats, 2013; Bento et al., 2017). Heat stress also affects organization of microtubule in mitotic cell, elongation of spindle, and formation of microtubule aster in mitotic cells (Hemantaranjan et al., 2014). All these injuries eventually lead to inhibition of plant growth, starvation, reduced ion flux, production of toxic compounds, and reactive oxygen species (ROS) (Hasanuzzaman et al., 2013; Filewod and Thomas, 2014; Brunel-Muguet et al., 2015; Driedonks et al., 2015; Bento et al., 2017).

The reproductive phase of crop plants is more sensitive to heat stress and it causes impaired pollen tube growth, reduced ovule viability, inconsistency in stigmatic and style positions, reduced number of pollen grains by the stigma, disturbed fertilization processes, proembryo, and unfertilized embryo, so all these effects diminish reproductive development in plant organ through abortion of flower and fruits (Hasanuzzaman et al., 2013; Liu et al., 2013). Heat stress-induced yield reduction was documented in different crops including in rice (Fahad et al., 2016a,b), in wheat (Liu et al., 2014; Barlow et al., 2015), in barley (Rollins et al., 2013), in soybean (Djannaguirman et al., 2011), in maize (Gabaldon-Leal et al., 2016; Daaryanto et al., 2016), and in chickpea (Devasirvatham and Tan, 2018).

The crop improvement through conventional breeding has marginal success, due to the complexity of stress tolerance traits. The transgenic approach is now being popularly used to develop transgenic crops tolerant to abiotic stresses (Yamaguchi and Blumwald, 2005), including heat stress. Therefore, deciphering the molecular mechanisms by which plants perceive and transduce stress signals to cellular machinery to initiate adaptive responses is an essential prerequisite for identification of the key genes and pathways to engineer heat stress-tolerant crops (Ray et al., 2009; Heidarvand and Amiri, 2010; Sanchez et al., 2011). The aim of plant biologists, breeders, and farmers is to improve the current germplasm and develop new cultivars that will be able to withstand and perform at the highest possible level under stressful conditions. The crop improvement can be achieved either through traditional breeding or by genetic engineering and recent genome editing techniques (Surabhi et al., 2019). To achieve this goal knowledge on crops response to heat stress at both cellular and organism level is required (Mittler and Blumwald, 2010; Bokszczanin et al., 2013). The genome wide analysis conveyed that crop plants affects with various types of environmental stresses, which are mainly activated by the induction or repression of many genes encoding with proteins that involved in numerous biological processes (Nakashima et al., 2012).

Numerous candidate genes involved in heat stress tolerance have been identified and functionally validated, which are generally classified into two types: functional and regulatory genes (Shinozaki et al., 2003). Functional genes encode important enzymes and metabolic proteins, such as heat shock proteins (HSPs), late embryogenesis abundant (LEA) proteins and detoxification enzymes, which directly protects cells from heat stress. The regulatory genes encode various regulatory proteins including TFs, protein kinases, and protein phosphatases, which regulate signal transduction and gene expression under heat stress.

The TFs play a crucial role in the conversion of stress signal perception to stress-responsive gene expression by interacting with *cis*-acting elements present in the promoter region of various target stress-responsive genes in the signal transduction processes. TFs activate the signaling cascade of the whole network of genes that act together in enhancing plant tolerance to the harsh environmental conditions (Akhtar et al., 2012), including heat stress. In plant genomes, approximately 7% of the coding sequences are assigned to TFs and many of these often belong to large gene families compared with animals and yeasts, such as the heat stress transcription factors (HSFs) family (Baniwal et al., 2004; Udvardi et al., 2007). The transcriptional reprogramming at higher temperatures is controlled by the activity of the HSFs (Fragkostefanakis et al., 2015). In recent years, several studies on transcription factors under heat stress revealed its significant role in enhancing heat stress tolerance in different crop plants (Grover et al., 2013). Thus, genetically modifying the expression of certain TFs (regulatory genes) can greatly influence crop plants tolerance to heat stress, because it mimics or enhances stress signals to regulate many downstream stress-responsive genes at a time (Yang et al., 2011).

2. Plant heat stress and activation of transcription factors

As a sessile organism, plants have produced various defense mechanisms to fight against unfavorable environmental conditions. In many abiotic stress signaling pathways, the phosphorylation and dephosphorylation of proteins mediated by protein kinase and phosphatases is an important array of the stress mechanism. Transcriptome analysis shows that genes encoding for heat stress transcription factors (HSFs), osmolytes, heat shock proteins (HSPs), which are constituted major downstream players in high-temperature response by affecting the membrane fluidity and enzyme involved in ROS homeostasis (Frank et al., 2009). Recently, many studies have paid attention to some regulatory genes including TFs, which play an essential role in the field of heat stress by regulating a large spectrum of downstream heat stress-responsive genes or overexpressing of some heat stress transcription factor can lead to acquire thermotolerance in plants. The regulation of plant TFs mainly put attention to four levels, including transcriptional, posttranscriptional, translational, and posttranslational levels (Fragkostefanakis et al., 2015; Surabhi, 2018).

TFs are activated or suppressed by specifically binding to *cis*-element in the promoter of stress-responsive genes and regulate their transcription (Danquah et al., 2014; Hernandez-Garcia and Finer, 2014). Under heat stress plants produce more amount of ROS such as H_2O_2, O_2^-, and O_2 which are mainly detoxified by various types of detoxifying or ROS scavenging enzymes and heat shock proteins, which is mainly generated by heat stress transcription factors (Baxter et al., 2014; Qu et al., 2013). Many TFs have been found and cover a large portion of the plant genome. These are categorized according to their DNA domain presence (Erpen et al., 2017).

3. Different types of plant heat stress transcription factors

With the availability of plant genomes and various omics tools such as genomics, transcriptomics, and proteomics have made a tremendous increase in crop productivity with higher yield and improved qualities against multiple environmental stresses (Liu et al., 2014). TFs are good candidates for genetic-engineering to breed stress-tolerant crops (Erpen et al., 2017). The major transcription factors under heat stress include, Hsfs (Hu et al., 2018; Chidambaranathan et al., 2018), WRKY (He et al., 2016), MYB

(Zhao et al., 2017; Casaretto et al., 2016), NAC (Guo et al., 2015; Fang et al., 2015), bZIP (Ramakrishna et al., 2018; Geng et al., 2018), zinc finger protein (Agarwal and Khurana, 2017; Hsu et al., 2014), AP2/ERF (Wu et al., 2015; Kang et al., 2011), DREB (Wu et al., 2018; Song et al., 2018b), ERF (Cheng et al., 2013), and bHLH (Wang et al., 2018a, b,c,d, 2019; Koini et al., 2009), brassinosteriod TFs (Yin et al., 2018; Sahni et al., 2016), which have been found to play a major role in controlling heat stress response pathways in different model and agricultural crops are described briefly in this chapter (Table 1).

3.1 Heat shock factors

The heat shock factors (Hsfs) play a central role in the regulation of heat-inducible genes at the time of heat acclimation process (Liu and Charng, 2013; Fragkostefanakis et al., 2015; Yabuta, 2016). Hsfs are defined as important molecular chaperon and gives protection to plant cells against heat stress by interacting with different type of *cis*-acting elements present in the promoter region of stress-responsive genes (Hu et al., 2009; Anckar and Sistonen, 2011; Liu et al., 2011; Akhtar et al., 2012; Guo et al., 2016; Jiang et al., 2018; Wang et al., 2018b). Plants Hsfs have unique feature and develop tens of members Hsfs families with various expression and function as compared to nonplant Hsfs (Scarf et al., 2012; Ohama et al., 2017). Various types of Hsfs have been found in different plants, but first *HSF* gene was cloned from tomato (Scarf et al., 1990). The plant Hsfs show a multiplicity due to gene duplication and genome duplication at different points of evolution, followed by extensive gene loss (Scarf et al., 2012; Chidambaranathan et al., 2018). Plant Hsfs present a well-conserved modular structure and mode of action. Almost all members of the Hsfs family share a several highly conserved domains including a common core structure comprised of DNA-binding domains (DBDs), hydrophobic heptad regions or oligomerization domains (ODS or HR-A/B), nuclear localization signal (NLS), C-terminal activation domains (CTADs), a leucine-rich nuclear export signal (NES), and activator motifs (AHA motifs) (Scarf et al., 2012; Chidambaranatham et al., 2016; Guo et al., 2016; Yabuta, 2016; Jiang et al., 2018; Lavania et al., 2018). Hsfs are classified into three distinct types, including classes HsfA (A1, A2, A3, A4, A5, A6, A7, A8, A9), HsfB (B1, B2, B3, B4), and HsfC (C1, C2) according to their flexible linker and oligomerization domains (Scarf et al., 2012; Cheng et al., 2015; Huang et al., 2015; Yabuta, 2016; Wang et al., 2018b).

Table 1 Different transcription factor genes overexpressed through transgenic approach in different crop and model plants and their physiological, biochemical, reproductive, and yield responses under imposed heat stress.

TFs family	Gene	Source plant	Transgenic plant source	Temperature range/ duration	Stress responses key findings	References
HSFs	*SlHsfA3*	Wheat (*Triticum aestivum* L.)	*Arabidopsis* (*Arabidopsis thalaiana* L.)	37°C, 12h	In tomato, SlHsfA3 expression confirmed that it is a multistress-responsive gene. But in *Arabidopsis* ectopic overexpression of SlHsfA3 showed acquired and basal thermotolerance under 37°C heat stress	Li et al. (2013)
	TaHsf3	Wheat (*Triticum aestivum* L.)	*Arabidopsis* (*Arabidopsis thalaiana* L.)	42°C, 3h	A novel HSF class B2 gene, *TaHsf3* was overexpressed in *Arabidopsis* and it showed thermotolerance by regulating expression of *AtHSp70-1*. Survivability of transgenic plant exhibited 82% as compared to WT, which was shown only 25%. Under heat stress, chlorophyll content in transgenic was higher ($0.67\,mg\,g^{-1}$) than WT ($0.35\,mg\,g^{-1}$)	Zhang et al. (2013)
	TaHsfA2d	Wheat (*Triticum aestivum* L.)	*Arabidopsis* (*Arabidopsis thalaiana* L.)	42°C, 20min	Under heat stress, overexpression of *TaHsfA2d* in *Arabidopsis* exhibited better thermotolerance capacity with high biomass accumulation, better growth and number of rosette leaves. Also transgenic plants showed higher photosynthesis and faster seedling germination	Chauhan et al. (2018)
	CarHSFB2	Chickpea (*Cicer arietinum* L.)	*Arabidopsis* (*Arabidopsis thaliana* L.)	46°C, 36h	Overexpression of a class B HSF gene, *CarHSFB2* showed enhanced thermotolerance in transgenic *Arabidopsis* with survival rate of 95.5% as compared to WT plants	Ma et al. (2016)

	ZmHsf04	Maize (*Zea mays* L.)	*Arabidopsis* (*Arabidosis thaliana* L.)	44°C, 2h	Overexpression of *ZmHsf04* in transgenic *Arabidopsis* plants significantly enhanced thermotolerance with survival rate of 80% as compared to WT plants (survival rate of 25%)	Jiang et al. (2018)
	ZmHsf06	Wheat (*Triticum aestivum* L.)	*Arabidopsis* (*Arabidosis thaliana* L.)	BT-45°C, 1h AT-46°C, 2h	Under heat shock, overexpression of *ZmHsf06* in *Arabidopsis* enhanced range of thermotolerance and lead to various types of advantages to transgenic plants. Further, transgenic *Arabidopsis* plants showed germination rate up to 92%, root growth up to 8cm and enhanced antioxidant enzyme activity (SOD, CAT, POD) compared to WT plants	Li et al. (2015)
	FaHsf2c	Tall fescue (*Festuca arundinacea* Schreb. L.)	*Arabidopsis* (*Arabidopsis thaliana* L.)	45°C/40°C or 37°C/32°C 14h photoperiod for 42 days	Ectopic overexpression of *FaHsfA2c* gene in *Arabidopsis* plant showed heat tolerance by triggering heat protective gene expression. Overexpression of *FaHsfA2c* in *Arabidopsis* plant showed 70% of survival rate compared to WT (survival rate 45%) plants	Wang et al. (2017a,b,c,d)
	LlHSFA1	Easter lily (*Lilium longiflorum* L.)	*Arabidopsis* (*Arabidpsis thaliana* L.)	37°C, 1h	Under heat stress, transgenic *A. thaliana* overexpressing *LlHSFA1* showed increased thermotolerance and survival rate up to 55% as compared to WT plants which showed 31.8%	Gong et al. (2014)
	VpHSF1	Chinese wild grapevine (*Vitis pseudoreticulata* L.)	Tobacco (*Nicotiana tabacum* L.)	40°C/48°C-BT/ 2.5 and 3h 50°C-AT, 2h	Transgenic tobacco plants with overexpressing *VpHSF1* showed increased thermotolerance with survival rate of 61% compared to WT plants	Peng et al. (2012)

Continued

Table 1 Different transcription factor genes overexpressed through transgenic approach in different crop and model plants and their physiological, biochemical, reproductive, and yield responses under imposed heat stress—cont'd

TFs family	Gene	Source plant	Transgenic plant source	Temperature range/ duration	Stress responses key findings	References
	TaHSFA6f	Wheat (*Triticum aestivum* L.)	Wheat (*Triticum aestivum* L.)	45°C, 2h	*TaHSF6f* was acted as positive regulator of several classes of heat protection genes and increased thermotolerance in wheat. *TaHsfA6f* is constitutively expressed in green organs but markfully upregulated when wheat plant subjected to r heat stress. Transgenic wheat plant overexpressing *TaHSFA6f* showed increased thermotolerance by regulating some *HSP* genes, *TaGAAP*, and *TaRof*	Xue et al. (2015)
	TaHsfC2a-B	Wheat (*Triticum aestivum* L.)	Wheat (*Triticum aestivum* L.)	43–45°C, 2h	Under 43°C heat stress, overexpression of *TaHsfC2a-B* in the root and leaves of transgenic wheat plant showed better growth, increased thermotolerance with survival rate of 90% while only 15% survival rate in WT plants	Hu et al. (2018)
	TaHsfA6e	Wheat (*Triticum aestivum* L.)	Wheat (*Triticum aestivum* L.)	38°C, 2h	The expression of HSF family subclass A6 gene, *TaHsfA6e* TF and their targets (HSP70, HSP90, and HSP17) showed upregulation under HS, which has a significant role in heat stress tolerance of wheat at different growth stages	Kumar et al. (2018)
	CarHsfs	Chickpea (*Cicer arietinum* L.)	Chickpea (*Cicer arietinum* L.)	352°C, 24h	Under heat stress, the chickpea expressed regulation of some *Hsf* genes such as *CarHsfA2, A6,* and *B2* at pod developmental stage, which conferred heat tolerance in chickpea plant	Chidambaranathan et al. (2018)

	BcHsfA1	Mustard (*Brassica campestris* L.)	Tobacco (*Nicotiana tabacum* L.)	42°C, 0–5 h	The transgenic tobacco lines with *BcHsfA1* overexpression showed more heat tolerance, better growth and less damage under HS compared to WT. It showed higher content of Chl, Chl-a, Chl-b, reduced amount of MDA content, higher soluble sugar content, lower comparative electrical conductivity compared to WT. Many heat stress defense genes were upregulated in transgenic tobacco plants compared to control	Zhu et al. (2018)
WRKY	*TaWRKY33*	Wheat (*Triticum aestivum* L.)	*Arabidopsis* (*Arabidopsis thaliana* L.)	45°C, 5 h	Overexpression of *TaWRKY33* in *Arabidopsis* transgenic lines exhibited thermotolerance capacity compared to WT. The transgenic lines expressed survival rate of 45% as compared to WT under heat exposure	He et al. (2016)
	ZmWRKY106	Maize (*Zea mays* L.)	*Arabidopsis* (*Arabidopsis thaliana L.*)	45°C, 5 h	*ZmWRKY106* overexpressing *Arabidopsis* lines exhibited improved thermotolerance and also had survival rate more than 30%	Wang et al. (2018a,b,c,d,e)
	CaWRKY40	Pepper (*Capsicum annum* L.)	Tobacco (*Nicotina tabacum* L.)	42°C, 24 or 48 h	Under heat stress, overexpressed *CaWRKY40* in tobacco plants evoked thermotolerance capacity by regulation if some transcripts such as; *NtHSP18, NtsmallHsp*, and *NtHsf2*	Dang et al. (2013)

Continued

Table 1 Different transcription factor genes overexpressed through transgenic approach in different crop and model plants and their physiological, biochemical, reproductive, and yield responses under imposed heat stress—cont'd

TFs family	Gene	Source plant	Transgenic plant source	Temperature range/ duration	Stress responses key findings	References
	WRKY39	*Arabidopsis* (*Arabidopsis thaliana* L.)	*Arabidopsis* (*Arabidopsis thaliana* L.)	45°C, 4h	Under heat stress *A. thaliana* exhibited 5.3-fold expression of *WRKY39*, which confirms the significant role toward heat stress. Further, overexpressing *35S:WRKY39* plants exhibited greater thermotolerance and 84%–95% seed germination as compared to WT	Li et al. (2010a,b)
	AtWRKY25/26/33	*Arabidopsis* (*Arabidopsis thaliana* L.)	*Arabidopsis* (*Arabidopsis thaliana* L.)	42–48°C, 4h	The function of *WRKY25*, *26*, and *33* explained as positive regulators in *Arabidopsis* plants under heat stress, which showed partial redundancy in thermotolerance and regulate the cooperation between the Hsps and MBF1c pathways that mediate responses to heat stress. After heat treatment transgenic *Arabidopsis* seeds survivability, 81%–88% germination, newly emerging leaves	Li et al. (2010)
	WRKY25	*Arabidopsis* (*Arabidopsis thaliana* L.)	*Arabidopsis* (*Arabidopsis thaliana* L.)	45°C, 4h	The *WRKY25* overexpressing seeds evoked enhanced thermotolerance compared to WT seeds. It showed up regulation of HS-related genes and oxidative stress-responsive genes. Due to induced expression of all HS-related genes, transgenic seeds showed better seed germination under heat stress	Li et al. (2009)

	OsWRKY11	Rice (*Oryza sativa* L.) Cv. Sasanishiki	Rice (*Oryza sativa* L.) Cv. Sasanishiki	37°C/22°C, 2.5 days	When *OsWRKY11* cDNA was fused to the promoter of HSP101 of rice, the transgenic lines showed significant heat and drought tolerance, as indicated by the slower leaf-wilting and less-impaired survival rate of green parts of plants	Wu et al. (2009)
MYB	*OsMYB55*	Rice (*Oryza sativa* L.)	Maize (*Zea mays* L.)	40°C, 5 days	Overexpression of *OsMYB55* in maize improved maize plant tolerance to high-temperature stress. Transgenic maize plants showed less significant reduction in plant growth, chlorophyll content, plant height, and leaf damage under heat stress	Casaretto et al. (2016)
	TaMYB80	Wheat (*Triticum aestivum* L.)	*Arabidopsis* (*Arabiopsis thaliana* L.)	45°C, 1 h	Six MYB genes constitutively expressed in wheat under heat stress, out of them only *TaMYB80* conferred thermotolerance. Under heat stress, overexpressed *TaMYB80* plants exhibited 70% survival rate compared to WT plants	Zhao et al. (2017)
	LeAN2	Tomato (*Solanum lycopersicum* L.)	Tomato (*Solanum lycopersicum* L.)	40°C, 24 h	Transgenic tomato plants with increased R2-R3 MYB gene, *LeAN2* expression resulted in thermotolerance under heat stress. Overexpressed *LeAN2* in tomato plants indicated less accumulation of H_2O_2, O_2^-, ROS, less serious membrane damage, mitigate oxidative damage and both young seedling and grown plant exhibited better growth, which confers thermotolerance capacity of *LeAN2* under heat stress	Meng et al. (2015)

Continued

Table 1 Different transcription factor genes overexpressed through transgenic approach in different crop and model plants and their physiological, biochemical, reproductive, and yield responses under imposed heat stress—cont'd

TFs family	Gene	Source plant	Transgenic plant source	Temperature range/ duration	Stress responses key findings	References
NAC	*NAC019/RCF2*	*Arabidopsis* (*Arabidopsis thaliana* L.)	*Arabidopsis* (*Arabidopsis thaliana* L.)	37°C, 1 h	Transgenic *Arabidopsis* plants expressing *RCF2* or *NAC019* under the control of 35S promoter exhibited increased thermotolerance, reduced damage to whole plant and flowers by regulating five *HSFs* or four *HSPs* under heat stress	Guan et al. (2014)
	TaNAC2L	Wheat (*Triticum aestivum* L.)	*Arabidopsis* (*Arabidopsis thaliana* L.)	45°C, 2 h	The overexpression of *TaNAC2L* in *Arabidopsis* plants enhanced acquired heat tolerance by upregulating downstream stress-responsive genes, which induced *TaNAC2L* as a heat stress sensitive and heat stress tolerant genes	Guo et al. (2015)
	ANAC042 or *JUNGBRUNNEN1*	*Arabidopsis* (*Arabidopsis thaliana* L.)	*Arabidopsis* (*Arabidopsis thaliana* L.)	37°C, 90 min	A NAC transcription factor, *ANAC042* increased survivability up to 100% and thermotolerance under heat stress in *Arabidopsis thaliana* when overexpressed, in comparison to knock down lines	Shahnejat-Bushehri et al. (2012)
	SNAC3	Rice (*Oryza sativa* L.)	Rice (*Oryza sativa* L.)	42°C, 1–2 days	*SNAC3-OE* rice plant survived up to 40%, less extensive membrane lipid peroxidation and improved cell membrane stability, attenuated oxidative damage was found under heat stress treatment. Also *SNAC3* overexpressed in rice plant scavenged ROS through controlling the expression of downstream genes, which induced *SNAC3* a positive NAC regulator of heat resistance	Fang et al. (2015)

bZIP	*EcbZIP17*	Finger millet (*Eleusine coracana* L.)	Tobacco (*Nicotiana tabacum* L.)	42°C, 3 days	Under heat treatment, overexpressing *EcbZIP17* transgenic tobacco plant retained their green color, chlorophyll content and the plant survived after 3 days of heat treatment	Ramakrishna et al. (2018)
	TabZIP60	Wheat (*Triticum aestivum* L.)	*Arabidopsis* (*Arabidopsis thaliana* L.)	45°C, 2h	The *TabZIP60s* overexpression (*35S::TabZIP60s-MYC*) *Arabidopsis* lines examined phenotypic variations in response to heat stress, such as germination up to 74.8% and survival rate up to 48.2% compared to control, respectively, which indicates clearly that *TabZIP60* heat stress tolerance in *Arabidopsis*	Geng et al. (2018)
AP2/ERF	*FTL1/DDF1*	*Arabidopsis* (*Arabidopsis thaliana* L.)	*Arabidopsis* (*Arabidopsis thaliana* L.)	45°C, 12h	*FTL1/DDF1*overexpressed *Arabidopsis* plant enhanced tolerance to high temperature by showing 100% survival rate	Kang et al. (2011)
	AtDREB1A	*Arabidopsis* (*Arabidopsis thaliana* L.)	*Chrysanthemum* (*Chrysanthemum morifolium* L.)	45°C, 36h	The overexpression of *AtDREB1A* in *Chrysanthemum* enhanced thermotolerance and also enhanced tolerance of photosynthesis. Under HS overexpressing *AtDREB1A* in *Chrysanthemum* plants experienced 56.9% of recovery from heat stress injury, approximately 70% of survival rate, lower electrolyte leakage, enhanced photosynthetic activity by showing higher levels of CO_2 assimilation rate and Rubisco enzymatic activity	Hong et al. (2009)

Continued

Table 1 Different transcription factor genes overexpressed through transgenic approach in different crop and model plants and their physiological, biochemical, reproductive, and yield responses under imposed heat stress—cont'd

TFs family	Gene	Source plant	Transgenic plant source	Temperature range/ duration	Stress responses key findings	References
	CmDREB6	*Chrysanthemum* (*Chrysanthemum morifolium* L.)	*Chrysanthemum* (*Chrysanthemum morifolium* L.)	45°C, 24h	*CmDREB6* play an important role in heat stress tolerance, where it showed higher expression in leaves under HS. After 1-week recovery growth, two transgenic overexpressed *CmDREB6* lines exhibited survival rate up to 85% and 50%, respectively, compared to WT. The overexpression of *CmDREB6* also induced expression of transcript *CmHsfA4*, *CmHSP90*, *CmDREB6*, and also ROS scavenging gens *CmSOD* and *CmCAT*, which are enhanced tolerance under heat stress	Du et al. (2018)
	DREB2C (VOZ1)	*Arabidopsis* (*Arabidopsis thaliana* L.)	*Arabidopsis* (*Arabidopsis thaliana* L.)	44°C, 35min	The *voz1 voz2-2* plants exhibited higher thermotolerance by increasing the activity of *DREB2C*, which induced the expression of *HsfA3*	Song et al. (2018b)
	LiDREB2B	Lily (*Lilium longiflorum* L.)	*Arabidopsis* (*Arabidopsis thailiana* L.)	37–45°C, 3h	Overexpression of *LiDREB* in transgenic *Arabidopsis* plants showed better acquired thermotolerance after long-time recovery and higher survival rate after 7 days of heat stress treatment	Wu et al. (2018)
	DREB2A (DPB3-1)	*Arabidopsis* (*Arabidopsis thailiana* L.)	*Arabidopsis* (*Arabidopsis thailiana* L.)	45°C, 50min	Overexpression of *DPB3-1* in transgenic *Arabidopsis* plants formed a transcriptional complex with NF-YA and NF-YB that enhanced heat stress-inducible gene expression during heat stress by activating heat stress-inducible *DREB2A*. Transgenic *Arabidopsis* plant showed less growth retardation as compared to WT plants under heat stress	Sato et al. (2014)

	ERF1	*Arabidopsis* (*Arabidopsis thailiana* L.)	*Arabidopsis* (*Arabidopsis thailiana* L.)	45°C, 1 h	Overexpressing 35S:RNAi *Arabidopsis* plants resulted augmented thermotolerance by resulting half of the 35S: ERF1 plants survived and many heat shock-related genes such as *AtHsfA3* and *Hsps* as compared to *ERF1*:RNAi plants	Cheng et al. (2013)
BR TFs	*BZR1*	Tomato (*Solanum lycopersicum* L.)	Tomato (*Solanum lycopersicum* L.)	42°C/38°C/24 h	*FER2/3* genes as direct targets of *BZR1* mediate BR response and heat stress tolerance in tomato plants. After heat stress overexpressed *BZR1* tomato plants resulted in high thermotolerance, a less severe wilting phenotype, increase in photosystem-II and a decrease in electrolyte leakage by regulating the expression of heat stress responsive genes such as, *HSFA2, HSFA3, HSP70,* and *HSP90*	Yin et al. (2018)
	DWF4	*Arabidopsis* (*Arabidopsis thaliana* L.)	Rapeseed (*Brassica napus* L.)	45°C, 3 and 4 h	Ectopic overexpression of *AtDWF4* in *B. napus* seedling under heat treatment expressed survivability up to 77% and 55% as compared to WT and VC. Marker of thermotolerance such as *Hsp101*, *Hsp90* accumulated in much higher level than WT seedling	Kagale et al. (2007)
ZNF Novel zinc finger transcription factor	*TaZNF*	Wheat (*Triticum aestivum* L.)	*Arabidopsis* (*Arabidopsis thailiana* L.)	42°C, 2 h	Overexpression of *TaZnF* in *Arabidopsis thaliana* conferred improved tolerance to both basal and high-temperature stress. The transgenics were recovered and showed early flowering compared to WT. Enhanced growth and early recovery resulted in bigger plants with more yield	Agarwal and Khurana (2018)

The Hsfs are normally found in an inactive form inside the cytoplasm of eukaryotes during nonstressed conditions due to association with other heat-stressed proteins. But when it comes under heat stress condition, according to chaperon model all heat-stressed protein comes away from Hsfs to act as molecular chaperon. Then this released Hsfs lead to activate expression of HSPs and other heat-responsive genes through forming trimers and relocate to the nucleus (Hasanuzzaman et al., 2013; Driedonks et al., 2015; Volkov et al., 2006). Hsfs play a pivotal role in the regulation of heat-inducible genes (Liu and Charng, 2013; Fragkostefanakis et al., 2015; Yabuta, 2016). Transcriptional activation or transactivation acts through binding of Hsfs to heat shock element (HSE; 5′AGAANNTTCT-3′) in the promoter region of HS stress-inducible genes (Yabuta, 2016).

Overexpression of Hsfs in transgenic tomato (Li et al., 2013), soyabean (Guo et al., 2008), *Arabidopsis* (Li et al., 2015; Wang et al., 2016a,b,c; Jiang et al., 2018; Zhang et al., 2012), wheat (Chauhan et al., 2018; Hu et al., 2018; Kumar et al., 2018), rice (Guo et al., 2008), strawberry (Liao et al., 2016), lily (Gong et al., 2014), chickpea (Chidambaranathan et al., 2018), and grape (Li et al., 2015) caused upregulation of stress-associated gene and enhanced thermotolerance. The previous studies have shown that HsfA2 is important for thermotolerance in plants (Meiri and Breiman, 2009; Yoshida et al., 2008). Guo et al. (2008) have reported that a rice A2-type heat stress transcription factors *OsHsfA2e* conferred better thermotolerance in *Arabidopsis* plants. Overexpression of *OsHsfA2e* in *Arabidopsis* plants showed stem elongation, flowering after 1 week and exhibited better thermotolerance at heat treatment of 45°C. Overexpression of *OsHsfA2e* in transgenic *Arabidopsis* plants synthesized some heat-responsive gene to acquire thermotolerance capacity such as *HSP17.7-CII, HSP26.5-P, HSP70, HSP101, HSP17.6A-CI, HSP17.4-CIII* (Guo et al., 2008). The role of a subclass HsfA6 member of the Hsf family (TaHsfA6) was studied to investigate the thermotolerance mechanism in wheat grain (Xue et al., 2015). The constitutive expression of TaHsfA6f (is a transcriptional activator) in green organs (leaf, stem, and hull) of wheat acted as positive regulator of several classes of heat stress protection genes (*TaHSP*, *TaGAAP*, and *TaRof1*) and provided enhanced thermotolerance capacity to wheat under heat stress conditions (Xue et al., 2015) (Fig. 1).

The chickpea is one of the important pulse crops in India and it is most affected by heat stress at the reproductive stage of crop growth (Devasirvatham et al., 2013). A number of Hsf genes were specifically upregulated in response to heat stress during pod development stage in chickpea

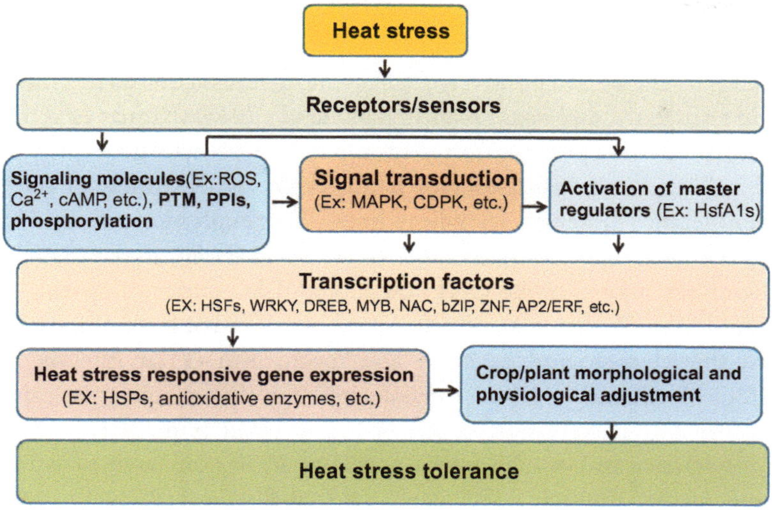

Fig. 1 Regulation of heat stress transcription factors in crop plants.

plant (Chidambaranathan et al., 2018). The expression of the *CarHsfs* was analyzed at both the seedling and pod stage. The in silico analysis showed three classes of *CarHsfs* genes (HsfA2, A6, and B2), and expressed in the early response to heat stress at seedling stage and also found to be upregulating during pod developmental stage (Chidambaranathan et al., 2018).

The ectopic overexpression of *SlHsfA3* in transgenic tomato showed increased thermotolerance by directly activating the expression of *SlHsp26.1-P* and *SlHsp21.5.ER* (Li et al., 2013). An HsfC member class TaHsfC2A showed that constitutive overexpression of *TaHsfC2a-B* positively upregulated heat protection genes (*TaHSP16.9b, TaHSP17.3, TaHSP26.6, TaHSP62.4, TaHSP70d, HSP101d, TaGalSyn, TaHSA32, TaRof1,* and *TaβAmy1*) in the roots and leaves of transgenic wheat and provide a high basal level of the heat protection capacity in the grains (Hu et al., 2018). Overexpression of *Triticum aestivum TaHsf3*, a class B2 HSF gene in *Arabidopsis* showed thermotolerance capacity of 75%–91% after imposing heat treatment. The overexpressing TaHsf3 upregulated the expression of *AtHSP70-1*, so it could prevent protein damage and stabilize plant membrane under extreme temperature (Zhang et al., 2013). Jiang et al. (2018) have focused on the classic member of *HsfA2* gene, ZmHsf04 in *Arabidopsis* plants for its function at the level of heat stress. Compared to wild type, transgenic overexpressing *ZmHsf04 Arabidopsis* plant significantly increased thermotolerance by upregulating the expression of heat-specific

HSP genes such as *AtHSP25.3-P, AtHsp18.2-CI, AtHsp70B,* and *AtHsp101.* Overexpression of a maize gene ZmHsf04 in *Arabidopsis* showed improved heat stress tolerance and positively regulate short-term acquired thermotolerance in transgenic *Arabidopsis* plants (Jiang et al., 2018). Kumar et al. (2018) have reported that a putative heat-responsive *TaHsfA6e* TF gene from wheat which belongs to subclass A6 of HSF family can be used as candidate genes for thermotolerance potential of wheat lines. The expression of *TaHSFA6e* upregulated the expression of HSP (*HSP70, HSP 90,* and *HSP17*) and biochemical parameters like CAT, GPX, and TAC under heat stress (30°C and 38°C, 2 h) (Kumar et al., 2018). Wang et al. (2018a,b,c,d,e) have investigated the physiological function and mechanism of an A2-type HSF, *FaHsf2C* of tall fescue (*Festuca arundinacea*) for thermotolerance capacity. The ectopic overexpression of *FaHsfA2c* revealed positive regulation of thermotolerance in both tall fescue and *Arabidopsis* compared to wild type by upregulating multiple HSP genes, such as *Hsp70, sHSPs,* and *Hsp101* (Wang et al., 2018a,b,c,d, 2019).

Recently, *Brachypodium distachyon,* commonly called purple false brome makes itself a promising model for studying their expression profiles under different abiotic stresses including HS due to its whole-genome sequence data (Wen et al., 2014). Wen et al. (2017) used the method of phytogenetic analysis, chromosomal localization, gene duplication, and investigated the involvement of *Hsfs* in *Brachypodium distachyon* under different abiotic stresses including heat stress. The whole-genome expression profile analysis exhibited that the *BdHsfs* and *BdHsp70s* play a role in attenuating the damage by HS through expressing highly in the root region. It was found that under HS of 45°C for 2 h, the expression level of both *BdHsfs* and *BdHsp70s* were high compared with normal HS condition (22°C, 2 h). Heat map analysis presented that both *BdHsfs* and *BdHsp70s* expressed highly up to 70% under HS as compared to other abiotic stresses. Moreover, a list of heat stress-related genes was upregulated such as; *Hsf02, Hsf03, Hsf15, cHsp70-2, CHsp70-6, cHsp-11, cHsp70-1, Bip-1, Hsp110-2,* and *Hsp110-8.* Also heat-induced expression profile revealed that *BdHsfs* expression under HS can acquire thermotolerance for severe HS challenge in plants (Wen et al., 2017).

3.2 WRKY

The plant-specific WRKY transcription factors form one of the largest TF families extensively distributed in plants. WRKY TFs are characterized by the presence of highly conserved WRKYGQK heptatide at the N-terminus and zinc finger-like motif and about 60 amino acid residues (Bakshi and Oelmüller, 2014). WRKY proteins are classified into three groups according to the number of WRKY domain and the feature of the zinc finger motif. Group I has two WRKY domains, groups II and III have one WRKY domain. Group I and II have the zinc finger-like motif C2H2, where group III contains a C2HC zinc finger-like motif (Li et al., 2010a,b). The WRKY TFs need high binding affinity to the W-box *cis*-element with a core sequence of TTGACC/T (Bakshi and Oelmüller, 2014; He et al., 2016; Wang et al., 2016a,b,c). The first WRKY protein (SPF1) was identified from sweet potato (Ishiguro and Nakamura, 1994). A number of WRKY TFs have been identified in different plants such as *Arabidopsis* (Li et al., 2010a,b; Ulker and Somssich, 2004), rice (Wu et al., 2005), soybean (Bencke-Malato et al., 2014), cotton (Dou et al., 2014), wheat (He et al., 2016; Wang et al., 2017a,b,c,d), tobacco (Dang et al., 2013), and maize (Wang et al., 2018a,d).

WRKY39, a group II WRKY protein characterized as a positive regulator of thermotolerance in *A. thaliana* was exposed to 42°C HS for 1 h, the transcript level increased up to 5.3-folds compared to nontreated plants. Transgenic plants constitutively overexpressing *WRKY39* showed more resistance to high temperature than wild-type plants. To analyze the function of WRKY39, full-length WRKY39 cDNA was transferred into *A. thaliana* plants using CaMV35S promoter. Both mutant wrky39 and transgenic *35S:WRKY39* were exposed to 45°C for 4 h. After 3 days of HS treatment *35S:WRKY39* lines showed a germination rate up to 84%–95%, whereas mutant wrky39 and wild-type plants were exhibited 47% and 63% germination, respectively. After 9-days surviving plants clearly exhibited the presence of newly emerging leaves while nonsurviving plants evoked with arrested leaf formation, complete etiolation. The degree of thermotolerance was greater in *35S:WRKY39* plants as compared to wrky39 and wild-type plants. On the basis of this assay, it is confirmed that the *WRKY39* positively co-regulate thermotolerance in *A. thaliana*, specifying that this gene has a vital role in thermotolerance (Li et al., 2010b).

Dang et al. (2013) have reported that a WRKY TF *CaWRKY40* transcriptionally responds to high temperature and regulates thermotolerance. Under 42°C at 48-h postheat treatment in pepper plant leaves exhibited the expression of *CaWRKY40* up to 5.5-folds. Overexpression of *CaWRKY40* in tobacco plants exposed to 100% humidity under 42°C heat treatment for 48 h revealed that out of 15 *CaWRKY40-OE* tobacco plants, 13 plants remained healthy and two of them died with necrosis, while in wild-type case 12 plants died due to necrosis out of 15 plants. The overexpression of *CaWRKY40* in tobacco plants under heat stress enhanced transcript level such as HR-associated genes, ET-biosynthesis-related genes, ROS detoxification-associated genes as well as heat shock genes compared with wild-type plants. The study clearly indicates that *CaWRKY40* can be used for enhancing heat stress tolerance in crop plants (Dang et al., 2013).

The *TaWRKY1* and *TaWRKY33* from group II and III of WRKY TFs of wheat plants were conferred tolerance to heat stress in transgenic *Arabidopsis* plants (He et al., 2016). Under a high temperature of 45°C for 2h, TaWRKY1 showed a maximum level of threefold as compared to *TaWRKY33* which expressed more than 33-fold. Meanwhile, the seed germination rates of transgenic *TaWRKY33* lines were high up to 90%, while transgenic *TaWRKY1* lines exhibited 82% of germination rate. The rate of water loss and survivability rate were 20.3% and more than 50% in transgenic *TaWRKY33* lines while transgenic *TaWRKY1* lines were presented 22.1% and 30%. Several stress-related genes were possibly upregulated by both *TaWRKY1* and *TaWRKY33*, but overexpressed *TaWRKY33* regulated transcripts at a high level compared to *TaWRKY1*. All these data suggested that both the transcription factor *TaWRKY33* and *TaWRKY1* showed positive role in thermotolerance, but most probably, *TaWRKY33* transgenic lines manifested enhanced tolerance to heat stress (He et al., 2016). Wang et al. (2018c) have identified a drought-induced gene *ZmWRKY106* as part of group WRKY II from maize, it involves in multiple abiotic stress response pathways and also acts as a positive regulator of heat stress by inducing thermotolerance in *Arabidopsis* plants. Under 45°C of HS for 2h, the transcript level of *ZmWRKY106* increased gradually up to 7.6-fold in transgenic plants compared to wild types. The survival rate of transgenic *Arabidopsis* plants overexpressing ZmWRKY106 was more than 30%, while WT plants exhibited less than 20% survival rate. Stress-related transcripts were high in transgenic lines compared to WT plants. These results showed that *ZmWRKY106* reveals better thermotolerance under HS (Wang et al., 2018a).

3.3 Myeloblastosis oncogene

The myeloblastosis oncogene (MYB) family is a large group TF identified in eukaryotes and is widely distributed in plants (Baldoni et al., 2015; Li et al., 2015). MYB proteins are a highly conserved domain for DNA binding, which contains one to four imperfect repeats of 51–53 amino acids.

MYB proteins are generalized in four groups according to their number of MYB domain repeats; IR-MYB (one repeat), R2R3MYB (two repeat), R1R2R3-MYB (three repeat), and 4R-MYB (four repeat) protein. Plant MYB transcription factors play numerous process during the plant life cycle including responses to environmental stresses such as salinity (Xiong et al., 2014), drought (Baldoni et al., 2015), cold (Yang et al., 2012), and oxidative stress (Chen et al., 2015). A very few plant MYB genes have been studied for thermotolerance in different crop plants such as *Arabidopsis* (Feng et al., 2004), tomato (Meng et al., 2015; Guo et al., 2015), maize (Casaretto et al., 2016), rice (El-kereamy et al., 2012), wheat (Zhao et al., 2017).

Studies of El-kereamy et al. (2012) have revealed that a R2R3-MYB TF *OsMYB55* elevates the vegetative growth and improve grain yield of rice growth under HS conditions. To study the role of *OsMYB55* in rice plants transgenic *Oryza sativa* cv. Kaybonnet plants were used and exposed to 39°C for 4 weeks. It was found that after long-time exposure to HS, *OsMYB55* overexpressed plants resulted in greater plant dry biomass, plant height, and leaf sheath length than wild type. Also, overexpression of *OsMYB55* enhanced amino acid metabolism which was crucial for plant tolerance to high-temperature conditions. Transgenic *OsMYB55* overexpressed rice plant increase the level of amino acid, glutamic acids, proline, GABA, and arginine. In conclusion, *OsMYB55* could help the plant to acquire thermotolerance and also to increase crop yield in many rice-growing regions (El-kereamy et al., 2012). An R2-R3-MYB gene, *LeAN2* is involved in enhancing thermotolerance of tomato plants under heat stress (Meng et al., 2015). Under 40°C heat stress treatment overexpression of *LeAN2* in transgenic tomato seedlings and grown plants exhibited better growth, less accumulation of ROS, alleviated photoinhibition of PSII with high thermotolerance as compared to wild-type tomato plants (Meng et al., 2015). Overexpression of *OsMYB55* in maize improves plant tolerance to high temperature (Casaretto et al., 2016). Under 42°C of HS for 5 days, overexpressed *OsMYB55* maize lines indicated a small reduction of dry mass, plant height, stem diameter, decrease in chlorophyll content, reduced leaf damage compared to WT plants. The reduced MDA accumulation in transgenic lines expressed that enhanced ROS scavenging ability under HS conditions (Casaretto et al., 2016). Transgenic *Arabidopsis* plants overexpressing *TaMYB80* of wheat showed increased tolerance to heat stress (Zhao et al., 2017). The survival rates of all transgenic lines were significantly higher than wild-type plants under heat stress of 45°C for 1 h (Zhao et al., 2017).

3.4 NAC

NACs are also a large group of TFs in plants (Puranik et al., 2012). NAC stands for NAM-ATAF1, 2-CUC2 (no apical meristem (NAM), *Arabidopsis thaliana* transcription activation factor (ATAF1/2) and cup-shaped cotyledon proteins (CUC2)) (Erpen et al., 2017). The typical feature of highly conserved NAC domain is present at the N-terminal region and a variable transcriptional regulatory region is present at the C-terminal domain (Wang et al., 2016c). The NAC family across the plant kingdom reflects their numerous functions including hormone signaling (Allu et al., 2016), grain nutrient remobilization (Ricachenevsky et al., 2013), plant development, senescence (Jensen and Skriver, 2014), cell wall biosynthesis (Nakano et al., 2015), and enhanced abiotic stress tolerance such as drought, salinity, and cold (Wang et al., 2013, 2014; Jin et al., 2017). To date, a lot of NAC TFs have been established in different plants related to heat stress such as in *Arabidopsis* (Guan et al., 2014; Shahnejat-Bushehri et al., 2012; Guo et al., 2015), wheat (Guo et al., 2015), and rice (Fang et al., 2015).

A NAC transcription factor, NAC019 get interacted with RCF2/CPL1 and result in severely reduced thermotolerance and reduced expression of heat-stress-responsive genes (Guan et al., 2014). Overexpression of both NAC019 and RCF2 in transgenic *Arabidopsis* plants expressed more heat tolerant, minimal damage to flowers and plant growth with reduced accumulation of HSFs (HsfA1b, Hsf6b, Hsf7a, HsfB1, and HsfC1) and HSPs (Hsp18, Hsp26.5, Hsp70B, and Hsp101 (Guan et al., 2014). The wheat (*Triticum aestivum* L.) NAC transcription factor gene *TaNAC2L* expressed thermotolerance in transgenic *Arabidopsis thaliana* under heat stress of 38°C (Guo et al., 2015). The TaNAC2L transgenic lines showed an enhanced thermotolerance to heat stress compared to wild type by expressing some stress-responsive gene at a high level such as *AtCYP18-1, RD17, HSP26.5, LEA, HSP70, AtHsfA3* (Guo et al., 2015). Shahnejat-Bushehri et al. (2012) reported that NAC transcription factors JUNGBRUNNENI (also called ANAC042) positively involved in regulating of thermomemory-related expression and enhanced heat stress tolerance in *Arabidopsis thaliana*. The overexpression of *JUB1* in *Arabidopsis* transgenic lines enhanced survival rate at 37°C and also showed some thermomemory expressive genes such as HSP18.2 and HSP21 compared to wild-type (Shahnejat-Bushehri et al., 2012). Another NAC transcription factor gene, *SNAC3*, was analyzed for the involvement of enhancing heat tolerance in rice under heat stress of 42°C. The plants with SNAC3-OE survived up to 40% and whereas wild-type plants, exhibited 20%

survival rate. SNAC3-OE plants evoked high heat resistance due to lower accumulation of H_2O_2 and MDA and also capability to scavenge ROS under heat stress by inducing ROS-associated genes (Fang et al., 2015).

3.5 Basic leucine zipper

The basic leucine zipper (bZIP) family contains a conserved bZIP domain. The bZIP domain is of 60–80 amino acid residues, which contains basic region at N-terminus for DNA binding and nuclear localization and also leucine-rich motif at C-terminus for specific recognition and dimerization (Wang et al., 2018c). Now many members of bZIP TF family have been identified related to heat stress tolerance in different crop plants such as in rice (Hsu et al., 2014), strawberry (Wang et al., 2017b), finger millet (Ramakrishna et al., 2018), and model plant *Arabidopsis* (Geng et al., 2018; Gao et al., 2008). Ramakrishna et al. (2018) have identified a novel endoplasmic reticulum (ER) membrane-tethered bZIP transcription factor, *EcbZIP17* from finger millet and induced it to produce transgenic tobacco plant which showed better thermotolerance under heat stress. The overexpressed *EcbZIP17* plants under imposed heat stress (42°C) was evaluated, and the transgenic plant exhibited green leaves per plant, increased seedling weight, green color disc, and retention of chlorophyll content compared to wild-type plants (Ramakrishna et al., 2018). Recently, Geng et al. (2018) have reported that constitutive expression of the spliced form of *TabZIP60* from wheat enhanced heat tolerance in transgenic *Arabidopsis*. A bZIP28 belongs to a putative bZIP-type MTTF (membrane-tethered transcription factors) in *Arabidopsis* which contributes to thermotolerance by upregulating heat-responsive genes (Gao et al., 2008).

3.6 Zinc finger protein

The zinc-binding protein is the largest studied essential transcription factor in eukaryotes and it showed variability in secondary structure and in functional diversity, it has been divided into six groups according to their sequence of zinc finger protein such as C_2H_2, C_3H, C_2C_2, A_{20}/AN_1, $C_3H_2C_3$, and C_3HC_4 (Jin et al., 2018). Zinc finger proteins are involved in regulation of plant growth (Wu et al., 2014), root development (Cui et al., 2017), seed germination (Bogamuwa and Jang, 2013), fruit ripening (Wang et al., 2016a,b,c) and also do play an important role in abiotic stresses such as drought (Yin et al., 2017), salinity (Zhang et al., 2016), and cold

(Iqbal et al., 2017). To date, many studies have been done on zinc finger protein in different plants such as *Arabidopsis* (Koguchi et al., 2017), wheat (Agarwal et al., 2017), and rice (Hsu et al., 2014). OsRZFP34, a ring zinc finger protein showed the highest heat stress responsiveness (Hsu et al., 2014). *TaZNF* is C4HC3-type zinc finger from wheat suggested as a positive regulator in mediating plant heat stress response in *Arabidopsis* (Agarwal et al., 2017). The transgenic lines overexpressing TaZNF were subjected to heat stress at 42°C for 2h showed faster and robust growth, increased thermotolerance, increased membrane stability, chlorophyll retention, higher photosynthetic rate as compared to wild types (Agarwal et al., 2017).

3.7 APETALA2/ethylene response factor

The APETALA2/ethylene response factor (AP2/ERF) represents a large family of transcription factors in plants. It contains a 60–70 amino acid residue of AP2 DNA-binding domain (Licausi et al., 2013). AP2/ERF subfamily has been classified into four major subfamilies according to DNA-binding domain and sequence similarities. These four subfamilies are DREB, ERF, and AP2 and related to AB13 and VP1 (RAV) (Owjia et al., 2017). AP2/ERF TFs involve in important biological processes including development, reproduction, and adaption to biotic and abiotic stresses (Licausi et al., 2013). AP2/ERF TFs may play a vital role in abnormal temperature stress response in tea plants (*Camellia sinenis*) subjected to 38°C (Wu et al., 2015). The *FTL1-1D/DDF1* enhances tolerance to high temperature in *Arabidopsis thaliana* plant (Kang et al., 2011). The survivability rate of *FTL1/DDF1* mutant plant showed 100% under heat stress at 45°C up to 12h treatment as compared to wild-type plants, which showed only 16% survival rate. The overexpression of *FTL1/DDF1* plants induce some heat stress transcription factor, *Hsf3* which leads to enhanced thermotolerance under heat stress (Kang et al., 2011).

The ERFs belong to the large AP2/ERF multigene family. These are known to be last downstream components of the ethylene transduction pathway. ERFs made up of 58–59 amino acids highly conserved DNA-binding domain which binds to multiple *cis*-acting elements also include the GCC box and DRE/CRT (dehydration responsive element/C-repeat) (Pirrello et al., 2012; Klay et al., 2018). ERFs from different plant species have been shown to play major roles in response to various abiotic stresses including heat stress such as *Arabidopsis* (Cheng et al., 2013). ERF1 had a significant role in heat stress tolerance of *Arabidopsis* under heat stress exposure to 45°C for 1 h. The *35S:ERF1* RNA interference (RNAi) knockdown plant found that many heat shock-inducible genes (*ATHSFA3, HSP101, HSP70,* and *HSP23.6-M*) were upregulated in *35S:ERF1* plants and downregulated in *ERF1* RNAi plants (Cheng et al., 2013).

3.8 Dehydration responsive element binding

Many transcription factors have been identified by involving HS-signaling pathway. Among all of these, dehydration responsive element binding (DREB) TFs are unique subfamily composed of two class of protein DREB1 and DREB2 of plant-specific APETALA2/ethylene responsive element binding factor (AP2/ERF) with a single AP2 domain that broadly participates with various stress response signaling pathways (Agarwal and Khurana, 2018; Zhao et al., 2014). Both class-1 and class-2 gene of DREB binds to dehydration responsive element (DRE) and also termed as C-repeat (CRT). This DRE/CRT contains a conserved core sequence (A/GCCGAC), which is present in the promoter of many heat stress-inducible genes and it is sufficient to induce gene transcription under heat stress (Song et al., 2018b; Karki et al., 2013). Many DREBs have been studied and characterized in various plants including grasses, crops, legumes, *Arabidopsis*, and their involvement in heat stress tolerance (Song et al., 2018b; Sato et al., 2014; Morimotoa et al., 2017; Kidokoro et al., 2015; Wu et al., 2015, 2018).

VOZ1 (vascular plant one-zinc finger 1) functions as a negative regulator of HS-inducible DREB2C signaling by blocking to the AP2 DNA-binding domain of DREB2C in *Arabidopsis* (Song et al., 2018b). It was analyzed that, *voz1 voz2-2* mutant plants with HS (44°C) showed increased activity of DREB2C, the expression of *HsfA3* with increased thermotolerance compared to wild-type plant of HS (22°C) (Song et al., 2018b). The DREB1-type transcription factors GmDREB1, activated transcription in *Arabidopsis* and soybean via DREs (dehydration responsive element) in response to heat stress (Kidokoro et al., 2015). The expression of GmDREB1of *Arabidopsis* is similar to the expression of DREB1A of soybean. The overexpression of GmDREB1 in transgenic *Arabidopsis* showed many heat stress-inducible genes such as dehydrin/LEA (late embryogenesis abundant), these genes might be involved in cellular protection and heat stress tolerance in both soybean and *Arabidopsis* (Kidokoro et al., 2015). A novel DREB2 gene (*LlDREB2B*) from lily

(*Lilium* spp.) was known to enhance thermotolerance of *Arabidopsis* under heat stress of 37–45°C (Wu et al., 2018). The overexpression of *LIDREB2B* could improve the induced expression of *AtHsfA3* in transgenic *Arabidopsis*, which is involved in heat stress response. The transgenic *Arabidopsis* showed many downstream genes such as *AtHsp101, AtHsp32, A22.0, AtGols1, AtAB15,* and *AtAPX2*, and some target genes of *AtDREB2A*, these genes might contribute stronger thermotolerance in the transgenic *Arabidopsis* plants (Wu et al., 2018). Sato et al. (2014) identified that a trimer comprising of NF-Y2, NF-YB3, and DPB3-1 may be important in the target selectivity to enhance *HsfA3* promoter transactivation by *DREB2A* in the adaptive response of plant under heat stress. First, DPB3-1 and NF-YB3 form a dimer in the nucleus and then form a trimer with NF-YA2. The overexpression of DPB3-1 in transgenic *Arabidopsis* showed overexpression of *DREB2A* with improved thermotolerance under 45°C (Sato et al., 2014). So far, functions of a limited number of DREB have been well characterized, but most of DREB family members have yet to be identified in response to heat stress tolerance in different crop plants.

3.9 Brassinosteroids transcription factors (BR TFs)

Brassinosteroids (BRs) are plant steroid hormones, which regulates different types of biological processes like fertility, root development, and various types of abiotic stresses including heat stress (Vardhini and Anjum, 2015). BRs get involved in heat stress tolerance by activating some transcription factors through a BR mediated signaling. At first, BR binds to the receptor-like kinase brassinosteroid intensive 1 (BRI1) at the cell surface and activates intracellular kinase activity. Activated BRI1 interacts with its co-receptor, BRI1-associated kinase 1 (BAK1) and initiates subsequent phosphorylation cascades. Brassinosteroid insensitive 2 (BIN2) a negative regulator of BR signaling is inactivated by BR, resulting in relief of downstream transcription factors brassinazole resistant (BZR1) and BRI1 EMS-suppressor 1(BES1 and also named BZR2). Active BZR1/BZR2 binds to *cis*-element called as E box (CANNTG) and directly regulates the BR responsive genes (Wu et al., 2016).

The role of Brassinosteroids transcription factors has been studied in various plants under heat stress such as in tomato (Yin et al., 2018) and *Brassica napus* (Sahni et al., 2016). When plants get exposed to heat stress BZR1/BES1translocates from the cytosol to nucleus and regulate thermomorphogenesis, while this regulation is dependent on BR synthesis (Ibanez et al., 2018). BR modulates various physiological and developmental processes such as cell division, elongation, vascular differentiation, seed germination, and also it increases resistance to abiotic stresses such as drought, salinity, and chilling (Wei and Li, 2016; Sharma et al., 2013; Zhang et al., 2014; Kim et al., 2012).

Yin et al. (2018) reported that BZR1 transcription factors participated in BR-induced heat stress tolerance in tomato plant by regulating RBOH1 (respiratory burst oxidase homologue 1)-dependent reactive oxygen species signaling and feronia (*FER*) homologs genes (*FER2* and *FER3*) which activates defense mechanism against heat stress in tomato plants. Under heat stress (48°C), overexpressed BZR1 tomato plant resulted in a less severe wilting phenotype, higher photosynthetic activity, reduced electrolyte leakage. Overexpressed BZR1 tomato plants presented most binding site for *FER2* and *FER3*, which demonstrated that upregulation of BZR1-dependent BR signaling regulated expression FER, where FER involves in the production of apoplastic ROS, H_2O_2 accumulation and provide heat stress tolerance in tomato plants (Yin et al., 2018). Heat stress caused a reduction in vegetative growth and reproductive yield in *B. napus* (Young et al., 2004; Angadi et al., 2000), and exogenous supplement of BR enhanced thermotolerance in *B. napus* (Dhaubhadel et al., 1999; Kagale et al., 2007). Sahni et al. (2016) have demonstrated that overexpression of a brassinosteroid synthetic gene *DWF4* could increase crop yield and heat stress tolerance in *B. napus*. Transgenic *B. napus* seedling which overexpressed DWF4 survived up to 55% as compared to WT seedlings, where survival was only 15%. Transgenic *B. napus* plants were characterized with increased seed yield, higher root biomass, higher root length, and better tolerance to heat stress (Sahni et al., 2016).

3.10 Basic/helix-loop-helix

The basic/helix-loop-helix (bHLH) is a large superfamily transcription factors, identified in eukaryotes such as plant, fungi, and metazoan (Carretero-Paulet et al., 2010; Pires and Dolan, 2010). bHLH is characterized by one specific bHLH domain including one basic region and HLH region. Normally 17 amino acid substitute of basic domain presents at N-domain and contains *E*-box (CANNTG) for binding of bHLH TFs and HLH region participates in the formation of homodimers or heterodimers. bHLH transcription factors are in general divided into six groups (from A to F). Plant bHLH transcription factors participate in a number of biological processes, such as growth (Dong et al., 2014; Chen et al., 2017), development (Zhang et al., 2018; Chen et al., 2017), and different abiotic stress responses such as drought (Dong et al., 2014), cold (Feng et al., 2012), salinity (Li et al., 2010a,b), osmotic stress (Liu et al., 2014), and heat stress (Wang et al., 2018d; Koini et al., 2009).

Recent studies by Wang et al. (2018e) have performed gene ontology analysis which showed the involvement of *StbHLH* in response to abiotic stress, including heat stress in potato plants. In gene annotation, it was observed that *StbHLH45* genes were extremely expressed in potato plant under heat stress. Many genes were upregulated (StbHLH65, 76, and 79) or downregulated (StbHLH19, 60, and 78) in response to heat stress in over-expression of StbHLH45 potato plants (Wang et al., 2018d). A bHLH transcription factor phytochrome interacting factor 4 (PIF4) which mediated architectural adaption in *Arabidopsis* plant under high-temperature environment from 22°C to 28°C (Koini et al., 2009). When *Arabidopsis* plant went under high-temperature condition, it showed an increasing level of *Pif4* transcripts and auxin-responsive gene *IAA29*, which indicated PIF4 is an important regulator of auxin-mediated signaling pathway. Plant development with hypocotyls elongation, petiole elongation, and leaf hyponasty responses was not expressed in mutant *Pif4* plants, while *Pif4* containing *Arabidopsis* plants showed these entire characteristics under high temperature. The study demonstrated how PIF4 impacts plant architecture in response to high temperature (Koini et al., 2009).

3.11 MADS box

The MADS-box family members were originally developed as floral homeotic genes, and most enormously studied transcription factors in plants (Arora et al., 2007). MADS-box gene name stands for the yeast *minichromosome maintenance 1* (*MCM1*), *Arabidopsis agamous* (*AG*), Antirrhinum *deficiens* (*DEF*) and human serum response factor (*SRF*) (Agarwal and Khurana, 2019). The MADS-box family has been categorized into two main groups type-1 ARG 80/SRF (also called M-type) and type-II MEF2 (also called MIKC type) (Yin et al., 2016). The plant MADS-box TF contains a ~60 amino acid MADS-box domain known as DNA-binding domain at N-terminal region, ~30 AA of intervening region, ~70 AA of keratin like domain and a highly variable C-terminal region involved in transcription activation (Arora et al., 2007). MADS-box TFs are highly diversified with their function in plant growth, development (such as flowering and fruit ripening), and under abiotic stress conditions such as drought, salinity, and cold (Khong et al., 2015; Li et al., 2018; Chen et al., 2019; Wei et al., 2018).

The role of MADS-box transcription factors has been studied in different plants under heat stress such as in rice (Chen et al., 2019), petunia (Wang et al., 2019), and wheat (Agarwal and Khurana, 2019). Chen et al. (2019) have elucidated the involvement of MADS-box transcription factor in syncytial cellularization, regulates rice seed size and improved thermal resilience. Rice plants cv. Kitaake (ssp. Japonica) were used in this study and exposed to moderate (D/N temperature of 35°C/39°C) and severe (D/N temperature of 39°C/34°C) temperatures and maintained it for 24-, 48-, or 72-h after fertilization (HAF). The shriveling of seed occurred during severe heat stress under 72HAF. Several M-γ types I MADS-box gene had higher transcript abundance under moderate and severe heat stress compared to control seed. Out of three genes of MADS transcription factor (*OsMADS82, OsMADS87, OsMADS89*), *OsMADS87* expressed highly up to 90-fold under 35°C at 48 HAF during rice seed development in endosperm as well as pericarp. For further functional analysis of *OsMADS87*, transgenic rice plant was generated by overexpressing or with repressed expression via RNA interference (RNAi). Overexpression of *OsMADS87* in rice plants exhibited greater sensitivity to heat stress and also visualized reduced seed dimension. This data suggests that how *OsMADS87* regulates seed size and heat-induced deficiency during the syncytial stage accelerates endosperm cellularization (Chen et al., 2019). Wang et al. (2018b) have suggested that, *SOC1* and *SOC1*-like genes such as *FBP21* and *FBP22* of MADS-box TF from *Arabidopsis* and petunia plants control the chloroplast biogenesis under HS and produced green petals. Under 40°C for 12-h heat treatment, overexpression of *SOC1, FBP21*, and *FBP2* in *Arabidopsis* and tobacco plant showed flowering with green petals, accumulation of chlorophyll up to three- to sevenfold, wild-type plants expressed flowering with fading pink color petals. More genes were expressed regarding chloroplast biogenesis in transgenic plants compared to WT such as *Mg* protoporphyrin IX monomethyl ester cyclise (CHL27), pchlide oxidoreductase (POR), chlorophyllide-α oxygenase (CAO), HSP70, and HSP90. Their study suggests that use of MADS-box TF can influence chloroplast biogenesis and photosynthesis in higher plants under HS (Wang et al., 2019).

Studies of Agarwal and Khurana (2019) demonstrated that a type II (MIKC) *TaMADS* from wheat plant was involved under heat stress conditions in *Arabidopsis* plants and played an active role in response to thermal stress. Under 37°C for 2h of heat treatment, the expression of *TaMADS* was found in lemma, awn, glumes, and seed in transgenic *Arabidopsis* plants. Also, expression was observed high during heat stress and gets decreased after the recovery period. After 10-days recovery from 42°C HS for 2 and 4h treatment, overexpressed *TaMADS* in *Arabidopsis* plants recovered with good vegetative and healthier growth and also longer root than wild type. It was observed that transgenic *Arabidopsis* plants showed enhanced leaf number, leaf size, and rosette diameter. The photosynthetic parameters such as Fv/Fm, ETR, and YII can be used as an indicator of thermotolerance in plants. The transgenic lines showed better performance in response to all these photosynthetic parameters. Also, some heat stress markers like HSF and HSP101 was also found to be higher in transgenic plants. All these phenomena reveal that overexpression of *TaMADS* in *Arabidopsis* plants results in faster recovery, enhanced growth, increased yield, and better thermotolerance (Agarwal and Khurana, 2019).

4. Genetic engineering and transgenic approaches in the field of heat stress transcription factors

Rather using the technological solution, use of genetic or transgenic approach enables plant or crops to survive under increasing ambient temperature (Grover et al., 2013). Recent advancement of plant transcription factors deals with how transcription factors cope with abiotic stresses including heat stress (Wang et al., 2016a,b,c). There are some transcription factors which were recently discovered and functionally validated certain extent through molecular technology to know about its involvement in heat stress (Guo et al., 2016). Shockingly, very fewer investigations have been done on heat stress as compared to drought, salt, and cold tolerance in crop plants (Nadeem et al., 2018). Different transcription factors and its recent advancement in connection with varied stress tolerance have been described briefly later.

4.1 Synthetic transcription factors

Recent studies by Savo et al. (2016) have revealed that Earth's temperatures are increasing; there is an urgent need to produce crops that are more tolerant to heat and other abiotic stress. The field of plant synthetic biology has the potential to address these challenges by producing new and improved traits that function with unprecedented precision, in a way that traditional breeding methods would not be able to accomplish. Testing and quantitatively characterizing the behavior of synthetic parts are essential for modeling and predicting their function in engineered genetic circuits (Kassaw et al., 2018). These synthetic traits can be programmed to be active under specific conditions (e.g., "on-demand" heat stress tolerance that is turned "off" when the plants are placed in ambient temperature), providing precise temporal, spatial, and developmental control of gene expression, and reducing any metabolic burden associated with transgene overexpression. More recently, a deactivated form of the Cas9 protein (dCas9) from *Streptococcus pyogenes* has been used to engineer transcriptional activators and repressors (Lowder et al., 2015; Piatek et al., 2015). In this case, dCas9 is no longer able to introduce DNA breaks. Instead, dCas9's programmable DNA-binding specificity, determined by the guide RNA (gRNA) sequence, is used to build a synthetic transcription factor. Synthetic transcription factors are composed of a translational fusion between these DNA-binding domains and effector domains (repressor or activator). The DNA-binding domains provide the desired binding specificities capable of recognizing almost any predetermined DNA sequence, directing the molecular machinery to a specific promoter to activate or repress expression of a gene of interest. Application of synthetic biology tools in designing and validating novel heat stress synthetic TFs, which are more effective in enhancing heat stress tolerance in different crop plants is a viable option to the plant biologists.

In plants, heat stress responses always start from plasma membrane (Saidi et al., 2012) because, all proteins in the plasma membrane play a critical role in response to heat stress (Gao et al., 2012; Finka et al., 2012). Liu et al. (2014) have identified a membrane-bound RINGvgene (really interesting new gene variant) and *B. napus* thermal resistance gene 1 (*BnTR1*). *B. napus* is likely to be involved in a conserved thermal resistance mechanism in plants. To analyze *BnTR1* expression in response to heat stress, transgenic lines overexpressing *BnTR1* in *B. napus is* kept under 35°C for 6–9 days, and transgenic *O. sativa* were kept under 48°C. The transgenic *B. napus* showed survivability up to 9 days while nontransgenic (CK) lines died after 6 days. The transgenic *O. sativa* plants were unaffected and fringes appeared normal, pollen viability recorded up to 65%–90%, while CK plants were disturbed and fringes were short and sloppy. Both transgenic plants exhibited yield up to 8%–17% more than CK in comparison to grain weight up to 9.4 g and total grains per plant up to 963–1026 number. Under heat stress, upregulated *BnTR1* transgenic plants revealed expression of HsfA1a up to ninefold and Hsps up to 4- to 12-fold than in CK plants. Overall data seems that *BnTR1* is a novel yet conserved gene in the thermal resistance of plants (Liu et al., 2014).

4.2 Identification of TFs from thermotolerant grasses

There have been few findings which focus on heat stress response in perennial grass species. Based on the temperature categories, it is divided into two-types, warm-season perennial grass species ranges from 26°C to 35°C and cool-season perennial grass ranges from 15°C to 24°C, and compared to both, warm-season perennial grass shows better heat thermotolerance capacity than cool-season perennial grass (Beard, 1973). Wang et al. (2016a,b,c) have found a heat stress transcription factor A2 (*HsfA2*) gene *CtHsfA2b* from a warm-season grass species, African Bermudagrass (*Cynodon transvaalensis* Burtt-Davy), which was utilized as a candidate gene for improving thermo-tolerance in *Arabidopsis* plant. Under 45°C for 4h HS, all ectopic overexpressed *CtHsfA2b Arabidopsis* lines showed a survival rate from 50% to 76%, whereas WT exhibited 28% after 7 days of recovery from treatment. After returning to normal growth conditions, transgenic lines produced new leaves, whereas leaves of WT lines were wilted. Also, *CtHsfA2b Arabidopsis* lines exhibited higher photosynthetic rate, photochemical efficiency, chlorophyll content, and less electrolyte leakage compared to control lines. Many genes were highly upregulated regarding thermotolerance in transgenic plants such as*AtApx2*, *AtHsp18.1-CI*, *AtHsp22.0-ER*, *AtHsp25.3-P*, *AtHsp26.5-P*, *AtHsp70b*, and *AtHsp101-3* than WT. All these results clarified that *CtHsfA2b* may be a crucial HSF during increased temperature that influence heat tolerance mechanism. It can also be used for improving heat tolerance in cool-season grasses and crop improvement programs (Wang et al., 2016a,b,c). A warm-season turf grass (Zoysia grass; *Zyosia japonica* Steud.) is a good candidate for its tolerance phenotype to drought, high salinity, and cold stress (Hinton et al., 2012; Li et al., 2016).

The *ZjDREB1.4* showed temperature tolerance without growth inhibition in *Arabidopsis* plants. After 7 days of recovery from 60°C for 2h of heat stress, the overexpression of *ZjDREB1.4* in *Arabidopsis* plant showed significant thermotolerance with a higher survival rate than WT. Some important osmolytes like proline and soluble sugar accumulated higher level up to

30%–50% in the transgenic plants compared to WT, which are most important osmolytes under stress. The transgenic line with overexpression of *ZjDREB1.4* showed no growth inhibition and produced seeds. The study implies that *ZjDREB1.4* is useful in plant breeding for enhancing heat stress tolerance (Feng et al., 2019). Zhuang et al. (2018) have cloned a class C *Hsf*, *FaHsfC1b* from a perennial grass species tall fescue and overexpressed this gene in *Arabidopsis* plants under high-temperature conditions to examine positive phenotype and tolerance to heat stress conditions. Under 45°C of HS for 5 h, many heat stress-protective genes upregulated compared to the control line (WT). *FaHsfC1b* overexpression transgenic lines exhibited survival rate from 46% to 77% compared to WT (Zhuang et al., 2018).

4.3 Role of epigenetic regulation and miRNAs under heat stress

TFs are the core regulators of transcription during heat stress (HS). However, recent studies have evidenced that a multilayered regulatory system comprising not only TFs but also epigenetic regulators and small RNAs, has an important role in adaption to HS (Ohama et al., 2017). The miR398 represses reactive oxygen species (ROS) scavengers, copper/zinc superoxide disutase 1 (CSD1), CSD2, and copper chaperone for SOD1 (CCS). The downregulation of these genes results in the accumulation of ROS, which subsequently activate *HsfA1s*. Further, miR156 represses squamosa promoter binding-like (SPL) transcription factors (Ohama et al., 2017). TF-mediated epigenetic regulation may be a widely adopted mechanism in heat stress response (HSR) because human HSF1 also has the ability to modulate the chromatin landscape in response to HS (Miozzo et al., 2015). In addition, NF-Ys are reported to modulate transcription through their effects on the epigenetic status in mouse and *Arabidopsis* (Oldfield et al., 2014; Hou et al., 2014). More focus is required on synthetic biology aspects and use of synthetic TFs in heat stress response and tolerance in crop plants.

5. Conclusion

The TFs will provide more information on the complex regulatory networks involved in abiotic stress responses and the cross talk between different signaling pathways during the adaptation of plants to various abiotic stresses. The functional redundancy among different members of TFs may hinder the functional validation of an individual member of TFs. The holistic regulation mechanism of individual TFs including its downstream and upstream co-regulators, as well as their interactions, remains largely unknown. One transcription factor may regulate a huge array of target genes with the corresponding *cis*-elements on the promoters, whereas one gene with several types of *cis*-elements may be regulated by different families of TFs (Wang et al., 2016c). The TFs-based transgenic studies were mostly conducted on single stage (either vegetative or reproductive) under heat stress. It is required to carry out studies during different stages (vegetative and reproductive) of crop development under heat stress, to shed the lights on the functional role of individual TFs in stress protection at all stages of plant development. The TF genes constitutively overexpressed could improve heat stress tolerance in crop plants, but often lead to undesired effects in transgenic plants such as loss of root phenotype, stunting of growth, delay in flowering, and reduced yields. To address this issue in a better way, omics-based approaches should be applied extensively to understand the alterations of transcriptome, proteome, metabolome, and phenomics in crop plants to assign the functional role to each of the TFs in heat stress tolerance.

It is very important to identify commonly regulated different members of TF genes under multiple abiotic stresses and understand their cross talk between signaling pathways (Prasch and Sonnewald, 2015). These genes can be functionally validated using stress-inducible promoters for utilizing them effectively in improving crop plants tolerance to heat and other abiotic stresses, rather manipulating individual functional genes. It reduces the detrimental effects brought about as a result of overexpression of certain TFs in crop plants. Incorporation of heat stress response pathways in the vital vegetative development and reproductive stages in crops is an efficient strategy to effectively improve productivity in field conditions. In addition to current studies on TFs, more focus is required on synthetic biology aspects, epigenetic regulation, small RNAs (miRNAs), and integration of omics-approaches to develop novel TFs which can enhance heat stress tolerance in crop plants in a better way.

Acknowledgment

Work in the laboratory of GKS supported by the Forest and Environment Department, Government of Odisha, India, is gratefully acknowledged. The authors apologize for being unable to cite all relevant papers.

References

Agarwal, P., Khurana, P., 2017. Characterization of a novel zinc finger transcription factor (TaZnF) from wheat conferring heat stress tolerance in Arabidopsis. Cell Stress Chaperones 23, 253–267.

Agarwal, P., Khurana, P., 2018. Characterization of a novel zinc finger transcription factor (TaZnF) from wheat conferring heat stress tolerance in Arabidopsis. Cell Stress Chaperones 23 (2), 253–267.

Agarwal, P., Khurana, P., 2019. Overexpression of TaMADS from wheat promotes flowering by upregulating expression of floral promoters and provides protection against thermal stress. Plant Gene 17, 1–14.

Agarwal, P.K., Gupta, K., Lopato, S., Agarwal, P., 2017. Dehydration responsive element binding transcription factors and their applications for the engineering of stress tolerance. J. Exp. Bot. 68, 2135–2148.

Akhtar, M., Jaiswal, A., Taj, G., Jaiswal, J.P., Qureshu, M.I., Singh, N.K., 2012. DREB1/CBF transcription factors: their structure, function and role in abiotic stress tolerance in plants. J. Genet. 91, 385–395.

Allu, A.D., Bortman, Y., Xue, G.-P., Balazedeh, S., 2016. Transcription factor ANAC032 modulates JA/SA signalling in response to *Pseudomonal syringae* infection. EMBO Rep. 17, 1578–1589.

Anckar, J., Sistonen, L., 2011. Regulation of HSF1 function in the heat stress response: implications in ageing and disease. Annu. Rev. Biochem. 80, 1089–1115.

Angadi, S.V., Cutforth, H.W., Miller, P.R., McConkey, B.G., Entz, M.H., Brandt, S.A., Volkmar, K.M., 2000. Response of three Brassica species to high temperature stress during reproductive growth. Can. J. Plant Sci. 80, 693–701.

Arora, R., Agarwal, P., Ray, S., Singh, A.K., Singh, V.P., Tyagi, A.K., Kapoor, S., 2007. MADS-box gene family in rice: genome-wide identification, organization and expression profiling during reproductive development and stress. BMC Genomics 8, 1–21.

Bakshi, M., Oelmüller, R., 2014. WRKY transcription factors. Plant Signal. Behav. 9, 1–18.

Baldoni, E., Genga, A., Cominelli, E., 2015. Plant MYB transcription factors: their role in drought response mechanisms. Int. J. Mol. Sci. 16, 15811–15851.

Baniwal, S.K., Bharti, K., Chan, K.Y., Fauth, M., Ganguli, A., Kotak, S., et al., 2004. Heat stress response in plants: a complex game with chaperones andmore than twenty heat stress transcription factors. J. Biosci. 29, 471–487. https://doi.org/10.1007/BF02712120.

Barlow, K.M., Christy, B.P., O'Leary, G.J., Riffkin, P.A., Nuttall, N.G., 2015. Simulating the impact of extreme heat and frost events of wheat crop production: a review. Field Crop Res. 171, 109–119.

Baxter, A., Mittler, R., Suzuki, N., 2014. ROS as key players in plant stress signalling. J. Exp. Bot. 65, 1229–1240.

Beard, J., 1973. Turfgrass: Science and Culture. Prentice-Hall, Englewood Cliffs, NJ, pp. 54–131.

Bencke-Malato, M., Cabreira, C., Webke-Strohm, B., Bucker-Neto, L., Mancini, E., Osorio, M.B., Homrich, M.S., Turchetto-Zolet, A.C., De Carvalho, M.C., Stolf, R., Weber, R.L., Westergaard, G., Casragnaro, A.P., Abdelnoor, R.V., Marcelino-Guimaraes, F.C., Margis-Pinheiro, M., Bodanese-Zanettini, M.H., 2014. Genome-wide annotation of the soybean WRKY family and functional characterization of genes involved in response to Phakopsora pachyrhizi infection. BMC Plant Biol. 14, 1–18.

Bento, M., Pereira, S.G., Viegas, W., Silva, M., 2017. Durum wheat diversity for heat stress tolerance during inflorescence emergence is correlated to TdHSP101C expression in early developmental stages. PLoS One 12, 1–12.

Bita, C.E., Gerats, T., 2013. Plant tolerance to high temperature in a changing environment: scientific fundamentals and production of heat stress-tolerant crops. Front. Plant Sci. 4, 1–18.

Bogamuwa, S., Jang, J.C., 2013. The Arabidopsis tandem CCCH zinc finger proteins AtTZF4, 5 and 6 are involved in light-, abscisic acid- and gibberellic acid-mediated regulation of seed germination. Plant Cell Environ. 36, 1507–1519.

Bokszczanin, K.L., Solanaceae Pollen Thermotolerance Initial Training Network C, Fragkostefanakis, S., 2013. Perspectives on decipheringmechanisms underlying plant heat stress response and thermotolerance. Front. Plant Sci. 4, 315.
Brunel-Muguet, S., D'Hooghe, P., Bataillé, M.-P., Larré, C., Kim, T.-H., Trouverie, J., Avice, J.-C., Etienne, P., Dürr, C., 2015. Heat stress during seed filling interferes with sulfur restriction on grain composition and seed germination in oilseed rape (Brassica napus L.). Front. Plant Sci. 6, 1–12.
Carretero-Paulet, L., Galstyan, A., Roig-Villanova, I., Martínez-García, J.F., Bilbao Castro, J.R., Robertson, D.L., 2010. Genome-wide classification and evolutionary analysis of the bHLH family of transcription factors in Arabidopsis, poplar, Rice, Moss, and algae. Plant Physiol. 153, 1398–1412.
Casaretto, J.A., El-kereamy, A., Zeng, B., Stiegelmeyer, S.M., Chen, X., Bi, Y.-M., Rothstein, S.J., 2016. Expression of OsMYB55 in maize activates stress-responsive genes and enhances heat and drought tolerance. BMC Genomics 17, 1–15.
Chauhan, H., Khurana, N., Agarwal, P., Khurana, J.P., Khurana, P., 2018. A seed preferential heat shock transcription factor from wheat provides abiotic stress tolerance and yield enhancement in transgenic Arabidopsis under heat stress environment. PLoS One 8, 1–14.
Chen, T., Li, W., Hu, X., Guo, J., Liu, A., Zhang, B., 2015. A cotton MYB transcription factor, GbMYB5, is positively involved in plant adaptive response to drought stress. Plant Cell Physiol. 56, 917–929.
Chen, S., Yang, W., Jia, Q., Wang, W., Zhang, N., Wang, X., Wang, S., 2017. *Pleurotusostreatus*b HLH transcription factors regulate plant growth and development when expressed in Arabidopsis. J. Plant Interact. 12, 542–549.
Chen, R., Ma, J., Luo, D., Hou, X., Ma, F., Zhang, Y., Meng, Y., Zhang, H., Guo, W., 2019. CaMADS, a MADS-box transcription factor from pepper, plays an important role in the response to cold, salt, and osmotic stress. Plant Sci. 280, 164–174.
Cheng, M.-C., Liao, P.-M., Kuo, W.-W., Lin, T.-P., 2013. The Arabidopsis ETHYLENE RESPONSE FACTOR1 regulates abiotic stress-responsive gene expression by binding to different cis-acting elements in response to different stress signals. Plant Physiol. 162, 1566–1582.
Cheng, Q., Zhou, Y., Liu, Z., Zhang, L., Song, G., Guo, Z., Wang, W., Qu, X., Zhu, Y., Yang, D., 2015. An alternatively spliced heat shock transcription factor, OsHSFA2dI, functions in the heat stress-induced unfolded protein response in rice. Plant Biol. 17, 419–429.
Chidambaranathan, P., Jagannadham, P.T.K., Satheesh, V., Kohli, D., Basavarajappa, S.H., Chellapilla, B., Kumar, J., Jain, P.K., Srinivasan, R., 2018. Genome-wide analysis identifies chickpea (*Cicer arietinum*) heat stress transcription factors (*Hsfs*) responsive to heat stressat the pod development stage. J. Plant Res. 131, 525–542.
Cui, P., Liu, H., Ruan, S., Ali, B., Gill, R.A., Ma, H., Zheng, Z., Zhou, W., 2017. A zinc finger protein, interacted with cyclophilin, affects root development via IAA pathway in rice. J. Integr. Plant Biol. 59, 496–505.
Dang, F.-F., Wang, Y.-N., Yu, L., Eulgem, T., Lai, Y., Liu, Z.-Q., Wang, X., Qiu, A.-L., Zhang, T.-X., Lin, J., Chen, Y.-S., Guan, D.-Y., Cai, H.-Y., Mou, S.-L., He, S.-L., 2013. CaWRKY40, a WRKY protein of pepper, plays an important role in the regulation of tolerance to heat stress and resistance to *Ralstonia solanacearum infectionpce.* Plant Cell Environ. 36, 757–774.
Danquah, A., de Zelicourt, A., Colcombet, J., Hirt, H., 2014. The role of ABA and MAPK signaling pathways in plant abiotic stress responses. Biotechnol. Adv. 32, 40–52.
Devasirvatham, V., Tan, D.K.Y., 2018. Impact of high temperature and drought stresses on chickpea production. Agronomy 8, 1–9.

Devasirvatham, V., Gaur, P.M., Mallikarjuna, N., Raju, T.N., Trethowan, R.M., Tan, D.K.Y., 2013. Reproductive biology of chickpea response to heat stress in the field is associated with the performance in controleed environments. Field Crop Res. 142, 9–19.

Dhaubhadel, S., Chaudhary, S., Dobinson, K.F., Krishna, P., 1999. Treatment with 24-epibrassinolide, a brassinosteroid, increases the basic thermotolerance of Brassica napus and tomato seedlings. Plant Mol. Biol. 40, 333–342.

Djannaguirman, M., Prasad, P.V.V., Boyle, D.L., Schapaugh, W.T., 2011. High-temperature stress and soybean leaves: leaf anatomy and photosynthesis. Crop Sci. 51, 2125–2131.

Dong, Y., Wang, C., Han, X., Tang, S., Liu, S., Xia, X., Yin, W., 2014. A novel bHLH transcription factor PebHLH35 from *Populuseuphratica* confers drought tolerance through regulating stomatal development, photosynthesis and growth in Arabidopsis. Biochem. Biophys. Res. Commun. 450, 453–458.

Dou, L., Zhang, X., Pang, C., Song, M., Wei, H., Fan, S., Yu, S., 2014. Genome-wide analysis of the *WRKY* gene family in cotton. Mol. Genet. Genomics 289 (6), 1103–1121.

Driedonks, N., Xu, J., Peters, J.L., Park, S., Rieu, I., 2015. Multi-level interactions between heat shock factors, heat shock proteins, and the redox system regulate acclimation to heat. Front. Plant Sci. 6, 1–9.

Du, X., Li, W., Sheng, L., Deng, Y., Wang, Y., Zhang, W., Yu, K., Jiang, J., Fang, W., Guan, Z., Chen, F., Chen, S., 2018. Over-expression of chrysanthemum *CmDREB6* enhanced tolerance of chrysanthemum to heat stress. BMC Plant Biol. 18, 178.

El-kereamy, A., Bi, Y.-M., Ranathunge, K., Beatty, P.H., Good, A.G., Rothstein, S.J., 2012. The rice R2R3-MYB transcription factor OsMYB55 is involved in the tolerance to high temperature and modulates amino acid metabolism. PLoS One 7, 1–16.

Erpen, L., Devi, H.S., Grosser, J.W., Dutt, M., 2017. Potential use of the DREB/ERF, MYB, NAC and WRKY transcription factors to improve abiotic and biotic stress in transgenic plants. Plant Cell Tiss. Org. Cult. 132, 1–25.

Fahad, S., Hussain, S., Saud, S., Hassan, S., Tanveer, M., Ihsaan, M.Z., 2016a. A combined application of biochar and phosphorus alleviates heat-induced adversities on physiochemical, agronomical and quality attributes of rice. Plant Physiol. Biochem. 101, 191–198.

Fahad, S., Hussain, S., Saud, S., Khan, F., Hassan, S., Amanullah, J., et al., 2016b. Exogenously applied plant growth regulators affect heat-stressed rice pollens. J. Agron. Crop Sci. 202, 139–150.

Fang, Y., Liao, K., Du, H., Xu, Y., Song, H., Li, X., Xiong, L., 2015. A stress-responsive NAC transcription factor SNAC3 confers heat and drought tolerance through modulation of reactive oxygen species in rice. J. Exp. Bot. 66, 6803–6817.

Feng, C., Andreasson, E., Maslak, A., Mock, H.P., Mattsson, O., Mundy, J., 2004. Arabidopsis MYB68 in development and responses to environmental cues. Plant Sci. 167, 1099–1107.

Feng, X.-M., Zhao, Q., Zhao, L.-L., Qiao, Y., Xie, X.-B., Li, H.-F., Yao, Y.-X., You, C.-X., Hao, Y.-J., 2012. The cold-induced basic helix-loop-helix transcription factor gene MdCIbHLH1 encodes an ICE-like protein in apple. BMC Plant Biol. 12, 1–14.

Feng, W., Li, J., Long, S., Wei, S., 2019. A DREB1 gene from zoysia grass enhances Arabidopsis tolerance to temperature stresses without growth inhibition. Plant Sci. 278, 20–31.

Filewod, B., Thomas, S.C., 2014. Impacts of a spring heat wave on canopy processes in a northern hard wood forest. Glob. Chang. Biol. 20, 360–371.

Finka, A., Quendet, A.F.H., Maathuis, F.J.M., Saidi, Y., Goloubinoff, P., 2012. Plasma membrane cyclic nucleotide gated calcium channels control land plant thermal sensing and acquired thermotolerance. Plant Cell 24, 3333–3348.

Fragkostefanakis, S., Röth, S., Schleiff, E., Scharf, K.D., 2015. Prospects of engineering thermotolerance in crops through modulation of heat stress transcription factor and heat shock protein networks. Plant Cell Environ. 38, 1881–1895.
Frank, G., Pressman, E., Ophir, R., Althan, L., Shaked, R., Freedman, M., Shen, S., Firon, N., 2009. Transcriptional profiling of maturing tomato (Solanum lycopersicum L.) microspores reveals the involvement of heat shock proteins, ROS scavengers, hormones, and sugars in the heat stress response. J. Exp. Bot. 60, 3891–3908.
Gabaldon-Leal, C., Webber, H., Otegul, M.E., Slafer, G.A., Ordonez, R.A., Gaiser, T., Lorite, I.J., Ruiz-Ramos, M., Ewert, F., 2016. Modelling the impact of heat stress on maize yield formation. Field Crop Res. 198, 226–237.
Gao, H., Brandizzi, F., Benning, C., Larkin, R.M., 2008. A membrane-tethered transcription factor defines a branch of the heat stressresponse in *Arabidopsis thaliana*. Proc. Natl. Acad. Sci. U. S. A. 105, 16398–16403.
Gao, F., Han, X., Wu, J., Zheng, S., Shang, Z., Sun, D., Zhou, R., Li, B., 2012. A heat-activated calcium-permeable channel-Arabidopsis cyclic nucleotide-gated ion channel 6- is ivolved in heat shock responses. Plant J. 70, 1056–1069.
Geng, X., Zanga, X., Lia, H., Liua, Z., Zhaoc, A., Liud, J., Penga, H., Yaoa, Y., Hua, Z., Nia, Z., Suna, Q., Xina, M., 2018. Unconventional splicing of wheat TabZIP60 confers heat tolerance in transgenic Arabidopsis. Plant Sci. 274, 252–260.
Gong, B., Yi, J., Wu, J., Sui, J., Ali Khan, M., Wu, Z., Zhong, X., Seng, S., He, J., Yi, M., 2014. LlHSFA1, a novel heat stress transcription factor in lily (*Lilium longiflorum*), can interact with LlHSFA2 and enhance the thermo tolerance of transgenic *Arabidopsis thaliana*. Plant Cell Rep. 33, 1519–1533.
Grover, A., Mittal, D., Negi, M., Lavania, D., 2013. Generating high temperature tolerant transgenic plants: archivents and challenges. Plant Sci. 205, 38–47.
Guan, Q., Yue, X., Zeng, H.Y., Zhu, J., 2014. The protein phosphatase RCF2 and its interacting partner NAC019 are critical for heat stress–responsive gene regulation and thermotolerance in Arabidopsis. Plant Cell 26, 438–453.
Guo, L., Chen, S., Liu, Y., Ni, L., Zhang, K., Zhang, L., 2008. Isolation of heat shock factor Hsf1a-binding sites in vivo revealed variations of heat shock elements in *Arabidopsis thaliana*. Plant Cell Physiol. 49, 1306–1315.
Guo, W., Zhang, J., Zhang, N., Xin, M., Peng, H., Hu, Z., Ni, Z., Du, J., 2015. The wheat NAC transcription factor TaNAC2L is regulated at the transcriptional and post-translational levels and promotes heat stress tolerance in transgenic Arabidopsis. PLoS One 10, 1–11.
Guo, M., Liu, J.-H., Ma, X., Luo, D.-X., Gang, Z.-H., Lu, M.-H., 2016. The plant heat transcription factors (HSFs): structure, regulation, and function in response to abiotic stresses. Front. Plant Sci. 7, 1–13.
Hasanuzzaman, M., Nahar, K., Alam, M.M.D., Roychowdhary, R., Fujita, M., 2013. Physiological, biochemical, and molecular mechanisms of heat stress tolerance in plants. Int. J. Mol. Sci. 14, 9643–9684.
He, G.-H., Xu, J.-Y., Wang, Y.-X., Liu, J.-M., Li, P.-S., Chen, M., Ma, Y.-Z., Xu, Z.-S., 2016. Drought-responsive WRKY transcription factor genes TaWRKY1 and TaWRKY33 from wheat confer drought and/or heat resistance in Arabidopsis. BMC Plant Biol. 16, 1–16.
Heidarvand, L., Amiri, R.M., 2010. What happens in plant molecularresponses to cold stress? Acta Physiol. Plant. 32, 419–431. https://doi.org/10.1007/s11738009-0451-8.
Hemantaranjan, A., Bhanu, A.N., Singh, M.N., Yadav, D.K., Patel, P.K., Singh, R., Katiyar, D., 2014. Heat stress responses and thermotolerance. Adv. Plants Agric. Res. 1, 62–70.
Hernandez-Garcia, C.M., Finer, J.J., 2014. Identification and validation of promoters and cis-acting regulatory elements. Plant Sci. 217–218, 109–119.

Hinton, J.D., Livingston, D.P., Miller, G.L., Peacock, C.H., Tuang, T., 2012. Freeze tolerance of nine zoysiagrass cultivars using natural cold acclimation and freeze chambers. Hort Sci. 47, 112–115.

Hong, B., Ma, C., Yang, Y., Wang, T., Yamaguchi-Shinozaki, K., Gao, J., 2009. Overexpression of AtDREB1A in chrysanthemum enhances tolerance to heat stress. Plant Mol. Biol. 70, 231–240.

Hou, X., et al., 2014. Nuclear factor Y-mediated H3K27me3 demethylation of the SOC1 locus orchestrates flowering responses of Arabidopsis. Nat. Commun. 5, 4601.

Hsu, K.-H., Liu, C.-C., Wu, S.-J., Kuo, Y.-Y., Lu, C.-A., Wu, C.-R., Lian, P.-J., Hong, C.-J., Ke, Y.-T., Huang, J.-H., Yeh, C.-H., 2014. Expression of a gene encoding a rice RING zinc-finger protein, OsRZFP34, enhances stomata opening. Plant Mol. Biol. 86, 125–137.

Hu, W., Hu, G., Han, B., 2009. Genome-wide survey and expression profiling of heat shock proteins and heat shock factors revealed overlapped and stress specific response under abiotic stresses in rice. Plant Sci. 176, 583–590.

Hu, X.-J., Chen, D., Mlntyre, C.-L., Dreccer, M.-F., Zhang, Z.-B., Drenth, J., Kalaipandian, S., Chang, H., Xue, G.-P., 2018. Heat shock factor C2a serves as a protective mechanism for heat protection in developing grains in wheat via an ABA mediated regulatory pathway. Plant Cell Environ. 41, 79–98.

Huang, X.Y., Tao, P., Li, B.Y., Wang, W.H., Yue, Z.C., Lei, J.L., Zhong, X.M., 2015. Genome-wide identification, classification, and analysis of heat shock transcription factor family in Chinese cabbage (*Brassica rapapekinensis*). Genet. Mol. Res. 14, 2189–2204.

Ibanez, C., Delker, C., Martinez, C., Burstenbinder, K., Janitza, P., Lippmann, R., et al., 2018. Brassinosteroids dominate hormonal regulation of plant thermomorphogenesis via BZR1. Curr. Biol. 28, 303–310.

IPCC, 2012. Managing the risks of extreme events and disasters to advance climate change adaptation. In: Field, C.B., Barros, V., Stocker, T.F., Qin, D., Dokken, D.J., Ebi, K.L., Midgley, P.M. (Eds.), A Special Report of Working Groups I and II of the Intergovernmental Panel on Climate Change. Cambridge University Press, Cambridge, UK.

Iqbal, F., Sadique, S., Batool, F., Sarwar, M.B., Rashid, B., Shahid, M.N., Shahid, A.A., Husnain, T., 2017. Zinc finger transcription factor induces the drought, salt and cold stress tolerance in transgenic cotton. Indian J. Biotechnol. 16, 333–340.

Ishiguro, S., Nakamura, K., 1994. Characterization of a cDNA encoding a novel DNA-binding protein, SPF1, that recognizes SP8 sequences in the 5′ upstream regions of genes coding for sporamin and beta-amylase from sweet potato. Mol. Gen. Genet. 244, 563–571.

Jensen, M.K., Skriver, K., 2014. NAC transcription factor gene regulatory and protein-protein interaction networks in plant stress responses and senescence. IUBMB Life 66, 156–166.

Jiang, Y., Zheng, Q., Chen, L., Liang, Y., Wu, J., 2018. Ectopic overexpression of maize heat shock transcription factor gene ZmHsf04 confers increased thermo and salt-stress tolerance in transgenic Arabidopsis. Acta Physiol. Plant. 40, 1–12.

Jin, C., Li, K.Q., Xu, X.Y., Zhang, H.P., Chen, H.X., Chen, Y.H., Hao, J., Wang, Y., Huang, X.S., Zhang, S.L., 2017. A novel NAC transcription factor, PbeNAC1, of Pyrus betulifolia confers cold and drought tolerance via interacting with PbeDREBs and activating the expression of stress-responsive genes. Front. Plant Sci. 8, 1049.

Jin, Y.-M., Piao, R., Yan, Y.-F., Chen, M., Wang, L., He, H., Liu, X., Gao, X.-A., Jiang, W., Lin, X.-F., 2018. Overexpression of a new zinc finger protein transcription factor OsCTZFP8 improves cold tolerance in rice. Int. J. Genomics 2018, 1–13.

Kagale, S., Divi, U.K., Krochko, J.E., Keller, W.A., Krishna, P., 2007. Brassinosteroid confers tolerance in Arabidopsis thaliana and Brassica napus to a range of abiotic stresses. Planta 225 (2), 353–364.

Kang, H.-G., Kim, B., Jeong, H., Choi, S.H., Kim, E.K., Lee, H.-Y., Lim, P.O., 2011. Overexpression of FTL1/DDF1, an AP2 transcription factor, enhances tolerance to cold drought and heat stresses in Arabidopsis thaliana. Plant Sci. 180, 634–641.

Karki, A., Horvath, D.P., Sutton, F., 2013. Induction of DREB2A pathway with repression of E2F, jasmonic acid biosynthetic and photosynthesis pathways in cold acclimation-specific freeze-resistant wheat crown. Funct. Integr. Genomics 13, 57–65.

Kassaw, T.K., Donayre-Torres, A.J., Antunes, M.S., Morey, K.J., Medford, J.I., 2018. Engineering synthetic regulatory circuits in plants. Plant Sci. 273, 13–22.

Khong, G.N., Pati, P.K., Richaud, F., Parizot, B., Bidzinski, P., Mai, C.D., et al., 2015. OsMADS26 negatively regulates resistance to pathogens and drought tolerance in Rice. Plant Physiol. 169, 2935–2949.

Kidokoro, S., Watanabe, S., Ohori, T., Moriwaki, T., Maruyama, K., Mizoi, J., Htwe, N.M.P.S., Fujita, Y., Sekita, S., Shinozaki, K., Yamaguchi-Shinozaki, K., 2015. Soybean DREB1/CBF-type transcription factors function in heat and drought as well as cold stress-responsive gene expression. Plant J. 81, 505–518.

Kim, T.W., Michniewicz, M., Bergmann, D.C., Wang, Z.Y., 2012. Brassinosteroid regulates stomatal development by GSK3-mediated inhibition of a MAPK pathway. Nature 482, 419–422.

Klay, I., Gouia, S., Liu, M., Mila, I., Khoudi, H., Bernadac, A., Bouzayen, M., Pirrello, J., 2018. Ethylene response factors (ERF) are differentially regulated by different abiotic stress types in tomato plants. Plant Sci. 274, 137–145.

Koguchi, M., Yamasaki, K., Hirano, T., Sato, M.H., 2017. Vascular plant one-zinc-finger protein 2 is localized both to the nucleus and stress granules under heat stress in Arabidopsis. Plant Signal. Behav. 12, 1–8.

Koini, M.A., Alvey, L., Allen, T., Tilley, C.A., Harberd, N.P., Whitelam, G.C., Franklin, K.A., 2009. High temperature-mediated adaptations in plant architecture require the bHLH transcription factor PIF4. Curr. Biol. 19, 408–413.

Kumar, R.R., Goswami, S., Singh, K., Dubey, K., Rai, G.K., Singh, B., Singh, S., Grover, M., Mishra, D., Kumar, S., Bakshi, S., Rai, A., Pathak, H., Chinnusamy, V., Praveen, S., 2018. Characterization of novel heat-responsive transcription factor (TaHSFA6e) gene in regulation of heat shock proteins (HSPs) – a key member of heat stress tolerance network of wheat. J. Biotechnol. 279, 1–12.

Lavania, D., Dhingra, A., Grover, A., 2018. Analysis of transactivation potential of rice (*Oryza sativa* L.) heat shock factors. Planta 247, 1267–1276.

Li, S., Fu, Q., Huang, W., Yu, D., 2009. Functional analysis of an Arabidopsis transcription factor WRKY25 in heat stress. Plant Cell Rep. 28, 683–693.

Li, S., Zhou, X., Chen, L., Huang, W., Diqiu Yu, D., 2010. Functional characterization of *Arabidopsis thaliana* WRKY39 in heat stress. Mol. Cells 29, 475–483.

Li, F., Guo, S., Zhao, Y., Chen, D., Chong, K., Xu, Y., 2010a. Overexpression of a homopeptide repeat-containing bHLH protein gene (*OrbHLH001*) from Dongxiang Wild Rice confers freezing and salt tolerance in transgenic Arabidopsis. Plant Cell Rep. 29, 977–986.

Li, S., Zhou, X., Chen, L., Huang, W., Yu, D., 2010b. Functional characterizationof *Arabidopsis thaliana* WRKY39 in heat stress. Mol. Cells 29, 475–483.

Li, Z., Zhang, L., Wang, A., Xu, X., Li, J., 2013. Ectopic overexpression of SlHsfA3, a heat stress transcription factor from tomato, confers increased thermotolerance and salt hypersensitivity in germination in transgenic Arabidopsis. PLoS One 8, 1–16.

Li, Z., Tian, Y., Zhao, W., Xu, J., Wang, L., Peng, R., Yao, Q., 2015. Functional characterization of a grape heat stress transcription factor VvHsfA9 in transgenic Arabidopsis. Acta Physiol. Plant. 37, 1–10.

Li, S., Jin, H., Zhang, Q., 2016. The effect of exogenous spermidine concentration on polyamine metabolism and salt tolerance in zoysiagrass (*Zoysia japonica* Steud) subjected to short-term salinity stress. Front. Plant Sci. 7, 1–12.

Li, S., Xu, H., Ju, Z., Cao, D., Zhu, H., Fu, D., Grierson, D., Qin, G., Luo, Y., Zhu, B., 2018. The *RIN-MC* fusion of MADS-box transcription factors has transcriptional activity and modulates expression of many ripening genes. Plant Physiol. 176, 891–909.

Liao, W.-Y., Lin, L.-F., Jheng, J.-L., Wang, C.-C., Yang, J.-H., Chou, M.-L., 2016. Identification of heat shock transcription factor genes involved in thermotolerance of octoploid cultivated strawberry. Int. J. Mol. Sci. 17, 2130.

Licausi, F., Ohmi-Takagi, M., Perata, P., 2013. APETALA2/ethylene responsive factor (AP2/ERF) transcription factors: mediators of stress responses and developmental programs. New Phytol. 199, 639–649.

Liu, H.C., Charng, Y.Y., 2013. Common and distinct functions of Arabidopsis class A1 and A2 heat shock factors in diverse abiotic stress responses and development. Plant Physiol. 163, 276–290.

Liu, H.C., Liao, H.T., Charng, Y.Y., 2011. The role of class A1 heat shock factors (HSFA1) in response to heat and other stresses in Arabidopsis. Plant Cell Environ. 34, 738–751.

Liu, Y.-H., Offler, C.E., Ruan, Y.-L., 2013. Regulation of fruit and seed response to heat and drought by sugars as nutrients and signals. Front. Plant Sci. 4, 1–12.

Liu, Z.-B., Wang, J.-M., Yang, F.-X., Yang, L., Yue, Y.-F., Xiang, J.-B., Gao, M., Xiong, F.-J., LV, D., Wu, X.-J., Liu, N., Zhang, X., Li, X.-F., Yang, Y., 2014. A novel membrane-bound E3 ubiquitin ligase enhances the thermal resistance in plants. Plant Biotechnol. J. 12, 93–104.

Lowder, L.G., Zhang, D., Baltes, N.J., Paul, J.W., Tang, X., Zheng, X., Voytas, D.F., Hsieh, T.F., Zhang, Y., Qi, Y., 2015. A CRISPR/Cas9 toolbox for multiplexed plant genome editing and transcriptional regulation. Plant Physiol. 169, 971–985.

Ma, H., Wang, C., Yang, B., Cheng, H., Wang, Z., Mijiti, A., Ren, C., Qu, G., Zhang, H., Ma, L., 2016. *CarHSFB2*, a class B heat shock transcription factor, is involved in different developmental processes and various stress responses in Chickpea (*Cicer arietinum* L.). Plant Mol. Biol. Rep. 34, 1–14.

Meiri, D., Breiman, A., 2009. Arabidopsis ROF1 (FKBP62) modulates thermotolerance by interacting with HSP90 and affecting the accumulation of HsfA2-regulated HSPs. Plant J. 59, 387–399.

Meng, X., Wang, J.-R., Wang, G.-D., Liang, X.-Q., Li, X.-D., Meng, Q.-W., 2015. An R2R3-MYB gene, LeAN2, positively regulated the thermo-tolerance in transgenic tomato. J. Plant Physiol. 175, 1–8.

Miozzo, F., et al., 2015. HSFs, stress sensors and sculptors of transcription compartments and epigenetic landscapes. J. Mol. Biol. 427, 3793–3816.

Mittler, R., Blumwald, E., 2010. Genetic engineering for modern agriculture:challenges and perspectives. Annu. Rev. Plant Biol. 61, 443–462.

Morimotoa, K., Ohamaa, N., Kidokoroa, S., Mizoia, J., Takahashib, F., Todakaa, D., Mogamia, J., Satoa, H., Qinc, F., Kimb, J.-K., Fukaod, Y., Fujiwarad, M., Shinozakib, K., Yamaguchi-Shinozakia, K., 2017. BPM-CUL3 E3 ligase modulates thermotolerance by facilitating negative regulatory domain-mediated degradation of DREB2A in Arabidopsis. Proc. Natl. Acad. Sci. U. S. A. 114, E8520–E8536.

Nadeem, M., Li, J., Wang, M., Shah, L., Lu, S., Wang, X., Ma, C., 2018. Unravelling field crops sensitivity to heat stress: mechanisms, approaches, and future prospects. Agronomy 8, 1–34.

Nakano, Y., Yamaguchi, M., Endo, H., Rejab, N.A., Ohtani, M., 2015. NAC MYB-based transcriptional regulation of secondary cell wall biosynthesis in land plants. Front Plant Sci. 6, 288.

Nakashima, K., Takasaki, H., Mizoi, J., Shinozaki, K., Yamaguchi-Shinozaki, K., 2012. NAC transcription factors in plant abiotic stress responses. Biochim. Biophys. Acta 1819, 97–103.

Ohama, N., Sato, H., Shinozaki, K., Yamaguchi-Shinozaki, K., 2017. Transcriptional regulatory network of plant heat stress response. Trends Plant Sci. 22 (1), 53–65.

Oldfield, A., et al., 2014. Histone-fold domain protein NF-Y promotes chromatin accessibility for cell type-specific master transcription factors. Mol. Cell 55, 708–722.

Owjia, H., Hajiebrahimia, A., Seradjb, H., Hemmatia, S., 2017. Identification and functional prediction of stress responsive AP2/ERF transcription factors in *Brassica napus* by genome-wide analysis. Comput. Biol. Chem. 71, 32–56.

Peng, S.B., Zhu, Z.G., Zhao, K., Shi, J.L., Yang, Y.Z., He, M.Y., Wang, Y.J., 2012. A novel heat shock transcription factor, VpHsf1, from Chinese wild *Vitis pseudoreticulata* is involved in biotic and abiotic stresses. Plant Mol. Biol. Rep. 31, 240–247.

Piatek, A., Ali, Z., Baazim, H., Li, L., Abulfaraj, A., Al-Shareef, S., Aouida, M., Mahfouz, M.M., 2015. RNA-guided transcriptional regulation in planta via synthetic dCas9-based transcription factors. Plant Biotechnol. J. 13, 578–589.

Pires, N., Dolan, L., 2010. Origin and diversification of basic-helix-loop-helix proteins in plants. Mol. Biol. Evol. 27, 862–874.

Pirrello, J., Prasad, B.C.N., Zhang, W., Chen, K., Mila, I., Zouine, M., Latché, A., Pech, J.C., Ohme-Takagi, M., Regad, F., Bouzayen, M., 2012. Functional analysis and binding affinity of tomato ethylene response factors provide insight on the molecular bases of plant differential responses to ethylene. BMC Plant Biol. 190, 1–15.

Puranik, S., Sahu, P.P., Srivastava, P.S., Prasad, M., 2012. NAC proteins: regulation and role in stress tolerance. Trends Plant Sci. 17 (6), 369–381.

Prasch, C.M., Sonnewald, U., 2015. Signaling events in plants: stress factors in combination change the picture. Environ. Exp. Bot. 114, 4–14. https://doi.org/10.1016/j.envexpbot.2014.06.020.

Qu, A.L., Ding, Y.F., Jiang, Q., Zhu, C., 2013. Molecular mechanisms of the plant heat stress response. Biochem. Biophys. Res. Commun. 432, 203–207.

Ramakrishna, C., Singh, S., Raghavendrarao, S., Padaria, J.C., Mohanty, S., Sharma, T.R., Solanke, A.U., 2018. The membrane tethered transcription factor EcbZIP17 from finger millet promotes plant growth and enhances tolerance to abiotic stresses. Sci. Rep. 8, 2148.

Ray, S., Dansana, P.K., Bhaskar, A., Giri, J., Kapoor, S., Khurana, J.P., et al., 2009. Emerging trends in functional genomics for stress tolerance in crop plants. In: Hirt, H. (Ed.), Plant Stress Biology. Wiley-VCH Verlag GmbH & Co.KGaA, Weinheim, pp. 37–63.

Ricachenevsky, F.K., Menguer, P.K., Sperotto, R.A., 2013. NAC king on heaven's door: how important are NAC transcription factors for leaf senescence and Fe/Zn remobilization to seeds? Font. Plant Sci. 4, 1–7.

Rollins, J.A., Habte, E., Templer, S.E., Colby, T., Schmidt, J., von Korff, M., 2013. Leaf proteome alterations in the context of physiological and morphological responses to drought and heat stress in barley (Hordeum vulgare L.). J. Exp. Bot. 64, 3201–3212.

Sahni, S., Prasad, B.D., Liu, Q., Grbic, V., Sharpe, A., Singh, S.P., Krishna, P., 2016. Overexpression of the brassinosteroid biosynthetic gene *DWF4* in *Brassica napus*simultaneously increases seed yield and stress tolerance. Sci. Rep. 6, 1–14.

Saidi, Y., Finka, A., Muriset, M., Bromberg, Z., Weiss, Y.G., Maathuis, F.J.M., Goloubinoff, P., 2012. The heat shock response in moss plant is regulated by specific calcium-permeable channels in the plasma membrane. Plant Cell 21, 2829–2843.

Sanchez, D.H., Pieckenstain, F.L., Szymanski, J., Erban, A., Bromke, M., Hannah, M.A., et al., 2011. Comparative functional genomicsof salt stress in related model and cultivated plants identifies andovercomes limitations to translational genomics. PLoS One. 6: e17094. https://doi.org/10.1371/journal.pone.0017094.

Sato, H., Mizoi, J., Tanaka, H., Maruyama, K., Qin, F., Osakabe, Y., Morimoto, K., Ohori, T., Kusakabe, K., Nagata, M., Shinozaki, K., Yamaguchi-Shinozaki, K., 2014. Arabidopsis DPB3-1, a DREB2A interactor, specifically enhances heat-induced gene expression by forming a heat stress-specific transcriptional complex with NF-Y subunit. Plant Cell 26, 4954–4973.

Savo, V., Lepofsky, D., Benner, J.P., Kohfeld, K.E., Bailey, J., Lertzman, K., 2016. Observations of climate change among subsistence-oriented communities around the world. Nat. Clim. Chang. 6, 462–473.
Scarf, K.D., Rose, S., Zott, W., Schoffl, F., Nover, L., Scoff, F., 1990. Three tomato genes code for heat stress transcription factor with a region of remarkable homology to the DNA-binding domain of the yeast HSF. EMBO J. 9, 4495–4501.
Scarf, K.D., Berberich, T., Ebersberger, I., Novel, L., 2012. The plant heat stress transcription factor (Hsf) family: structure, function and evolution. Biochim. Biophys. Acta 1819, 104–119.
Shahnejat-Bushehri, S., Mueller-Roeber, B., Balazadeh, S., 2012. Arabidopsis NAC transcription factor JUNGBRUNNEN1 affects thermo memory associated genes and enhances heat stress tolerance in primed and unprimed conditions. Plant Signal. Behav. 7, 1518–1521.
Sharma, I., Ching, E., Saini, S., Bhardwaj, R., Pati, P.K., 2013. Exogenous application of brassinosteroid offers tolerance to salinity by altering stress responses in rice variety Pusa Basmati-1. Plant Physiol. Biochem. 69, 17–26.
Shinozaki, K., Yamaguchi-Shinozakiy, K., Sekiz, M., 2003. Regulatorynetwork of gene expression in the drought and cold stress responses. Curr. Opin. Plant Biol. 6, 410–417. https://doi.org/10.1016/S1369-5266(03)00092-X.
Song, C., Lee, J., Kim, T., Hong, J.C., Lim, C.O., 2018b. VOZ1, a transcriptional repressor of DREB2C, mediates heat stress responses in *Arabidopsis*. Planta 247, 1439–1448.
Surabhi, G.K., 2018. Update in root proteomics with special reference to abiotic stresses: achievements and challenges. J. Proteins Proteom. 9 (1), 31–55.
Surabhi, G.K., Badajena, B., Sahoo, S.K., 2019. Genome editing and abiotic stress tolerance in crop plants. In: Wani, S. (Eds) Recent Approaches in Omics for Plant Resilience to Climate Change. Springer Nature, Switzerland AG, pp. 35–56.
Udvardi, M.K., Kakkar, K., Wandrey, M., Montanari, O., Murray, J., Riankaja, A., 2007. Legume transcription factors: global regulators of plant development and response to the environment. Plant Physiol. 144, 538–549.
Ulker, B., Somssich, I.-E., 2004. WRKY transcription factors: from DNA binding towards biological function. Curr. Opin. Plant Biol. 7, 491–498.
Vardhini, B.V., Anjum, N.A., 2015. Brassinosteroids make plant life easier under abiotic stresses mainly by modulating major components of antioxidant defense system. Front. Environ. Sci. 2, 1–16.
Volkov, R.A., Panchuk, I.I., Mullineaux, P.M., Schoffl, F., 2006. Heat stress induced H_2O_2 is required for effective expression of heat shock genes in Arabidopsis. Plant Mol. Biol. 61, 733–746.
Wang, J.Y., Wang, J.P., Yuan, H., 2013. A populuseuphratica NAC protein regulating Na+/K+ homeostasis improves salt tolerance in Arabidopsis thaliana. Gene 521, 265–273.
Wang, D., Fan, J., Heckathorn, S.A., 2014. Acclimation of photosynthetic tolerance to acute heat stress at elevated CO_2 and N. Plant Sci. 226, 162–171.
Wang, D., Heckathron, S.A., Mainali, K., Tripathee, R., 2016a. Timing effects of heat-stress on plant ecophysiological characteristics and growth. Front. Plant Sci. 7, 1–11.
Wang, X., Huang, W., Yang, Z., Liu, J., Huang, B., 2016b. Transcriptional regulation of heat shock proteins and ascorbate peroxidase by CtHsfA2b from African bermudagrass conferring heat tolerance in Arabidopsis. Sci. Rep. 6, 1–12.
Wang, H., Wang, H., Shao, H., Tang, X., 2016c. Recent advances in utilizing transcription factors to improve plant abiotic stress tolerance by transgenic technology. Front. Plant Sci. 7, 67. https://doi.org/10.3389/fpls.2016.00067.
Wang, J., Tao, F., An, F., Zou, Y., Tian, W., Chen, X., Xu, X., Hu, X., 2017a. Wheat transcription factor TaWRKY70 is positively involved in high-temperature seedling-plant resistance to *Puccinia striiformis*f. sp. *Tritici*. Mol. Plant Pathol. 18, 649–661.

Wang, X., Huang, W., Liu, J., Yang, Z., Huang, B., 2017b. Molecular regulation and physiological functions of a novel FaHsfA2c cloned from tall fescue conferring plant tolerance to heat stress. Plant Biotechnol. J. 15, 237–248.
Wang, X.-L., Chen, X., Yang, T.-B., Cheng, Q., Cheng, Z.-M., 2017c. Genome-wide identification of bZIP family genes involved in drought and heat stresses in strawberry (*Fragaria vesca*). Int. J. Genomics 2017, 1–14.
Wang, J., Tao, F., Tian, W., Guo, Z., Chen, X., Xu, X., Shang, H., Hu, X., 2017d. The wheat WRKY transcription factors TaWRKY49 and TaWRKY62 confer differential high-temperature seedling-plant resistance to *Puccinia striiformis f.* sp. tritici. PLoS One 12, 1–23.
Wang, C.-T., Ru, J.-N., Liu, Y.-W., Li, M., Zhao, D., Yang, J.-F., Fu, J.-D., Xu, Z.-S., 2018a. Maize WRKY transcription factor ZmWRKY106 confers drought and heat tolerance in transgenic plants. Int. J. Mol. Sci. 19, 1–15.
Wang, X., Shi, X., Chen, S., Ma, C., Xu, S., 2018b. Evolutionary origin, gradual accumulation and functional divergence of heat shock factor gene family with plant evolution. Front. Plant Sci. 71, 1–11.
Wang, Y., Zhang, Y., Zhou, R., Dossa, K., Yu, J., Li, D., Liu, A., Mmadi, M.A., Zhang, X., You, J., 2018c. Identification and characterization of the bZIP transcription factor family and its expression in response to abiotic stresses in sesame. PLoS One 13, 1–21.
Wang, R., Zhao, P., Kong, N., Lu, R., Pei, Y., Huang, C., Ma, H., Chen, Q., 2018d. Genome-wide identification and characterization of the potato bHLH transcription factor family. Genes 9, 1–20.
Wang, Z., Shen, Y., Yang, X., Pan, Q., Ma, G., Bao, M., Zheng, B., Duanmu, D., Lin, R., Larkin, R.M., Ning, G., 2019. Overexpression of particulars MADS-box transcription factors in heat-stressed plants induces chloroplast biogenesis in petals. Plant Cell Environ. 42 (5), 1545–1560.
Wei, Z., Li, J., 2016. Brassinosteroids regulate root growth, development, and Symbiosis. Mol. Plant 9, 86–100.
Wei, M., Wang, Y., Pan, R., Li, W., 2018. Genome-wide identification and characterization of MADS-box family genes related to floral organ development and stress resistance in *Heveabrasiliensis* Müll. Arg. Forests 9, 1–16.
Wen, F., Zhu, H., Li, P., Jiang, M., Mao, W., Ong, C., Chu, Z., 2014. Genome-wide evolutionary characterization and expression analyses of *WRKY* family genes in *Brachypodium distachyon*. DNA Res. 21 (3), 327–339.
Wen, F., Wu, X., Li, T., Jia, M., Liu, X., Li, P., Zhou, X., Ji, X., Yue, X., 2017. Genome-wide survey of heat shock factors and heat shock protein 70s and their regulatory network under abiotic stresses in *Brachypodium distachyon*. PLoS One 12, 1–21.
Wu, K.L., Guo, Z.J., Wang, H.H., Li, J., 2005. The WRKY family of transcription factors in rice and Arabidopsis and their origins. DNA Res. 28, 9–26.
Wu, X., Shiroto, Y., Kishitani, S., Ito, Y., Toriyama, K., 2009. Enhanced heat and drought tolerance in transgenic rice seedlings overexpressing OsWRKY11 under the control of HSP101 promoter. Plant Cell Rep. 28, 21–30.
Wu, W., Cheng, Z., Liu, M., Yang, X., Qiu, D., 2014. C3HC4-type RING finger protein NbZFP1 is involved in growth and fruit development in *Nicotiana benthamiana*. PLoS One 9, 1–9.
Wu, Z.-J., Li, X.-H., Liu, Z.-W., Li, H., Wang, Y.-X., Zhuang, J., 2015. Transcriptome-based discovery of AP2/ERF transcription factors related to temperature stress in tea plant (*Camellia sinensis*). Funct. Integr. Genomics 6, 741–752.
Wu, X.X., Chen, J.L., Xu, S., Zhu, Z.W., Zha, D.S., 2016. Exogenous 24-epibrassinolide alleviates zinc-induced toxicity in eggplant (*Solanum melongena* L.) seedlings by regulating the glutathione ascorbate-dependent detoxification pathway. J. Hortic. Sci. Biotechnol. 91, 412–420.

Wu, Z., Liang, J., Zhang, S., Zhang, B., Zhao, Q., Li, G., Yang, X., Wang, C., He, J., Ki, M., 2018. A canonical DREB2-type transcription factor in lily is post-translationally regulated and mediates heat stress response. Front. Plant Sci. 9, 1–16.

Xiong, H., Li, J., Liu, P., Duan, J., Zhao, Y., Guo, X., Li, Y., Zhang, H., Ali, J., Li, Z., 2014. Overexpression of OsMYB48-1, a novel MYB-related transcription factor, enhances drought and salinity tolerance in rice. PLoS One 9, 1–13.

Xue, G.-J., Drenth, J., Mclntyre, L., 2015. TaHsfA6f a transcriptional activator that regulates a suite of heat stress protection gene in wheat (*Triticum aestivum* L.) including previously unknown Hsf targets. J. Exp. Bot. 66, 1025–1039.

Yabuta, Y., 2016. Functions of heat shock transcription factors involved in response to photooxidative stresses in *Arabidopsis*. Biosci. Biotechnol. Biochem. 80, 1254–1263.

Yamaguchi, T., Blumwald, E., 2005. Developing salt-tolerant cropplants: challenges and opportunities. Trends Plant Sci. 10, 615–620. https://doi.org/10.1016/j.tplants.2005.10.002.

Yang, W., Liu, X.D., Chi, X.J., Wu, C.A., Li, Y.Z., Song, L.L., et al., 2011. Dwarfapple MbDREB1 enhances plant tolerance to low temperature, drought, andsalt stress via both ABA-dependent and ABA-independent pathways. Planta 233, 219–229. https://doi.org/10.1007/s00425-010-1279-6.

Yang, A., Xiaoyan, D.X., Zhang, W.H., 2012. A R2R3-type MYB gene, OsMYB2, is involved in salt, cold, and dehydration tolerance in rice. J. Exp. Bot. 63, 2541–2556.

Yin, W., Hu, Z., Hu, J., Zhu, Z., Yu, X., Cui, B., Chen, G., 2016. Tomato (*Solanum lycopersicum*) MADS-box transcription factor SlMBP8 regulates drought, salt tolerance and stress-related genes. Plant Growth Regul. 83, 55–68.

Yin, M., Wang, Y., Zhang, L., Li, J., Quan, W., Yang, L., Wang, Q., Chan, Z., 2017. The Arabidopsis Cys2/His2 zinc finger transcription factor ZAT18 is a positive regulator of plant tolerance to drought stress. J. Exp. Bot. 68, 2991–3005.

Yin, Y., Qin, K., Song, X., Zhang, Q., Zhou, Y., Xia, X., Yu, J., 2018. BZR1 transcription factor regulates heat stress tolerance through FERONIA receptor-like kinases-mediated reactive oxygen species signaling in tomato. Plant Cell Physiol. 59, 2239–2254.

Yoshida, T., Sakuma, Y., Todaka, D., Maruyama, K., Qin, F., 2008. Functional analysis of an Arabidopsis heat-shock transcription factors HsfA3 in the transcriptional cascade downstream of the DREB2A stress-regulatory system. Biochem. Biophys. Res. Commun. 368, 515–521.

Young, L.W., Wilen, R.W., Bonham-Smith, P.C., 2004. High temperature stress of *Brassica napus* during flowering reduces micro- megagametophyte fertility, induces fruit abortion and disrupts seed production. J. Exp. Bot. 55, 485–495.

Zhang, S., Xu, Z.S., Li, P., Yang, L., Wei, Y., Chen, M., Li, L., Zhang, G., Ma, Y., 2012. Overexpression of TaHSF3 in transgenic Arabidopsis enhances tolerance to extreme temperatures. Plant Mol. Biol. Rep. 31, 688–697.

Zhang, S., Xu, Z.-S., Li, P., Yang, L., Wei, Y., Chen, M., Li, L., Zhang, G., Ma, Y., 2013. Overexpression of TaHSF3 in transgenic Arabidopsis enhances tolerance to extreme temperatures. Plant Mol. Biol. Rep. 31, 688–697.

Zhang, C., Bai, M.Y., Chang, K., 2014. Brassinosteroid-mediated regulation of agronomic traits in rice. Plant Cell Rep. 33, 683–696.

Zhang, A., Liu, D., Hua, C., Yan, A., Liu, B., Wu, M., et al., 2016. The Arabidopsis gene zinc finger protein 3(ZFP3) is involved in salt stress and osmotic stress response. PLoS One 11, 1–16.

Zhang, T., Lv, W., Zhang, H., Ma, L., Li, P., Ge, L., Li, G., 2018. Genome-wide analysis of the basic HelixLoop-Helix (bHLH) transcription factor family in maize. BMC Plant Biol. 18, 1–14.

Zhao, T., Xia, H., Liu, J., Ma, F., 2014. The gene family of dehydrationresponsive element-binding transcription factors in grape (*Vitisvinifera*): genome-wide identification and

analysis, expressionprofiles, and involvement in abiotic stress resistance. Mol. Biol. Rep. 41, 1577–1590.
Zhao, Y., Tian, X., Wang, F., Zhang, L., Xin, M., Hu, Z., Yao, Y., Ni, Z., Sun, Q., Peng, H., 2017. Characterization of wheat MYB genes responsive to high temperatures. BMC Plant Biol. 17, 1–14.
Zhu, X., Wang, Y., Liu, Y., Zhou, W., Yan, B., Yang, J., Shen, Y., 2018. Overexpression of BcHsfA1 transcription factor from *Brassica campestris* improved heat tolerance of transgenic tobacco. PLoS ONE. 13(11), e0207277.
Zhuang, L., Cao, W., Wang, J., Yu, J., Yang, Z., Huang, B., 2018. Characterization and functional analysis of FaHsfC1b from *Festuca arundinacea* conferring heat tolerance in Arabidopsis. Int. J. Mol. Sci. 19, 2702.

Further reading

Abd El-Daim, I.-A., Bejai, S., Meijer, J., 2014. Improved heat stress tolerance of wheat seedlings by bacterial seed treatment. Plant Soil 379, 337–350.
Bai, Y., Kissoudis, C., Visser, R.G.F., Linden, G.V., 2018. Plant behaviour under combined stress: tomato responses tocombined salinity and pathogen stress. Plant J. 93, 781–793.
Bechtold, U., Field, B., 2018. Molecular mechanisms controlling plant growth during abiotic stress. J. Exp. Bot. 69, 2753–2758.
Chen, W.R., Zheng, J.S., Li, Y.Q., Guo, W.D., 2012. Effects of high temperature on photosynthesis, chlorophyll fluorescence, chloroplast ultrastructure, and antioxidant activities in fingered citron. Russ. J. Plant Physiol. 59, 732–740.
Chen, C., Begcy, K., Liu, K., Folsom, J.J., Wang, Z., Zhang, C., Walia, H., 2014. Heat stress yields a unique MADS box transcription factor in determining seed size and thermal sensitivity. Plant Physiol. 161, 606–622.
Chwialkowska, K., Nowakowsk, U., Mroziewicz, A., Szarejke, I., Kwasniewski, M., 2016. Water deficiency conditions differently modulate the methylome of roots and leaves in barley (Hordeum vulgare L.). J. Exp. Bot. 67, 1109–1121.
Cvetkovic, J., Muller, K., Baier, M., 2017. The effect of cold priming on the fitness of Arabidopsis thaliana accessions under natural and controlled conditions. Sci. Rep. 7, 1–11.
Daryanto, S., Wang, L., Jacinthe, P.-A., 2016. Global synthesis of drought effects on maize and wheat production. PLoS One 11, 1–15.
Distéfano, A.M., Martin, M.V., Córdoba, J.P., Bellido, A.M., D'Ippólito, S., Colman, S.L., Soto, D., Roldán, J.A., Bartoli, C.G., Zabaleta, E.J., Fiol, D.F., Stockwell, B.R., Dixon, S.J., Pagnussat, G.C., 2017. Heat stress induces ferroptosis-like cell death in plants. J. Cell Biol. 216, 463–476.
Dong, C., Ma, Y., Zheng, D., Wisniewski, M., Cheng, Z.-M., 2018. Meta-analysis of the effect of overexpression of dehydration-responsive element binding family genes on temperature stress tolerance and related responses. Front. Plant Sci. 9, 1–15.
Fahad, S., Bajwa, A.A., Najir, U., Anjum, S.A., Farooq, A., Zohaib, A., Sadia, S., Nasim, W., Adkins, S., Saud, S., Ihsan, M.Z., Alharby, H., Wu, C., Wang, D., Huang, J., 2017. Crop production under drought and heat stress: plant responses and management options. Front. Plant Sci. 8, 1–15.
Faraji, A., Latifi, N., Soltani, A., Rad, A.H.S., 2000. Effect of high temperature stress and supplemental irrigation on flower and pod formation in two canola (*Brassica napus*. L) cultivars at Mediterranean climate. Asian J. Plant Sci. 7, 343–351.
Feng, B., Liu, P., Li, G., Dong, S.T., Wang, F.H., Kong, L.A., Zhang, J.W., 2014. Effect of heat stress on the photosynthetic characteristics in flag leaves at the grain-filling stage of different heat-resistant winter wheat varieties. J. Agron. Crop Sci. 200, 143–155.

Folsom, J.J., Begcy, K., Hao, X., Wang, D., Walia, H., 2014. Rice fertilizationindependent endosperm1 regulates seed size under heat stress by controlling early endosperm development. Plant Physiol. 165, 238–248.
Frich, P., Alexander, L.V., Della-Marta, P., Gleason, B., Haylock, M., Tank, A.M.G.K., et al., 2002. Observed coherent changes in climatic extremes during the second half of the twentieth century. Clim. Res. 19, 193–212.
Gill, S.S., Anjum, N.A., Gill, R., Jha, M., Tuteja, N., 2015. DNA damage and repair in plants under ultraviolet and ionizing radiations. Sci. World J. 2015, 1–11.
Gupta, B., Huang, B., 2014. Mechanism of salinity tolerance in plants: physiological, biochemical and molecular characterization. Int. J. Genomics 2014, 1–18.
Hwang, J.E., Kim, Y.J., Shin, M.H., Hyun, H.J., Bohnert, H.J., Park, H.J., 2018. A comprehensive analysis of the Korean fir (*Abieskoreana*) genes expressed under heat stress using transcriptome analysis. Sci. Rep. 8, 1–11.
International Brachypodium I, 2010. Genome sequencing and analysis of the model grass *Brachypodium distachyon*. Nature 463, 763–768.
IPCC, 2007. Climate change, the physical science basis. In: Solomon, S., Qin, D., Manning, M., Chen, Z., Marquis, M., Averyt, K.B., Tignor, M., Miller, H.L. (Eds.), Contribution of Working Group I to the Fourth Assessment Report of the Intergovernmental Panel on Change. Cambridge University Press, Cambridge, United Kingdom and New York.
IPCC, 2013. Climate Change 2013: The Physical Science Basis. Contribution of Working Group I to the Fifth Assessment Report of the Intergovernmental Panel on Climate Change. Cambridge University Press, Cambridge.
Ling-ling, D., Ya-ning, G., Evans, O., Chao-you, P., Heng-ling, W., Mei-zhen, S., Shu-li, F., Shu-Xun, Y., 2014. Identification and expression analysis of group III WRKY transcription factors in cotton. J. Integr. Agric. 15, 2469–2480.
Liu, J.-H., Peng, T., Dai, W., 2014a. Critical cis-acting elements and interacting transcription factors: key players associated with abiotic stress responses in plants. Plant Mol. Biol. Report. 32, 303–317.
Liu, W., Tai, H., Li, S., Gao, W., Zhao, M., Xie, X., Li, W.-X., 2014b. bHLH122 is important for drought and osmotic stress resistance in Arabidopsis and in the repression of ABA catabolism. New Phytol. 201, 1192–1204.
Lyman, N.B., Jagadish, K.S.V., Nalley, L.L., Dixon, B.L., Siebenmorgen, T., 2013. Neglecting rice milling yield and quality under estimates economic losses from high temperature stress. PLoS One. 8, e72157.
Mahmood, Q., Bilal, M., Jan, S., 2014. Herbicides, pesticides and plant tolerance: an overview. In: Ahmad, P., Rasool, S. (Eds.), Emerging Technologies and Management of Crop Stress Tolerance. In: vol. 2014. Academic Press, USA, pp. 423–448.
Meena, K.K., Sorty, A.M., Bitla, U.M., Choudhary, K., Gupta, P., Pareek, A., Singh, D.P., Prabha, R., Sahu, P.K., Gupta, V.K., Singh, H.B., Krishanani, K.K., Minhas, P.S., 2017. Abiotic stress responses and microbe mediated mitigation in plants: the omics strategies. Front. Plant Sci. 8, 1–25.
Mishra, S.K., Tripp, J., Winkelhaus, S., Tschiersch, B., Theres, K., Nover, L., Scharf, K.-D., 2002. In the complex family of heat stress transcription factors, HsfA1 has a unique role as master regulator of thermotolerance in tomato. Genes Dev. 16, 1555–1567.
Mittal, D., Enoki, Y., Lavania, D., Singh, A., Sakurai, H., Grover, A., 2011. Binding affinities and interaction of different heat shock element types and heat shock factors in rice (*Oryza sativa* L.). FEBS J. 278, 3076–3085.
Salehi-Lisar, S.Y., Bakhshayeshan-Agdam, H., 2016. Drought stress in plants: causes, consequences, and tolerance. In: Hossain, M., Wani, S., Bhattacharjee, S., Burritt, D., Tran, L.S. (Eds.), *Drought Stress Tolerance in Plants*, Vol. 1. Springer, Cham.
Salvucci, M.E., Deridder, B.P., Portis Jr., A.R., 2006. Effects of activase level and isoform on the thermotolerance of photosynthesis in Arabidopsis. J. Exp. Bot. 57, 3793–3799.

Song, C., Lee, J., Kim, T., Hong, J.C., Lim, C.O., 2018a. VOZ1, a transcriptional repressor of DREB2C, mediates heat stressresponses in *Arabidopsis*. Planta 47 (6), 1439–1448.
Suzuki, N., Rivero, R.M., Shulaev, V., Blumwald, E., Mittler, R., 2014. Abiotic and biotic stress combinations. New Phytol. 203, 32–43.
Szymanska, R., Susak, I., Orzechowska, A., Kruk, J., 2017. Physiological and biochemical responses to high light and temperature stress in plants. Environ. Exp. Bot. 139, 165–177.
Wahid, A., Gelani, S., Asharf, M., Foolad, M.R., 2007. Heat tolerance in plants: an overview. Environ. Exp. Bot. 61, 199–223.
Weng, L., Zhao, F., Li, R., Xu, R., Chen, K., Xiao, H., 2014. The zinc finger transcription factor SlZFP2 negatively regulates abscisic acid biosynthesis and fruit ripening in tomato. Plant Physiol. 163, 931–949.
Yamaguchi-Shinozaki, K., 2011. Arabidopsis HsfA1 transcription factors function as the main positive regulators in heat shock-responsive gene expression. Mol. Gen. Genomics. 286, 321–332.
Yang, J., Kloepper, J.W., Ryu, C., 2009. Rhizosphere bacteria help plants tolerate abiotic stress. Trends Plant Sci. 14, 1–4.
Zhaoa, C., Liub, B., Piao, S., Wanga, X., Lobelli, D.B., Huangj, Y., Huanga, M., et al., 2017. Temperature increase reduces global yields of major crops in four independent estimates. Proc. Natl. Acad. Sci. U. S. A. 101, 9971–9975.

CHAPTER 11

Transcription factors and their roles in phosphorus stress tolerance in crop plants

Anuj Kumar[a], Vijay Gahlaut[b], and M. Nagaraju[c]
[a]Advanced Centre for Computational and Applied Biotechnology, Uttarakhand Council for Biotechnology, Dehradun, Uttarakhand, India
[b]Department of Plant Molecular Biology, University of Delhi, Delhi, India
[c]Department of Genetics, Osmania University, Hyderabad, Telangana, India

1. Introduction

Phosphorus (P) is an essential mineral element required for normal plant growth and development. This indispensable mineral macro-element plays a central role in wide range of biological processes including nucleic acid synthesis, respiration, carbohydrate metabolism, and nitrogen fixation, glycolysis, photosynthesis, signal transduction, energy transfer, membrane synthesis and stability, enzyme activation/inactivation, and redox reactions (Abel et al., 2002; Vance et al., 2003; Niu et al., 2012, 2013; Kumar et al., 2018a). Mainly plants absorb P from the soil through their roots, in the form of phosphate (Pi) (Vance et al., 2003; Balyan et al., 2016). Further, soil Pi exists primarily in the forms of organic molecules and Ca (calcium), Fe (iron) and (Al) aluminum salts, which are very difficult for the roots to absorb (Raghothama and Karthikeyan, 2005). It decreases the bioavailability of Pi, leading to an available Pi amount in the soil that is far lower than the requirement for sustainable plant growth (Schachtman et al., 1998). In order to improve crop yield, a huge quantity of Pi fertilizers is often utilized to overcome the problem of Pi deficiency and maintain the stable cellular Pi concentrations. However, this approach leads to water eutrophication and overexploitation of phosphate ore, which is nonrenewable. Resolving the above problem is very critical to environmental protection and sustainable development. Basically, to adapt to Pi-deficient environments, plants increase the absorption of Pi from the soil through undergo phenotypic changes in the root system architecture (RSA) including favored lateral root growth through increased lateral root density and length and reduced primary root growth mediated by reduced cell

Transcription Factors for Abiotic Stress Tolerance in Plants
https://doi.org/10.1016/B978-0-12-819334-1.00011-3

elongation (Williamson et al., 2001; Vance et al., 2003). In the past, a number of molecular mechanisms which define Pi-transport system have been proposed previously to overcome Pi stress and also to maintain Pi homeostasis (Luan et al., 2017). It has been well established that Pi transporters are the basic effectors involved in Pi uptake, transport and storage (Kumar et al., 2016). These molecular mechanisms are controlled by complex gene regulatory networks. As part of the central dogma of molecular biology, transcriptional activation/repression presents an early and key step for modulating gene expression. This is mainly dependent on TFs and has been demonstrated as the main level of regulation in model plants and crops encountering Pi stress (Bustos et al., 2010; Secco et al., 2013). To date, many TFs and miRNAs have been reported and functionally annotated to be involved in plant Pi-stress responses in plants (Yi et al., 2005; Chen et al., 2007; Liu et al., 2009; Shen et al., 2013; Wang et al., 2014a,b; Dai et al., 2016; Ruan et al., 2017; Gu et al., 2017). TFs and miRNAs are playing a critical role in modulating pathways associated with stress-sensory and precise responses to environmental stresses including Pi stress (Deinlein et al., 2014; Gu et al., 2017; Kumar et al., 2016b). Various TF families responses to biotic and abiotic stresses, including WRKY, MYB, ZFP, and ARF have been most intensively studied for their functional roles in eliciting Pi-stress responses in plants (Wang et al., 2014a,b; Dai et al., 2016; Ruan et al., 2017; Gu et al., 2017). The availability of the whole genome sequences of model plants and crops has dramatically accelerated the attention from molecular biologist and plant physiologist to investigate their potential roles in producing stress-tolerant cultivars (Gahlaut et al., 2016; Wani et al., 2018; Kumar et al., 2016, 2018b). Multiple transgenics with improved tolerance against Pi starvation have been developed through classical genetic approaches and have already been reported in the previous studies. For last two decades, with the momentum the miRNAs (a class of noncoding RNAs) have gained the importance of genes-encoding TFs have given new height as they identified to be major regulatory targets of miRNAs for master modulation under abiotic stresses. In this chapter, we elaborate on the key TFs that contribute toward Pi-stress tolerance, and their regulatory networks, and potential utilization in developing crop varieties resilient to Pi stress.

2. TF families involved in Pi-stress tolerance

2.1 WRKY

The WRKY TFs are exclusively found in plants and form a large superfamily characterized by a highly conserved WRKY domain for DNA binding.

The approximately length of this conserved WRKY domain is 60 amino acid residues, which contains WRKYGQK consensus sequence at the N-terminus, followed by a C_2H_2 (C–X_{4-5}–C–X_{22-23}–H–X–H) or C_2HC (C–X_7–C–X_{23}–H–X–C) zinc-finger motif at the C-terminus (Eulgem et al., 2000; Rushton et al., 2010; Tripathi et al., 2014; Gahlaut et al., 2016; Ning et al., 2017). This highly conserved WRKY domain also referred to as the "signature sequences" at the N-terminus (Rushton et al., 1996; Maleck et al., 2000; Tripathi et al., 2014). It has been well established that the WRKY domain facilitates high-binding affinity for the *cis*-acting elements called W-box (TTGACC/T), which is found in several abiotic stress-responsive genes (Ulker and Somssich, 2004; Rushton et al., 2010; Gahlaut et al., 2016). Members of this family are well known to impart tolerance against abiotic and biotic stress (Zhu et al., 2013; Tripathi et al., 2014; Wani et al., 2018). According to the number of WRKY domain repeats and the features of their zinc-finger-like motif, the WRKY family members are divided into three different groups (I, II, and III) (Eulgem et al., 2000). Generally, a group I categorized by two WRKY domains, including a C_2H_2 zinc-finger structure, while group II and III typically contain a single functional domain, including a C_2H_2 and C_2H_2 zinc-finger motif, respectively (Eulgem et al., 2000; Rushton et al., 2010; Ning et al., 2017). Based on the phylogeny analysis, group II can be further divided into five subgroups (IIa-IIe) (Zhang and Wang, 2005; Rushton et al., 2008). The first foundation member of the family of WRKY TFs, namely the *SPF1* gene from sweet potato (*Ipomoea batatas*) was identified in 1994 (Ishiguro and Nakamura, 1994). Since then, several WRKY genes have been identified and characterized by different crop plants including *Oryza sativa*, *Zea mays*, and *Triticum aestivum* (Tripathi et al., 2014; Gahlaut et al., 2016; Wani et al., 2018). Whole-genome sequencing data availability of plants accelerates the systematic identification and characterization of WRKY-encoding genes in different plants. So far, genome-wide-based studies for WRKY TFs have been reported in *Arabidopsis thaliana*, *O. sativa*, *Z. mays*, *T. aestivum*, *Cucumis sativus*, *Brachypodium distachyon*, *Brassica rapa*, *Lotus japonicas*, *Vitis vinifera*, *Daucus carota*, *Manihot esculenta*, *Dendrobium officinale*, *kiwifruit*, *Dimocarpus longan*, *Arachis duranensis*, *Arachis hypogaea*, *Glycine max*, *Chenopodium quinoa*, and other plants (Wang et al., 2011; Wei et al., 2012; Xu et al., 2016a,b; Chen et al., 2016; Song et al., 2016a,b, 2018; Ning et al., 2017; Li et al., 2017a,b; He et al., 2017; Jing and Liu, 2018; Jue et al., 2018; Yue et al., 2019).

The functional members of WRKY family play a central role in various cellular and molecular processes including seed development, seed dormancy, seed germination, senescence, development, and biotic and abiotic stresses (Hinderhofer and Zentgraf, 2001; Robatzek and Somssich, 2001; Johnson et al., 2002; Sun et al., 2003; Zhang et al., 2004; Luo et al., 2005; Eulgem, 2006; Zhou et al., 2008a,b; Li et al., 2009; Popescu et al., 2009; Qiu and Yu, 2009; Tao et al., 2009; Wu et al., 2017). Here, we focus on the role of WRKY TFs in response to Pi stress. Multiple studies have shown that members of the WRKY family respond to Pi stress. In *Arabidopsis*, four WRKY TFs including *AtWRKY75*, *AtWRKY6*, *AtWRKY45*, and *AtWRKY42* have been reported as the key regulators involved in Pi starvation. *AtWRKY75* is the first member of WRKY TF family reported to be involved in regulating a Pi-stress response and root development in *Arabidopsis* (Devaiah et al., 2007a). It has been reported that *AtWRKY6* negatively regulates *PHO1* expression in response to low Pi stress (Chen et al., 2009). Wang et al. (2014b) showed that the overexpression of *AtWRKY45* involved in response to Pi stress by direct upregulation of phosphate transporter (*PHT1;1*) expression. Under Pi-sufficient condition, *AtWRKY42* modulated homeostasis by regulating the *PHO1* expression and *PHT;1* (Su et al., 2015). Transgenic of stress-responsive *OsWRKY74* (belonging to group III of the WRKY family) contributed in regulation to improved tolerance to Pi homeostasis and cross talk between Pi and Fe starvation, and cold stress in rice (Dai et al., 2016). However, apart from significant progress in functional annotations of WRKY TFs, to determine the specific biological significance of each and every member of this important family, a great deal of work is required.

2.2 MYB

The MYB family of TFs is large, functionally diverse, and exclusively found in all eukaryotes (Abe et al., 2003; Ambawat et al., 2013; Gahlaut et al., 2016). Members of MYB TF family harbors a highly conserved MYB DNA-binding domain (DBD) with 1–3 imperfect repeats, each acquire a helix-turn-helix structure of about 52aa residues (Stracke et al., 2001; Yanhui et al., 2006; Wani et al., 2018). Based on the number of MYB DBD in the protein structure, the whole family was divided into four major groups: 1R-MYB (MYB-related type), R2R3-MYB, R1R2R3-MYB, and 4R-MYB. These groups containing one, two, three, and four *MYB* repeats, respectively. Among them, the members of the R2R3-MYB group are more popular in higher plants (Dubos et al., 2010; Ambawat et al., 2013; Li et al., 2015b; Wang et al., 2017; Gahlaut et al., 2016). The first plant *MYB* gene (*C1*) from maize was identified and characterized in 1987 (Paz-Ares et al., 1987). MYB family members have been discovered and functionally characterized at the genome scale. To date, genome-wide identification and characterization of MYB TF family have been reported in wide range of plant species including such as *A. thaliana*, *Brassica napus*, kiwi, *Brachypodium distachyon*, *Setaria italica*, pear, sesame, *Phyllostachys edulis*, and cotton (Dubos et al., 2010; Muthamilarasan et al., 2014a,b; Cao et al., 2016; Mmadi et al., 2017; Hajiebrahimi et al., 2017; Li et al., 2017a,b; Chen et al., 2017; Yang et al., 2019; Huang et al., 2019). PlantTFDB (V4.0) contains 22,032 MYB genes from different plant species. The species with the most MYB genes include the following *Brassica napus* (489 members), *Gossypium hirsutum* (441 members), *Kalanchoe marnieriana* (440 members), *Glycine max* (430 members), and *Camelina sativa* (402 members) (Jin et al., 2017). Members of MYB TF family are well known to involve in wide arrange of biological process such as growth, development, and function of organs and specific cell types, metabolite biosynthesis, and abiotic stress responses (Ambawat et al., 2013; Baldoni et al., 2015; Wani et al., 2018).

The functions of plant MYB proteins, which are encoded by a large gene family, have been discovered and characterized by using classical genetic methods. Chlamydomonas phosphorus starvation resistant 1 (*CrPSR1*) was the first member of the MYB TF family reported to be involved in response to Pi starvation (Wykoff et al., 1999). It has been well studied that the *Arabidopsis PHOSPHATE STARVATION RESPONSE1 (AtPHR1)* and its homologue namely, *PHOSPHATE STARVATION RESPONSE1 (AtPHR2)* are positive regulators of Pi-acquisition and Pi-starvation-induced genes in *Arabidopsis* (Rubio et al., 2001; Todd et al., 2004). Zhou et al. (2008a,b) reported that *OsPHR2* positively regulates the Pi-starvation signaling and excessive Pi accumulation in shoots of rice. In *Arabidopsis*, expression of several MYB TFs such as *MYB108*, *PAP1*, *PAP2*, *At1g13300*, *At1g68670*, and *At3g25790* were shown to be induced, while *MYB111* protein was repressed in response to low-Pi stress (Nilsson et al., 2010; Li et al., 2015b). It has been reported that *OsMYB2*, an R2R3 MYB TF family member, plays a positive role in the modulation of primary root elongation under Pi starvation (Niu et al., 2013). In *Arabidopsis* and rice, the overexpression of *OsMYB2P-1* gene conferred the greater tolerance of transgenic plants and also activated the expression of *OsPT6*, *OsPT8*, and *OsPT10* genes under low-Pi stress (Dai et al., 2012). Overexpression of R2R3-type MYB TF OsMYB4P regulates Pi accumulation and expression of Pi-responsive transporter genes in rice crop (Yang et al., 2014). Gu et al.

(2017) identified and characterized the Pi stress-responsive *MYB1* gene (encoding an R2R3-type TF) in rice. In that study, the authors highlighted the role of *MYB1* as a regulator involved in both Pi-starvation signaling and gibberellic (GA) biosynthesis. Zhou et al. (2017) showed that *StMYB44* negatively regulates Pi transport by suppressing *StPHO1* expression in potato. *OsMYB5P* was recently demonstrated to modulate Pi-starvation responses by activating the expression of Pi transporter genes (Yang et al., 2018). However, further validation and functional characterization of MYB TF candidates are needed to elucidate their mode of molecular action during Pi starvation toward developing stress-tolerant crop genotypes.

2.3 Zinc finger

One of the largest plant TF families is zinc-finger proteins (ZFPs) TF family. ZFPs play a central role in regulating and modulating a wide range of plant growth and developmental processes and responses to environmental stresses. The family is characterized and functionally annotated because of the presence of the DBD referred to as the "zinc-finger (ZF) domain" (Takatsuji, 1999). It is well known that the zinc ion of the ZF domain is surrounded by cysteine (Cys) amino acids and/or the histidine (His) amino acids which help to stabilize the three-dimensional structure of ZF domain. Generally, the 3D structure of the ZF domain comprises a two-stranded antiparallel β-sheet and α-helix, which is essential for playing various biological roles of ZFPs (Li et al., 2013, 2015b). Based on the number of ZF domains in the protein structure and location of Cys and His residues, the whole family was divided into 10 different groups: C2H2, C2HC, C2HC5, C2C2, CCCH, C3HC4, C4, C4HC3, C6, and C8 (Moore and Ullman, 2003; Berg and Shi, 1996). Among them, the C2H2-ZFPs are more prevalent in plants (Liu et al., 2015). The C2H2-ZFP TFs also known as the TFIIIA-type ZFPs, can be denoted as X2-Cys-X(2–4)-Cys-X12-His-X(3–5)-His (Pabo et al., 2001). Since the first Petunia *ZFP* gene, was cloned and its function investigated (Takatsuji et al., 1992), a series of ZFP proteins has been identified and functionally annotated, by means of functional and comparative genomics and genome-wide studies, from different plant species such as *A. thaliana* (176 members), rice (189 members), foxtail millet (124 members), poplar (109), *Nicotiana tabacum* (118 members), and soybean (321 members) (Colasanti et al., 1998; Mukhopadhyay et al., 2004; Englbrecht et al., 2004; Agarwal et al., 2007; Muthamilarasan et al., 2014a,b; Liu et al., 2015; Minglei et al., 2016; Yuan et al., 2018). Furthermore, multiple

ZFP TFs in model plants and crops have been characterized and annotated in detail. The outcomes from previous studies have confirmed that they are play an important role in multiple physiological processes including trichome initiation, seed germination, floral organogenesis, and primary miRNA biogenesis (Dinneny et al., 2006; Joseph et al., 2014; Liu et al., 2015; Yang et al., 2006; Zhou et al., 2013; Li et al., 2010). Despite their central role in different biological processes, few of *ZFP* genes have been characterized and annotated in terms of Pi-starvation responses. Overexpression of *AtZAT6*, a member of the C2H2-ZFP TF family, mediates the Pi homeostasis through control of root architecture in *Arabidopsis* (Devaiah et al., 2007a,b). Ding et al. (2016) showed that *TaZAT8* plays important role in plant development and positively modulates Pi-stress resistance by regulating Pi acquisition and cellular ROS detoxicity in wheat. A working model of signal transduction and future utilization of TFs for developing Pi-stress cultivars has been apprehensively depicted in Fig. 1.

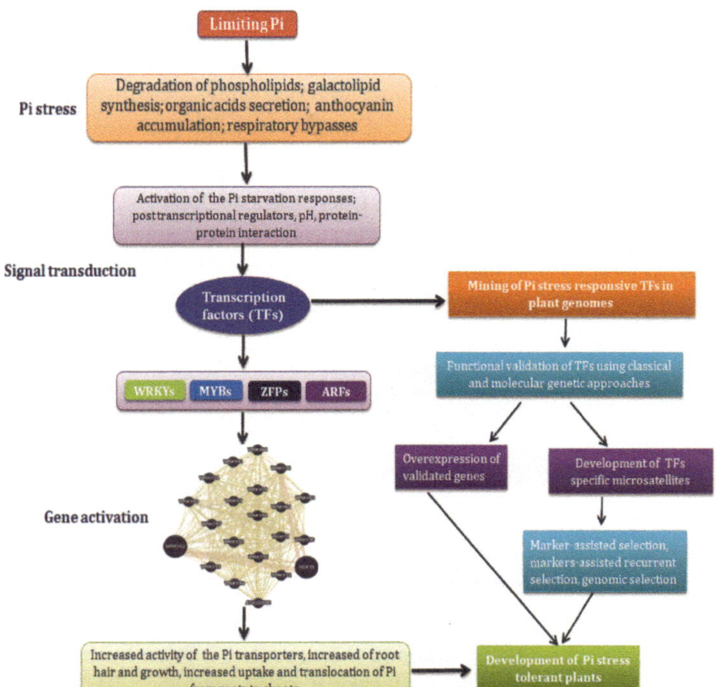

Fig. 1 A working model of Pi-stress-responsive TFs followed by gene expression in plants; their utilization in the development of Pi-stress-tolerant cultivars using marker-assisted breeding program is also shown.

2.4 ARF

As central regulator for many biological processes such as lateral root initiation, shoot elongation, embryogenesis, vascular growth, tropic development, and flowering and tissue architecture, auxin is an important phytohormone for plant growth and crop productivity (Teale et al., 2006; Domagalska and Leyser, 2011; Xiao et al., 2018). It is well known that ARFs are the important components of an auxin signaling pathway (Domagalska and Leyser, 2011; Xiao et al., 2018). Members of the ARF TF family regulate the auxin-responsive genes through directly binding to the auxin-responsive elements (Guilfoyle and Hagen, 2007). The promoters of ARFs harbor one or several copies of signature designated as TGTCTC or its variants, which is known as auxin-responsive element (AuxRE) (Guilfoyle and Hagen, 2007; Shen et al., 2015). Structurally, the proteins of ARF TF family are specified by three conserved domains: an N-terminal B3-type DBD; a variable middle region (MR); and two C-terminal dimerization domains (CTDs) (Ulmasov et al., 1999; Gray et al., 2001; Tiwari et al., 2003). The primary function of DBD is to recognize the AuxRE, while MR activates or represses the regulation of auxin-responsive genes depending on its residue combinations (Tiwari et al., 2003; Guilfoyle and Hagen, 2007, Guilfoyle, 2015). A CTD, similar to PB1 domain of the auxin/indole-3-acetic acid protein, is well characterized for its vital role in the formation of homo- and heterodimers between ARFs and Aux/IAA proteins (Kim et al., 1997; Gray et al., 2001; Xiao et al., 2018). The first *ARF* gene (*ARF1*) was identified and characterized in *Arabidopsis* in 1997 (Ulmasov et al., 1997). Since then numerous *ARF* genes have been cloned and functionally characterized form various models and crop plant species (Wang et al., 2007, 2012; Ha et al., 2013; Xiao et al., 2018). As an increasing number of genome sequences have been released, a variety of ARF TF family proteins have been identified and analyzed in wide range of plants species including *Arabidopsis* (23 members), tomato (22 members), maize (31 members), rice (25 members), soybean (51 members), cotton (56 members), poplar (39 members), *Cucumis sativus* (15 members), *Citrus sinensis* (19 members), *Medicago truncatula* (24 members), *Brachypodium distachyon* (19 members), *Musa acuminata* (47 members), *Salvia miltiorrhiza* (25 members), *Eucalyptus grandis* (17 members), and *V. vinifera* (17 members) (Guilfoyle and Hagen, 2007; Kalluri et al., 2007; Liu et al., 2007, 2018; Wang et al., 2007, 2012; Shen et al., 2013, 2015; Ha et al., 2013; Liu and Hu, 2013; Zouine et al.,

2014; Yu et al., 2014; Wan et al., 2014; Li et al., 2015b; Hu et al., 2015; Xu et al., 2016a,b; Xiao et al., 2018).

Several ARF TF family members are involved in regulating the response to abiotic stress including Pi starvation. Wang et al. (2014b) showed that *OsARF12*, an ARF TF family member required for auxin and Pi homeostasis responses in rice. It has been demonstrated that *OsARF16* is required for the cytokinin response and negatively regulates the Pi uptake and Pi signaling through cytokinin-mediated inhibition in rice (Shen et al., 2012, 2014). In maize, *ZmARF31* plays central roles in mediating tolerance to Pi starvation and root architecture regulation (Wu et al., 2016). Recently, two members of ARF namely, *auxin response factor 7 (ARF7)* and *auxin response factor 19 (ARF19)* genes were reported to plant resistance against Pi starvation through positive regulation of *PHR1* gene in *Arabidopsis* root (Huang et al., 2018). There are multiple reports on TFs which are involved in Pi-stress tolerance and have been systematically examined (Table 1).

3. Role of TFs in developing Pi-stress-tolerance plants

Plants respond and mediate Pi starvation through adaptive modifications such as altered root architecture (López-Bucio et al., 2003), elevated Pi activity (Lipton et al., 1987), secretion of organic acids (Marschner, 1995), and upregulation of Pi transporters (Raghothama and Karthikeyan, 2005), which help plants to overcome the Pi starvation by enhancing their ability to mobilize, acquire, and utilize Pi more efficiently (Rausch and Bucher, 2002). These adaptive mechanisms are regulated by complex molecular mechanisms, which are under the control of TFs are the molecular determinants, play an important role in Pi deficiency state. Cloning and alteration of the TFs expression to achieve desirable quality is a feasible approach to improve crop tolerance to stress (Century et al., 2008). *Arabidopsis* TFs including *MYB62*, *ZAT6*, *bHLH32*, *PTF1*, *WRKY75*, and rice *OsWRKY74* are the major candidates play an important role in the regulation of Pi homeostasis (Yi et al., 2005; Chen et al., 2007; Devaiah et al., 2007a,b, 2009; Dai et al., 2016).

The *PHR1* (Rubio et al., 2001), *WRKY75* (Devaiah et al., 2007a), *ZAT6* (Devaiah et al., 2007b), and MYB62 (Devaiah et al., 2009) were some important TFs involved in Pi-starvation signaling. The *PHR1*, an MYB TF regulates the Pi-starvation responses in both vascular plants and algae (Rubio et al., 2001). *Arabidopsis phr1* mutant exhibits impaired

Table 1 A list of TFs involved in Pi stress tolerance in plants.

TF family	TF genes	Plant	Function in stress	References
WRKY	*AtWRKY45*	*Arabidopsis*	Pi starvation	Wang et al. (2014a)
	AtWRKY75/6/45/42	*Arabidopsis*	Pi deficiency	Chen et al. (2009)
	OsWRKY74	Rice	Pi starvation	Dai et al. (2016)
MYB	*AtPHR1/2*	*Arabidopsis*	Pi starvation	Rubio et al. (2001), Todd et al. (2004)
	AtMYB108	*Arabidopsis*	Pi starvation	Nilsson et al. (2010), Li et al. (2015b)
	AtMYB111	*Arabidopsis*	Repressed in response to low-Pi stress	Nilsson et al. (2010), Li et al. (2015b)
	OsMYB2	Rice	Pi starvation	Niu et al. (2013)
	OsMYB4P	Rice	Pi starvation	Yang et al. (2014)
	OsMYB1	Rice	Pi starvation and gibberellic (GA) biosynthesis	Gu et al. (2017)
ZFP	*AtZAT6*	*Arabidopsis*	Pi starvation	Devaiah et al. (2007a,b)
	TaZA8	Wheat	Pi starvation and cellular ROS detoxicity	Ding et al. (2016)
ARF	*AtARF7/9*	*Arabidopsis*	Pi starvation	Huang et al. (2018)
	OsARF12	Rice	Pi starvation	Wang et al. (2014b)
	OsARF16	Rice	Cytokinin response and Pi homeostasis	Shen et al. (2012, 2014)
	ZmARF31	Maize	Pi starvation	Wu et al. (2016)

Pi-starvation responses, with lower anthocyanin, starch, and sugars, altered issuance of Pi between root and shoot, and also decreases the response of Pi deficiency responsive genes (Nilsson et al., 2007). The *OsPHR1* and *OsPHR2* are participated in the Pi-starvation signaling while the overexpression of *OsPHR2* accumulates Pi in shoots under Pi starvation (Zhou et al., 2008a,b). The rice *OsPHR2* and *AtPHR1* of *Arabidopsis* MYB TFs play a key role in Pi signaling (Wu et al., 2013). The *PHR1* binds to P1BS elements "GNATATNC" in the promoters of Pi-starvation-responsive genes (Bustos et al., 2010) and regulate their expressions under Pi-starvation state. The *Arabidopsis* PHR like gene overexpression enhanced the Pi uptake, which provides a platform for the breeding of crop plants toward more efficient Pi acquisition and assimilation (Nilsson et al., 2007). In rice, members of PHR family such as *OsPHR1*, *OsPHR2*, *OsPHR3*, and *OsPHR4* were functionally evaluated under Pi-starvation state (Ruan et al., 2016) and overexpression *OsPHR1*, *OsPHR2*, and *OsPHR3* resulted to enhance Pi accumulation in the shoot (Guo et al., 2015). The *OsPHR3*-transgenic lines showed enhanced tolerance to Pi deficient stress and showed normal growth under optimal Pi conditions (Guo et al., 2015). A total of 35 *GmPHR* genes were identified in soybean, of the *GmPHR25* mediates the Pi homeostasis (Xue et al., 2017). The *ZmPHR1* regulates the several metabolic genes under Pi starvation in *Arabidopsis* (Wang et al., 2013a,b). Overexpression of R2R3 MYB TF, *OsMYB2P-1* in *Arabidopsis* and rice upregulate the Pi-responsive genes, *O. sativa* UDP-sulfoquinovose synthase, *OsIPS1*, *OsPAP10*, *OsmiR399a*, and *OsmiR399j* and suppressed the *OsPHO2* under both Pi-sufficient and Pi-deficient conditions and exhibited enhanced tolerance to Pi deficiency, and suppression of *OsMYB2P-1* in rice become more sensitive to Pi starvation (Dai et al., 2012).

The *AtWRKY75*, *AtWRKY6*, *AtWRKY45*, and *AtWRKY42* of *Arabidopsis* WRKY TFs play a major role in Pi-starvation response (Devaiah et al., 2007a; Chen et al., 2009; Wang et al., 2014b; Su et al., 2015). The *WRKY75* suppression in *Arabidopsis* leads to impaired Pi-starvation responses which confirm its positive regulatory mechanism (Devaiah et al., 2007a), while *AtWRKY6* targets the PHOSPHATE1 (*PHO1*) and negatively regulates Pi-starvation response (Chen et al., 2009), whereas *WRKY45* alters the expression of PHT1;1 and regulates Pi homeostasis (Wang et al., 2014b). The *WRKY42*, a homologue of *WRKY6* in *Arabidopsis* (Eulgem et al., 2000), maintains the Pi homeostasis by controlling *PHO1* and *PHT1;1* (Su et al., 2015). Overexpression of *OsWRKYP74* in rice mediates the Pi-starvation conditions by targeting Pi-starvation-responsive genes such as *OsPT3*, *OsPT4*, and *OsPT10*, which may account for the Pi uptake in roots and modulate the architecture of root system in transgenics (Dai et al., 2016).

The *ZAT6*, *MYB62*, and *AtERF070* TFs upregulate or reduce the *PSI* genes which are participated in Pi sensing, translocation, transport, and mobilization (Devaiah et al., 2007b, 2009; Ramaiah et al., 2014). The *ARF7* and *ARF19* target the *PHR1*a key regulatory component of the Pi starvation and is positively regulated by auxin signaling in *Arabidopsis* roots (Huang et al., 2018). Heterologous expression of wheat C2H2-ZFP TF *TaZAT8* in tobacco exhibited tolerance to Pi starvation by detoxification of ROS and acquisition of Pi under deficiency state by regulating the *NtPT1* and *NtPT2* (Ding et al., 2016).

4. Role of miRNAs in Pi stress tolerance

MicroRNAs are endogenous noncoding RNAs, mature miRNAs are single-stranded and 20–24 nucleotides long in their length. MiRNAs have emerged as master regulators of plant development, and adaptive responses to stresses (both biotic and abiotic) via posttranscriptional regulation of gene expression (Mallory and Vaucheret, 2006; Sunkar et al., 2012; Ferdous et al., 2015). The miRNAs also involved in the regulation of important nutrient metabolic pathways and their homeostasis (phosphate, copper, nitrogen, and sulfate) in plants (Fujii et al., 2005; Chiou et al., 2006; Yamasaki et al., 2007; Kuo and Chiou, 2011; Liang et al., 2015). The first miRNA that has been demonstrated to be upregulated during Pi starvation and rapidly decreased upon Pi readdition is miRNA399 (Fujii et al., 2005). It was shown that transgenic *Arabidopsis* plants overexpressing the miR399 gene showed an increased Pi uptake (Fujii et al., 2005; Chiou et al., 2006). The miR399 targets three Pi transporter (PHT1;7), but only a ubiquitin-binding enzyme E2 encoded by the PHO2/UCB24 gene was validated (Allen et al., 2005). During normal Pi conditions, PHO2/UCB24 (E2 ligase) involved in the ubiquitination of the Pi/H+ transporters, avoiding their trafficking to cell membranes (Park et al., 2014). PHO2/UCB24 also involved in posttranslational inhibition of PHO1 (Liu et al., 2012). Therefore, by repressing *PHO2*, miR399 contributes to the accumulation of both the PHO2 downstream genes *PHT1* and *PHO1* during Pi-starvation conditions. Additionally, miR399 can be regulated by endogenous TM (eTM)-like lncRNA *induced by phosphate starvation1 (IPS1)* (Fig. 2). *IPS1* has a sequence

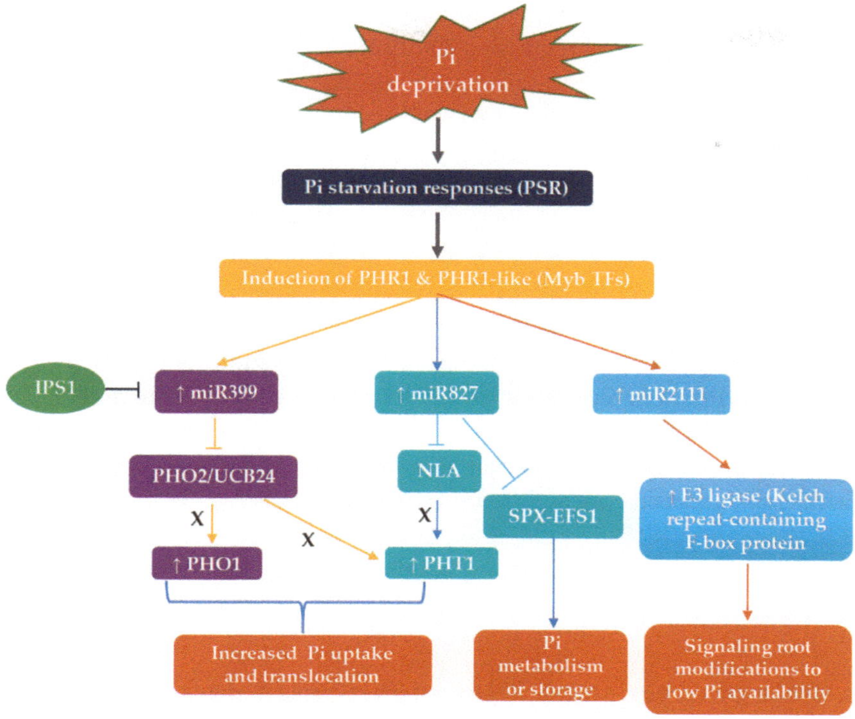

Fig. 2 The miRNAs pathway involved in plant Pi-starvation singling. Pi starvation activates the mechanism known as "Pi-starvation responses" (PSR), which leads to the induction of PHR1 and PHR1-like proteins (MYBTFs). The increased levels of these MYB TFs related to the upregulation of miR399, miR827, and miR2111. The miR399 represses phosphate2/UCB24 (PHO2) and miR827 represses NLA, while miR2111 regulates an E3 ligase. Repression of PHO2/UCB24 and NLA leads to PHT1 and PHO1 accumulation and consequently increase the Pi uptake and translocation. Additionally, miR399 also regulated by endogenous TM (eTM)-like lncRNA IPS1.

complementary to miR399 and acts as target mimic for miR399, allowing the expression of the PHO2 gene, this adds another level of regulation in Pi-starvation responses via miR399 (Franco-Zorrilla et al., 2007).

Another important microRNA that was induced in response to Pi deficiency is miRNA827. This miRNA plays an important role in regulating Pi homeostasis in a NO_3^--dependent manner by downregulating its target gene *nitrogen limitation adaptation* (*NLA*) which is an E3 RING ubiquitin enzyme (Hsieh et al., 2009; Kant et al., 2011). *NLA* is involved in *PHT1* ubiquitination (Hsieh et al., 2009). In rice, miR827 target SPX and MFS (major facilitator superfamily) domains containing genes which were involved in the Pi metabolism or storage (Lin et al., 2010), and participated in the transport of Pi by regulating the Pi transporters (Yao et al., 2014).

Under Pi-starvation conditions, the miR828 and its target *Trans-Acting SiRNA* gene 4 (*TAS4*) upregulated and produced the tasi-RNAs known as TAS4-Si81 (−) (Hsieh et al., 2009). This TAS4-Si81(−) targets the MYB TFs (*PAP1/MYB75*, *PAP2/MYB90*, and *MYB113*) and these were a key component of anthocyanin pigments biosynthesis pathway (Rajagopalan et al., 2006) and it well-known fact that plants increase the accumulation of anthocyanin pigments under Pi. Another important miRNA that induces in response to P starvation is miR2111. It is reported that miR2111 targets the gene At3g27150, which encodes a Kelch domain-containing F-box protein and this gene was involved in ubiquitin-mediated protein degradation (Hsieh et al., 2009).

Other important miRNAs that were upregulated under Pi starvation in different plant species are as follows: miR156, miR162, miR172, miR396, miR778, and miR1511 (Fujii et al., 2005; Zhao et al., 2012; Wang et al., 2013). However, their role in the regulation of Pi homeostasis in plants was not fully studied.

5. Role of other small RNAs in Pi-stress tolerance

Other classes of small RNAs (in addition to miRNAs i.e., transposons, tRNA-derived small RNAs) have been shown to be involved in Pi-stress tolerance (Hsieh et al., 2009). For example, a class of small RNAs named *smRPi1LTR*in *Arabidopsis* was upregulated in response to P starvation (Hsieh et al., 2009). The *smRPi1LTR* is generated from an intergenic (IG) region containing an array of truncated *Copia95* retrotransposon-derived long terminal repeats (LTRs) that are upregulated starvation (Hsieh et al., 2009). Another class of small RNAs are tRNA-derived and 19 nucleotides in length identified in *Arabidopsis*. These tRNA-derived mRNAs are present profusely in the root tissues under P-replete conditions, further upon Pi starvation, their level is more induced (Hsieh et al., 2009). However, the biological function of these transposons and tRNA-derived smRNAs in plants need further scientific exploration.

6. Conclusions and future perspectives

In order to cope with environmental stresses including Pi stress, multiple genes are regulated, and their function facilitates the stress tolerance in plants.

Understanding plant response mechanisms to Pi stress at the molecular level is crucial as it is exploiting them to improve stress tolerance leading to improve crop productivity. This chapter highlighted the role of key plant TFs belongs to WRKY, MYB, ZFP, and ARF multigene families. Members of these families play a vital role in providing tolerance against multiple abiotic stresses including Pi stress through *cis*-elements, DBD, DNA methylation, and other additional factors. Applications of forward and reverse genetic methods can be applied to TFs in order to produce transgenic plants with higher tolerance to Pi stress. Further, functional annotations of these TFs are a powerful tool for improving tolerance against Pi stress. Furthermore, considering TFs as candidate genes in molecular breeding programs, the field trials are required to validate the transgenic plants with focusing on their growth and eventually will lead breeders to develop crop varieties resilient to Pi stress.

Acknowledgment

AK would like to thank Shabir H. Wani for inviting him to write a book article.

Authors' contributions

AK wrote the entire initial draft with VG and MN focusing on miRNAs and other small RNAs. AK and VG proofread the manuscript. All authors read and approved the book article.

References

Abe, H., Urao, T., Ito, T., Seki, M., Shinozaki, K., Yamaguchi-Shinozaki, K., 2003. *Arabidopsis* AtMYC2 (bHLH) and AtMYB2 (MYB) function as transcriptional activators in abscisic acid signaling. Plant Cell 15, 63–78.

Abel, S., Ticconi, C.A., Delatorre, C.A., 2002. Phosphate sensing in higher plants. Physiol. Plant. 115, 1–8.

Agarwal, P., Arora, R., Ray, S., Singh, et al., 2007. Genome-wide identification of C2H2 zinc-finger gene family in rice and their phylogeny and expression analysis. Plant Mol. Biol. 65, 467–485.

Allen, E., Xie, Z.X., Gustafson, A.M., Carrington, J.C., 2005. microRNA-directed phasing during trans-acting siRNA biogenesis in plants. Cell 121, 207–221.

Ambawat, S., Sharma, P., Yadav, N.R., Yadav, R.C., 2013. MYB transcription factor genes as regulators for plant responses: an overview. Physiol. Mol. Biol. Plants 19, 307–321.

Baldoni, E., Genga, A., Cominelli, E., 2015. Plant MYB transcription factors: their role in drought response mechanisms. Int. J. Mol. Sci. 16, 15811–15851.

Balyan, H.S., Gahlaut, V., Kumar, A., et al., 2016. Nitrogen and phosphorus use efficiencies in wheat: physiology, phenotyping, genetics, and breeding. Plant Breed. Rev. 40, 167–214.

Berg, J.M., Shi, Y., 1996. The galvanization of biology: a growing appreciation for the roles of zinc. Science 271, 1081–1085.
Bustos, R., Castrillo, G., Linhares, F., et al., 2010. A central regulatory system largely controls transcriptional activation and repression response to phosphate starvation in Arabidopsis. PLoS Genet. 6, e1001102.
Cao, Y., Han, Y., Li, D., Lin, Y., Cai, Y., 2016. MYB transcription factors in Chinese Pear (*Pyrus bretschneideri* Rehd.): genome-wide identification, classification, and expression profiling during fruit development. Front. Plant Sci. 7, 577–591.
Century, K., Reuber, T.L., Ratcliffe, O.J., 2008. Regulating the regulators: the future prospects for transcription-factor-based agricultural biotechnology products. Plant Physiol. 147, 20–29.
Chen, Z.H., Nimmo, G.A., Jenkins, G.I., Nimmo, H.G., 2007. BHLH32 modulates several biochemical and morphological processes that respond to Pi starvation in *Arabidopsis*. Biochem. J. 405, 191–198.
Chen, Y.-F., Li, L.-Q., Xu, Q., Kong, Y.-H., Wang, H., Wu, W.-H., 2009. The WRKY6 transcription factor modulates PHOSPHATE1 expression in response to low Pi stress in Arabidopsis. Plant Cell 21, 3554–3566.
Chen, M., Tan, Q., Sun, M., Li, D., Fu, X., Chen, X., Xiao, W., Li, L., Gao, D., 2016. Genome-wide identification of WRKY family genes in peach and analysis of WRKY expression during bud dormancy. Mol. Gen. Genomics 291, 1319–1332.
Chen, S.K., Niu, X., Guan, Y.X., Li, H.F., 2017. Genome-wide analysis and expression profiles of the MYB genes in *Brachypodiumdistachyon*. Plant Cell Physiol. 58, 1777–1788.
Chiou, T.J., Aung, K., Lin, S.I., Wu, C.C., Chiang, S.F., Su, C.L., 2006. Regulation of phosphate homeostasis by microRNA in Arabidopsis. Plant Cell 18, 412–421.
Colasanti, J., Yuan, Z., Sundaresan, V., 1998. The indeterminate gene encodes a zinc finger protein and regulates a leaf-generated signal required for the transition to flowering in maize. Cell 93, 593–603.
Dai, X., Wang, Y., Yang, A., Zhang, W.H., 2012. OsMYB2P-1, an R2R3 MYB transcription factor, is involved in the regulation of phosphate-starvation responses and root architecture in rice. Plant Physiol. 159, 169–183.
Dai, X., Wang, Y., Zhang, W.H., 2016. OsWRKY74, a WRKY transcription factor, modulates tolerance to phosphate starvation in rice. J. Exp. Bot. 67, 947–960.
Deinlein, U., Stephan, A.B., Horie, T., et al., 2014. Plant salt-tolerance mechanisms. Trends Plant Sci. 19, 371–379.
Devaiah, B.N., Karthikeyan, A.S., Raghothama, K.G., 2007a. WRKY75 transcription factor is a modulator of phosphate acquisition and root development in *Arabidopsis*. Plant Physiol. 143, 1789–1801.
Devaiah, B.N., Nagarajna, V.K., Raghothama, K.G., 2007b. Phosphate homeostasis and root development in *Arabidopsis* are synchronized by the zing finger transcription factor ZAT6. Plant Physiol. 145, 147–159.
Devaiah, B.N., Madhuvanthi, R., Karthikeyan, A.S., Raghothama, K.G., 2009. Phosphate starvation responses and gibberellic acid biosynthesis are regulated by the MYB62 transcription factor in *Arabidopsis*. Mol. Plant 2, 43–58.
Ding, W., Wang, Y., Fang, W., Gao, S., Li, X., Xiao, K., 2016. TaZA8, a C2H2-ZFP type transcription factor gene in wheat, plays critical role in mediating tolerance to Pi deprivation trough regulating P acquisition, ROS homeostasis and root system establishment. Physiol. Plant. 158, 297–311.
Dinneny, J.R., Weigel, D., Yanofsky, M.F., 2006. *NUBBIN* and *JAGGED* define stamen and carpel shape in *Arabidopsis*. Development 133, 2285.
Domagalska, M.A., Leyser, O., 2011. Signal integration in the control of shoot branching. Nat. Rev. Mol. Cell Biol. 12, 211–221.

Dubos, C., Stracke, R., Grotewold, E., Weisshaar, B., Martin, C., Lepiniec, L., 2010. MYB transcription factors in *Arabidopsis*. Trends Plant Sci. 15, 573–581.
Englbrecht, C.C., Schoof, H., Bohm, S., 2004. Conservation, diversification and expansion of C2H2 zinc finger proteins in the Arabidopsis thaliana genome. BMC Genomics 5, 39.
Eulgem, T., 2006. Dissecting the WRKY web of plant defense regulators. PLoS Pathog. 2, e126.
Eulgem, T., Rushton, P.J., Robatzek, S., Somssich, I.E., 2000. The WRKY superfamily of plant transcription factors. Trends Plant Sci. 5, 199–206.
Ferdous, J., Hussain, S.S., Shi, B., 2015. Role of microRNAs in plant drought tolerance. Plant Biotechnol. J. 13, 293–305.
Franco-Zorrilla, J.M., Valli, A., Todesco, M., Mateos, I., Puga, M.I., Rubio-Somoza, I., Leyva, A., Weigel, D., García, J.A., Paz-Ares, J., 2007. Targetmimicry provides a new mechanism for regulation of microRNA activity. Nat. Genet. 39, 1033–1037.
Fujii, H., Chiou, T.J., Lin, S.I., Aung, K., Zhu, J.K., 2005. A miRNA involved in phosphate-starvation response in Arabidopsis. Curr. Biol. 15, 2038–2043.
Gahlaut, V., Jaiswal, V., Kumar, A., Gupta, P.K., 2016. Transcription factors involved in drought tolerance in developing drought tolerant cultivars with emphasis on wheat (*Triticum aestivum* L.). Theor. Appl. Genet. 129, 2019–2042.
Gray, W.M., Kepinski, S., Rouse, D., Leyser, O., Estelle, M., 2001. Auxin regulates SCF (TIR1)-dependent degradation of AUX/IAA proteins. Nature 414, 271–276.
Gu, M., Zhang, J., Li, H., Meng, D., Li, R., Dai, X., Wang, S., Liu, W., Qu, H., Xu, G., 2017. Maintenance of phosphate homeostasis and root development are coordinately regulated by MYB1, and R2R3-type MYB transcription factor in rice. J. Exp. Bot. 68, 3603–3615.
Guilfoyle, T.J., 2015. The PB1 domain in auxin response factor and aux/IAA proteins: a versatile protein interaction module in the auxin response. Plant Cell 27, 33–43.
Guilfoyle, T.J., Hagen, G., 2007. Auxin response factors. Curr. Opin. Plant Biol. 10, 453–460.
Guo, M.N., Ruan, W.Y., Li, C.Y., 2015. Integrative comparison of the role of the PHOSPHATE RESPONSE1 subfamily in phosphate signaling and homeostasis in rice. Plant Physiol. 168, 1762-U1134.
Ha, C.V., Le, D.T., Nishiyama, R., 2013. The auxin response factor transcription factor family in soybean: genome-wide identification and expression analyses during development and water stress. DNA Res. 20, 511–524.
Hajiebrahimi, A., Owji, H., Hemmati, S., 2017. Genome-wide identification, functional prediction, and evolutionary analysis of the R2R3-MYB superfamily in *Brassica napus*. Genome 60, 797–814.
He, C., Teixeira da Silva, J.A., Tan, J., Zhang, J., Pan, X., Li, M., Luo, J., Duan, J., 2017. A genome-wide identification of the WRKY family genes and a survey of potential WRKY target genes in *Dendrobium officinale*. Sci. Rep. 7, 9200.
Hinderhofer, K., Zentgraf, U., 2001. Identification of a transcription factor specifically expressed at the onset of leaf senescence. Planta 213, 469–473.
Hsieh, L.C., Lin, S.I., Shih, A.C., Chen, J.W., Lin, W.Y., Tseng, C.Y., Li, W.H., Chiou, T.J., 2009. Uncovering small RNA-mediated responses to phosphate deficiency in Arabidopsis by deep sequencing. Plant Physiol. 151, 2120–2132.
Hu, W., Zuo, J., Hou, X.W., Yan, Y., Wei, Y.X., Liu, J.H., Li, M.Y., Xu, B.Y., Jin, Z.Q., 2015. The auxin response factor gene family in banana: geneome-wide identification and expression analysis during development, ripening, and abiotic stress. Front. Plant Sci. 6, 742.
Huang, K.L., Ma, G.J., Zhang, M.L., et al., 2018. The *ARF7* and *ARF19* transcription factors positevly regulate *PHOSPAHTE STARVATION RESPONSE 1* in arabidopsis root. Plant Physiol. 178, 413–427.

Huang, J., Guo, Y., Sun, Q., Zeng, W., Li, J., Li, X., Xu, W., 2019. Genome-wide identification of R2R3-MYB transcription factors regulating secondary cell wall thickening in cotton fiber development. Plant Cell Physiol. 60, 687–701.

Ishiguro, S., Nakamura, K., 1994. Characterization of a cDNA encoding a novel DNA-binding protein, SPF1, that recognizes SP8 sequences in the 5′ upstream regions of genes coding for sporamin and beta-amylase from sweet potato. Mol. Gen. Genet. 244, 563–571.

Jin, J.P., Tian, F., Yang, D.C., Meng, Y.Q., Kong, L., Luo, J.C., Gao, G., 2017. PlantTFDB 4.0: toward a central hub for transcription factors and regulatory interactions in plants. Nucleic Acids Res. 45 (D1), D1040–D1045.

Jing, Z., Liu, Z., 2018. Genome-wide identification of WRKY transcription factors in kiwifruit (Actinidia spp.) and analysis of WRKY expression in responses to biotic and abiotic stresses. Genes Genomics 40, 429–446.

Johnson, C.S., Kolevski, B., Smyth, D.R., 2002. Transparent TESTA GLABRA2, a trichome and seed coat development gene of Arabidopsis, encodes a WRKY transcription factor. Plant Cell 14, 1359–1375.

Joseph, M.P., Papdi, C., Kozma-Bognar, L., et al., 2014. The *Arabidopsis* zinc finger protein3 interferes with abscisic acid and light signaling in seed germination and plant development. Plant Physiol. 165, 1203–1220.

Jue, D., Sang, X., Liu, L., Shu, B., Wang, Y., Liu, C., Xie, J., Shi, S., 2018. Identification of WRKY gene family from Dimocarpuslongan and its expression analysis during flower induction and abiotic stress responses. Int. J. Mol. Sci. 19, E2169.

Kalluri, U.C., Difazio, S.P., Brunner, A.M., Tuskan, G.A., 2007. Genome-wide analysis of Aux/IAA and ARF gene families in *Populus trichocarpa*. BMC Plant Biol. 7, 59.

Kant, S., Peng, M., Rothstein, S.J., 2011. Genetic regulation by NLA and microRNA827 for maintaining nitrate-dependent phosphate homeostasis in Arabidopsis. PLoS Genet. 7, e1002021.

Kim, J., Harter, K., Theologis, A., 1997. Protein–protein interactions among the aux/IAA proteins. Proc. Natl. Acad. Sci. U. S. A. 94, 11786–11791.

Kumar, A., Kumar, S., Kumar, U., Suravajhala, P., Gajula, M.N.V.P., 2016a. Functional and structural insights into novel DREB1A transcription factors in common wheat (*Triticum aestivum* L.): a molecular modeling approach. Comput. Biol. Chem. 64, 217–226.

Kumar, A., Rathore, M., Singh, K.P., Tyagi, P.K., Sharma, N., 2016b. In silico chromosomal mapping and functional annotation of TaPHT1;1 a high affinity phosphate transporter gene in wheat (*Triticum aestivum* L.). Online J. Bioinform. 17, 180–191.

Kumar, A., Sharma, M., Kumar, S., Tyagi, P., Wani, S.H., Gajula, M.N.V.P., Singh, K.P., 2018a. Functional and structural insights into candidate genes associated with nitrogen and phosphorus nutrition in wheat (*Triticum aestivum* L.). Int. J. Biol. Macromol. 118, 76–91.

Kumar, A., Batra, R., Gahlaut, V., et al., 2018. Genome-wide identification and characterization of gene family for RWP-RK transcription factors in wheat (*Triticum aestivum* L.). PLoS One 13, e0208409.

Kuo, H.F., Chiou, T.J., 2011. The role of microRNAs in phosphorus deficiency signaling. Plant Physiol. 156, 1016–1024.

Li, S., Fu, Q., Huang, W., Yu, D., 2009. Functional analysis of an Arabidopsis transcription factor WRKY25 in heat stress. Plant Cell Rep. 28, 683–693.

Li, S., Zhou, X., Chen, L., Huang, W., Yu, D., 2010. Functional characterization of *Arabidopsis thaliana* WRKY39 in heat stress. Mol. Cells 29, 475–483.

Li, L., Krymskaya, L., Wang, J., Henley, J., et al., 2013. Genomic editing of the HIV-1 coreceptor CCR5 in adult hematopoietic stem and progenitor cells using zinc finger nucleases. Mol. Ther. 21, 1259–1269.

Li, F., Fan, G., Lu, C., et al., 2015b. Genome sequences of cultivated uplannd cotton (*Gossypium hirsutum* TM-1) provides insights into geneome evolution. Nat. Biotechnol. 33, 524–530.

Li, W.B., Ding, Z.H., Ruan, M.B., Yu, X.L., Peng, M., Liu, Y.F., 2017a. Kiwifruit R2R3-MYB transcription factors and contribution of the novel *AcMYB75* to red kiwifruit anthocyanin biosynthesis. Sci. Rep. 7, 16861–16875.

Li, D., Liu, P., Yu, J., Wang, L., Dossa, K., Zhang, Y., Zhou, R., Wei, X., Zhang, X., 2017b. Genome-wide analysis of WRKY gene family in the sesame genome and identification of the WRKY genes involved in responses to abiotic stresses. BMC Plant Biol. 17, 152.

Liang, G., Ai, Q., Yu, D., 2015. Uncovering miRNAs involved in crosstalk between nutrient deficiencies in Arabidopsis. Sci. Rep. 5, 11813.

Lin, S.I., Santi, C., Jobet, E., et al., 2010. Complex regulation of two target genes encoding SPX-MFS proteins by rice miR827 in response to phosphate starvation. Plant Cell Physiol. 51, 2119–2131.

Lipton, D.S., Blanchar, R.W., Blevins, D.G., 1987. Citrate, malate, and succinate concentration in exudates from P-sufficient and P-stressed *Medicago sativa* L. seedlings. Plant Physiol. 85, 315–317.

Liu, S.Q., Hu, L.F., 2013. Genome-wide analysis of the auxin response factor gene family in cucumber. Genet. Mol. Res. 12, 4317–4331.

Liu, P.P., Montgomery, T.A., Fahlgren, N., Kasschau, K.D., Nonogaki, H., Carrington, J.C., 2007. Repression of AUXIN RESPONSE FACTOR10 by microRNA160 is critical for seed germination and post-germination stages. Plant J. 52, 133–146.

Liu, H., Yang, H., Wu, C., Feng, J., Liu, X., Qin, H., Wang, D., 2009. Overexpressing HRS1 confers hypersensitivity to low phosphate-elicited inhibition of primary growth in Arabidopsis thaliana. J. Integr. Plant Biol. 51, 382–392.

Liu, T.Y., Huang, T.K., Tseng, C.Y., et al., 2012. PHO2-dependent degradation of PHO1 modulates phosphate homeostasis in Arabidopsis. Plant Cell 24, 2168–2183.

Liu, Q., Wang, Z., Xu, X., Zhang, H., Li, C., 2015. Genome-wide analysis of C2H2 zinc-finger family transcription factors and their responses to abiotic stresses in poplar (*Populustrichocarpa*). PLoS One 10, e0134753.

Liu, N., Dong, L., Deng, X., Liu, D., Liu, Y., Li, M., Hu, Y., Yan, Y., 2018. Genome-wide identification, molecular evolution, and expression analysis of auxin response factor (ARF) gene family in *Brachypodium distachyon* L. BMC Plant Biol. 18, 336.

López-Bucio, J., Cruz-Ramírez, A., Herrera-Estrella, L., 2003. The role of nutrient availability in regulating root architecture. Curr. Opin. Plant Biol. 6, 280–287.

Luan, M., Tang, R., Tang, Y., Tian, W., Hou, C., Zhao, F., Lan, W., Luan, S., 2017. Transport and homeostasis factors for sustainable crop production. J. Exp. Bot. 68, 3091–3105.

Luo, M., Dennis, E.S., Berger, F., Peacock, W.J., Chaudhury, A., 2005. MINISEED3 (MINI3), a WRKY family gene, and HAIKU2 (IKU2), a leucine-rich repeat (LRR) kinase gene, are regulators of seed size in Arabidopsis. Proc. Natl. Acad. Sci. U. S. A. 102, 17531–17536.

Maleck, K., Levine, A., Eulgem, T., Morgan, A., Schmid, J., Lawton, K.A., Dangl, J.L., Dietrich, R.A., 2000. The transcriptome of *Arabidopsis thaliana* during systemic acquired resistance. Nat. Genet. 26, 403–410.

Mallory, A.C., Vaucheret, H., 2006. Functions of microRNAs and related small RNAs in plants. Nat. Genet. 38, S31–S36.

Marschner, H., 1995. Mineral Nutrition of Higher Plants. Academic Press, London.

Minglei, Y., Jiangtao, C., Dawei, W., et al., 2016. Genome-wide identification and expression profiling of the C2H2-type zinc finger protein transcription factor family in tobacco. Yi Chuan 38, 337–349.

Mmadi, M.A., Dossa, K., Wang, L.H., et al., 2017. Functional characterization of the versatile MYB gene family uncovered their important roles in plant development and responses to drought and waterlogging in sesame. Genes 8, 362–380.

Moore, M., Ullman, C., 2003. Recent developments in the engineering of zinc finger proteins. Brief. Funct. Genomics Proteomics 1, 342–355.

Mukhopadhyay, A., Vij, S., Tyagi, A.K., 2004. Overexpression of a zinc-finger protein gene from rice confers tolerance to cold, dehydration, and salt stress in transgenic tobacco. Proc. Natl. Acad. Sci. U. S. A. 101, 6309–6314.

Muthamilarasan, M., Bonthala, V.S., Mishra, A.K., et al., 2014a. C2H2 type of zinc finger transcription factors in foxtail millet define response to abiotic stresses. Funct. Integr. Genomics 14, 531–543.

Muthamilarasan, M., Khandelwal, R., Yadav, C.B., Bonthala, V.S., Khan, Y., Prasad, M., 2014b. Identification and molecular characterization of MYB transcription factor superfamily in C4 model plant Foxtail Millet (*Setariaitalica* L.). PLoS One. 9e109920.

Nilsson, L., Muller, R., Nielsen, T.H., 2007. Increased expression of the MYB-related transcription factor, PHR1, leads to enhanced phosphate uptake in *Arabidopsis thaliana*. Plant Cell Environ. 30, 1499–1512.

Nilsson, L., Müller, R., Nielsen, T.H., 2010. Dissecting the plant transcriptome and the regulatory responses to phosphate deprivation. Physiol. Plant. 139, 129–143.

Ning, P., Liu, C., Kang, J., Lv, J., 2017. Genome-wide analysis of WRKY transcription factors in wheat (*Triticum aestivum* L.) and differential expression under water deficit condition. Peer J. 5, e3232.

Niu, Y.F., Chai, R.S., Dong, H.F., Wang, H., Tang, C.X., Zhang, Y.S., 2012. Effect of elevated CO_2 on phosphorus nutrition of phosphate-deficient *Arabidopsis thaliana* (L.) Heynh under different nitrogen forms. J. Exp. Bot. 64, 355–367.

Niu, Y.F., Chai, R.S., Jin, G.L., Wang, H., Tang, C.X., Zhang, Y.S., 2013. Responses of root architecture development to low phosphorus availability: a review. Ann. Bot. 112, 391–408.

Pabo, C.O., Peisach, E., Garant, R.A., 2001. Design and selection of novel Cys2His2 zinc finger proteins. Annu. Rev. Biochem. 70, 313–340.

Park, B.S., Seo, J.S., Chua, N.H., 2014. Nitrogen limitation adaptation recruits phosphate2 to target the phosphate transporter PT2 for degradation during the regulation of Arabidopsis phosphate homeostasis. Plant Cell 26, 454–464.

Paz-Ares, J., Ghosal, D., Wienand, U., Peterson, P., Saedler, H., 1987. The regulatory c1locus of *Zea mays* encodes a protein with homology to MYB oncogene products and with structural similarities to transcriptional activators. EMBO J. 6, 3553–3558.

Popescu, S.C., Popescu, G.V., Bachan, S., et al., 2009. MAPK target networks in Arabidopsis thaliana revealed using functional protein microarrays. Genes Dev. 23, 80–92.

Qiu, Y., Yu, D., 2009. Over-expression of the stress-induced OsWRKY45 enhances disease resistance and drought tolerance in Arabidopsis. Environ. Exp. Bot. 65, 35–47.

Raghothama, K.G., Karthikeyan, A.S., 2005. Phosphate acquisition. Plant Soil 274, 37–49.

Rajagopalan, R., Vaucheret, H., Trejo, J., Bartel, D.P., 2006. A diverse and evolutionarily fluid set of microRNAs in Arabidopsis thaliana. Genes Dev. 20, 3407–3425.

Ramaiah, M., Jain, A., Raghothama, K.G., 2014. Ethylene Response Factor070 regulates root development and phosphate starvation-mediated responses. Plant Physiol. 164, 1484–1498.

Rausch, C., Bucher, M., 2002. Molecular mechanisms of phosphate transport in plants. Planta 216, 23–37.

Robatzek, S., Somssich, I.E., 2001. A new member of the Arabidopsis WRKY transcription factor family, AtWRKY6, is associated with both senescence-and defence-related processes. Plant J. 28, 123–133.

Ruan, W., Guo, M., Ping, W., Yi, K., 2016. Phosphate starvation induced OsPHR4 mediates Pi-signaling and homeostasis in rice. Plant Mol. Biol. 93, 1–14.

Ruan, W.Y., Guo, M.N., Wu, P., Yi, K.K., 2017. Phosphate starvation induced OsPHR2 mediates Pi-signaling and homeostasis in rice. Plant Mol. Biol. 93, 327–340.

Rubio, V., Linhares, F., Solano, R., Martín, A.C., Iglesias, J., Leyva, A., Paz-Ares, J., 2001. A conserved MYB transcription factor involved in phosphate starvation signaling both in vascular plants and in unicellular algae. Genes Dev. 15, 2122–2133.

Rushton, P.J., Torres, J.T., Parniske, M., Wernert, P., Hahlbrock, K., Somssich, I.E., 1996. Interaction of elicitor-induced DNA-binding proteins with elicitor response elements in the promoters of parsley PR1 genes. EMBO J. 15, 5690–5700.

Rushton, P.J., Bokowiec, M.T., Han, S.C., Zhang, H.B., Brannock, J.F., Chen, X.F., Laudeman, T.W., Timko, M.P., 2008. Tobacco transcription factors: novel insights into transcriptional regulation in the Solanaceae. Plant Physiol. 147, 280–295.

Rushton, P.J., Somssich, I.E., Ringler, P., Shen, Q.J., 2010. WRKY transcription factors. Trends Plant Sci. 15, 247–258.

Schachtman, D.P., Reid, R.J., Ayling, S.M., 1998. Phosphorus uptake by plants: from soil to cell. Plant Physiol. 116, 447–453.

Secco, D., Jabnoune, M., Walker, H., et al., 2013. Spatio-temporal transcript profiling of rice roots and shoot in response to phosphate starvation and recovery. Plant Cell 25, 4285–4304.

Shen, C., Wang, S., Zhang, S., Xu, Y., Qian, Q., Qi, Y., Jiang, D., 2012. *OsARF16*, a transcription factor, is required for auxin and phosphate starvation response in rice (*Oryza sativa* L.). Plant Cell Environ. 36, 607–620.

Shen, C., Wang, S., Zhang, S., et al., 2013. OsARF16, a transcription factor, is required for auxin and phosphate starvation response in rice (*Oryza sativa* L.). Plant Cell Environ. 36, 607–620.

Shen, C., Yue, R., Yang, Y., et al., 2014. *OsARF16* is involved in cytokinin-mediated inhibition of phosphate transport and phosphate signaling in rice (*Oryza sativa* L.). PLoS One 9, e112906.

Shen, C., Yue, R., Sun, T., Zhang, L., Xu, L., Tie, S., Wang, H., Yang, Y., 2015. Genome-wide identification and expression analysis of auxin response factor gene family in *Medicago truncatula*. Front. Plant Sci. 6, 73.

Song, H., Wang, P., Hou, L., Zhao, S., Zhao, C., Xia, H., Li, P., Zhang, Y., Bian, X., Wang, X., 2016a. Global analysis of WRKY genes and their response todehydration and salt stress in soybean. Front. Plant Sci. 7, 9. https://doi.org/10.3389/fpls.2016.00009.

Song, H., Wang, P., Lin, J.Y., Zhao, C., Bi, Y., Wang, X., 2016b. Genome-wide identification and characterization of WRKY gene family in peanut. Front. Plant Sci. 7, 534. https://doi.org/10.3389/fpls.2016.00534.

Song, H., Guo, Z., Chen, T., Sun, J., Yang, G., 2018. Genome-wide identification of LRR-containing sequences and the response of these sequences to nematode infection in Arachis duranensis. BMC Plant Biol. 18, 279.

Stracke, R., Werber, M., Weisshaar, B., 2001. The R2R3–MYB gene family in *Arabidopsis thaliana*. Curr. Opin. Plant Biol. 4, 447–456.

Su, T., Xu, Q., Zhang, F.C., Chen, Y., Li, L.Q., Wu, W.H., Chen, Y.F., 2015. WRKY42 modulates phosphate homeostasis through regulating phosphate translocation and acquisition in Arabidopsis. Plant Physiol. 167, 1579–1591.

Sun, C., Palmqvist, S., Olsson, H., Borén, M., Ahlandsberg, S., Jansson, C., 2003. A novel WRKY transcription factor, SUSIBA2, participates in sugar signaling in barley by binding to the sugar-responsive elements of the iso1 promoter. Plant Cell 15, 2076–2092.

Sunkar, R., Li, Y.F., Jagadeeshwaran, G., 2012. Functions of microRNAs in plant stress responses. Trends Plant Sci. 17, 196–203.

Takatsuji, H., 1999. Zinc-finger proteins: the classical zinc finger emerges in contemporary plant science. Plant Mol. Biol. 39, 1073–1078.
Takatsuji, H., Mori, M., Benfey, P.N., Ren, L., Chua, N.H., 1992. Characterization of a zinc finger DNA-binding protein expressed specifically in *Petunia* petals and seedlings. EMBO J. 11, 241–249.
Tao, Z., Liu, H., Qiu, D., et al., 2009. A pair of allelic WRKY genes play opposite roles in rice-bacteria interactions. Plant Physiol. 151, 936–948.
Teale, W.D., Paponov, I.A., Palme, K., 2006. Auxin in action: signalling, transport and control of plant growth and development. Nat. Rev. Mol. Cell Biol. 7, 847–859.
Tiwari, S.B., Hagen, G., Guilfoyle, T., 2003. The roles of auxin response factor domains in auxin-responsive transcription. Plant Cell 15, 533–543.
Todd, C.D., Zeng, P., Huete, A.M.R., Hoyos, M.E., Polacco, J.C., 2004. Transcripts of MYB-like genes respond to phosphorus and nitrogen deprivation in Arabidopsis. Planta 219, 1003–1009.
Tripathi, P., Rabara, R.C., Rushton, P.J., 2014. A system biology prospective on the role of WRKY transcription factors in drought responses in plants. Planta 239, 255–266.
Ulker, B., Somssich, I.E., 2004. WRKY transcription factors: from DNA binding towards biological function. Curr. Opin. Plant Biol. 7, 491–498.
Ulmasov, T., Hagen, G., Guilfoyle, T.J., 1997. ARF1, a transcription factor that binds to auxin response elements. Science 276, 1865–1868.
Ulmasov, T., Hagen, G., Guilfoyle, T.J., 1999. Activation and repression of transcription by auxin-response factors. Proc. Natl. Acad. Sci. U. S. A. 96, 5844–5849.
Vance, C.P., Uhde-Stone, C., Allan, D.L., 2003. Phosphorus acquisition and use: critical adaptations by plants for securing a nonrenewable resource. New Phytol. 157, 423–447.
Wan, S., Li, W., Zhu, Y., Liu, Z., Huang, W., Zhan, J., 2014. Genome-wide identification, characterization and expression analysis of the auxin response factor gene family in *Vitis vinifera*. Plant Cell Rep. 33, 1365–1375.
Wang, D., Pei, K., Fu, Y., Sun, Z., Li, S., Liu, H., Tang, K., Han, B., Tao, Y., 2007. Genome-wide analysis of the auxin response factors (*ARF*) gene family in rice (*Oryza sativa*). Gene 394, 13–24.
Wang, Q., Wang, M., Zhang, X., Hao, B., Kaushik, S.K., Pan, Y., 2011. WRKY gene family evolution in Arabidopsis thaliana. Genetica 139, 973–983.
Wang, Y., Deng, D., Shi, Y., Miao, N., Bian, Y., Yin, Z., 2012. Diversification, phylogeny and evolution of auxin response factor (ARF) family: insights gained from analyzing maize *ARF* genes. Mol. Biol. Rep. 39, 2401–2415.
Wang, Y., Zhang, C., Hao, Q., et al., 2013. Elucidation of miRNAs-mediated responses to low nitrogen stress by deep sequencing of two soybean genotypes. PLoS One 8, e67423.
Wang, J., Sun, J.H., Miao, J., Guo, J.K., Shi, Z.L., He, M.Q., Chen, Y., Zhao, X.Q., Li, B., Han, F.P., Tong, Y.P., Li, Z.S., 2013a. A phosphate starvation response regulator Ta-PHR1 is involved in phosphate signalling and increases grain yield in wheat. Ann. Bot. 111, 1139–1153.
Wang, X.H., Bai, J.R., Liu, H.M., Sun, Y., Shi, X.Y., Ren, Z.Q., 2013b. Overexpression of a maize transcription factor ZmPHR1 improves shoot inorganic phosphate content and growth of Arabidopsis under low-phosphate conditions. Plant Mol. Biol. Report. 31, 665–677.
Wang, H., Xu, Q., Kong, Y.H., Chen, Y., Duan, J.Y., Wu, W.H., Chen, Y.F., 2014a. Arabidopsis WRKY45 transcription factor activates *PHOSPHATE TRANSPORTER 1;1* expression in response to phosphate starvation. Plant Physiol. 164, 2020–2029.
Wang, S., Zhang, S., Sun, C., Xu, Y., Chen, Y., Yu, C., Qian, Q., Jiang, D.A., Qi, Y., 2014b. Auxin response factor (*OsARF12*), a novel regulator for phosphate homeostasis in rice (*Oryza sativa*). New Phytol. 201, 91–103.

Wang, Y., Zhan, D.-F., Li, H.-L., Guo, D., Zhu, J.-H., Peng, S.-Q., 2017. Transcriptome-wide identification and characterization of MYB transcription factor genes in the laticifer cells of *Heveabrasiliensis*. Front. Plant Sci. 8, 1974.
Wani, S.H., Tripathi, P., Zaid, P., et al., 2018. Transcriptional regulation of osmotic stress tolerance in wheat (*Triticum aestivum* L.). Plant Mol. Biol. 97, 469–487.
Wei, K.F., Chen, J., Chen, Y.F., Wu, L.J., Xie, D.X., 2012. Molecular phylogenetic and expression analysis of the complete WRKY transcription factor family in maize. DNA Res. 19, 153–164.
Williamson, L.C., Ribrioux, S.P.C.P., Fitter, A.H., Leyser, H.M.O., 2001. Phosphate availability regulates root system architecture in Arabidopsis. Plant Physiol. 126, 875–882.
Wu, P., Shou, H., Xu, G., Lian, X., 2013. Improvement of phosphorus efficiency in rice on the basis of understanding phosphate signaling and homeostasis. Curr. Opin. Plant Biol. 16, 205–212.
Wu, F., Liu, Z., Xu, J., Gao, S., Lin, H., Liu, L., Liu, Y., Lu, Y., 2016. Molecular evolution and association of natural variation in *ZmARF31* with low phosphorus tolerance in maize. Front. Plant Sci. 7, 1076.
Wu, J., Chen, J., Wang, L., Wang, S., 2017. Genome-wide investigation of WRKY transcription factors involved in terminal drought stress response in common bean. Front. Plant Sci. 8, 380.
Wykoff, D.D., Grossman, A.R., Weeks, D.P., Usuda, H., Shimogawara, K., 1999. Psr1, a nuclear localized protein that regulates phosphorus metabolism in *Chlamydomonas*. Proc. Natl. Acad. Sci. U. S. A. 96, 15336–15341.
Xiao, G., He, P., Zhao, P., Liu, H., Zhang, L., Pang, C., Yu, J., 2018. Genome-wide identification of the GhARF gene family reveals that GhARF2 and GhARF18 are involved in cotton fibre cell initiation. J. Exp. Bot. 69, 4323–4337.
Xu, Z., Ji, A., Song, J., Chen, S., 2016a. Genome-wide analysis of auxin response factor gene family members in medicinal model plant *Salvia miltiorrhiza*. Biol. Open 5, 848–857.
Xu, H., Watanabe, K.A., Zhang, L., Shen, Q.J., 2016b. WRKY transcription factor genes in wild rice *Oryza nivara*. DNA Res. 23, 311–323.
Xue, Y.B., Xiao, B.X., Zhu, S.N., Mo, X.H., Liang, C.Y., Tian, J., Liao, H., Miriam, G., 2017. GmPHR25, a GmPHR member up-regulated by phosphate starvation, controls phosphate homeostasis in soybean. J. Exp. Bot. 68, 4951–4967.
Yamasaki, H., Abdel-Ghany, S.E., Cohu, C.M., Kobayashi, Y., Shikanai, T., Pilon, M., 2007. Regulation of copper homeostasis by micro-RNA in Arabidopsis. J. Biol. Chem. 282, 16369–16378.
Yang, L., Liu, Z., Lu, F., Dong, A., Huang, H., 2006. SERRATE is a novel nuclear regulator in primary microRNA processing in *Arabidopsis*. Plant J. 47, 841–850.
Yang, W.T., Baek, D., Yun, D.-J., Hwang, W.H., Park, D.S., Nam, M.H., Chung, E.S., Chung, Y.S., Yi, Y.B., Kim, D.H., 2014. Overexpression of OsMYB4P, an R2R3-type MYB transcriptional activator, increases phosphate acquisition in rice. Plant Physiol. Biochem. 80, 259–267.
Yang, W.T., Baek, D., Yun, D.-J., et al., 2018. Rice OsMYB5P improves plant phosphate acquisition by regulation of phosphate transporter. PLoS One 13, e0194628.
Yang, K., Li, Y., Wang, S., Xu, X., Sun, H., Zhao, H., Li, X., Gao, 2019. Genome-wide identification and expression analysis of the MYB transcription factor in moso bamboo (*Phyllostachys edulis*). Peer J. 6, e6242.
Yanhui, C., Xiaoyuan, Y., Kun, H., et al., 2006. The MYB transcription factor superfamily of Arabidopsis: expression analysis and phylogenetic comparison with the rice MYB family. Plant Mol. Biol. 60, 107–124.
Yao, Z., Tian, J., Liao, H., 2014. Comparative characterization of GmSPX members reveals that GmSPX3 is involved in phosphate homeostasis in soybean. Ann. Bot. 114, 477–488.

Yi, K., Wu, Z., Zhou, J., Du, L., Guo, L., Wu, Y., Wu, P., 2005. OsPTF1, a novel transcription factor involved in tolerance to phosphate starvation in rice. Plant Physiol. 138, 2087–2096.

Yu, H., Soler, M., Mila, I., et al., 2014. Genome-wide characterization and expression profiling of the *AUXIN RESPONSE FACTOR* (*ARF*) gene family in *Eucalyptus grandis*. PLoS One 9, e108906.

Yuan, S., Li, X., Li, R., Wang, L., Zhang, C., Chen, L., Hao, Q., Zhang, X., Chen, H., Shan, Z., Yang, Z., Chen, S., Qiu, D., Ke, D., Zhou, X., 2018. Genome-wide identification and classification of soybean C2H2 zinc finger proteins and their expression analysis in legume-rhizobium symbiosis. Front. Microbiol. 9, 126.

Yue, H., Chang, X., Zhi, Y., Wang, L., Xing, G., Song, W., Nie, X., 2019. Evolution and identification of the WRKY gene family in quinoa (*Chenopodium quinoa*). Genes (Basel) 10, E131.

Zhang, Y., Wang, L., 2005. The WRKY transcription factor superfamily: its origin in eukaryotes and expansion in plants. BMC Evol. Biol. 5, 1.

Zhang, Z.L., Xie, Z., Zou, X., Casaretto, J., Ho, T.H., Shen, Q.J., 2004. A rice WRKY gene encodes a transcriptional repressor of the gibberellin signaling pathway in aleurone cells. Plant Physiol. 134, 1500–1513.

Zhao, M., Tai, H., Sun, S., Zhang, F., Xu, Y., Li, W.X., 2012. Cloning and characterization of maize miRNAs involved in responses to nitrogen deficiency. PLoS One 7, e29669.

Zhou, J., Jiao, F., Wu, Z., Li, Y., Wang, X., He, X., Zhong, W., Wu, P., 2008a. *OsPHR2* is involved in phosphate-starvation signaling and excessive phosphate accumulation in shoots of plants. Plant Physiol. 146, 1673–1686.

Zhou, Q.Y., Tian, A.G., Zou, H.F., et al., 2008b. Soybean WRKY-type transcription factor genes, GmWRKY13, GmWRKY21, and GmWRKY54, confer differential tolerance to abiotic stresses in transgenic Arabidopsis plants. Plant Biotechnol. J. 6, 486–503.

Zhou, Z., Sun, L., Zhao, Y., et al., 2013. Zinc finger protein 6 (ZFP6) regulates trichome initiation by integrating gibberellin and cytokinin signaling in *Arabidopsis thaliana*. New Phytol. 198, 699–708.

Zhou, X., Zha, M., Huang, J., Imram, M., Zhang, C., 2017. StMYB44 negatively regulates phosphate transport by suppressing expression of PHOSPHATE1 in potato. J. Exp. Bot. 68, 1265–1281.

Zhu, X., Liu, S., Meng, C., Qin, L., Kong, L., Xia, G., 2013. WRKY transcription factors in wheat and their induction by biotic and abiotic stress. Plant Mol. Biol. Report. 31, 1053–1067.

Zouine, M., Fu, Y., Chateigner-Boutin, A.L., et al., 2014. Characterization of the tomato ARF gene family uncovers a multi-levels post-transcriptional regulation including alternative splicing. PLoS One 9, e84203.

Further reading

Li, C., Ng, C.K.-Y., Fan, L.-M., 2015a. MYB transcription factors, active players in abiotic stress signaling. Environ. Exp. Bot. 114, 80–91.

Lin, W.Y., Huang, T.K., Chiou, T.J., 2013. Nitrogen limitation adaptation, a target of microRNA827, mediates degradation of plasma membrane-localized phosphate transporters to maintain phosphate homeostasis in Arabidopsis. Plant Cell 25, 4061–4074.

CHAPTER 12

Major transcription factor families involved in nitrogen stress adaptation in plants

Lekshmy Sathee[a], M.G. Mallikarjuna[b], B. Jagadhesan[a], and S.K. Jha[b]
[a]Division of Plant Physiology, Indian Agricultural Research Institute, New Delhi, India
[b]Division of Genetics, Indian Agricultural Research Institute, New Delhi, India

1. Introduction

Nitrogen (N) is an essential macronutrient that impacts many aspects of plant growth and development. Besides its nutritional role, N and the downstream metabolites act as signaling molecules that regulate the expression of a wide range of genes and biological processes (Undurraga et al., 2017). In developing countries, the production of a high yield in crops is associated with the application of a large quantity of nitrogenous fertilizers, which is also increasing the cost of cultivation. Further, subsequent soil and groundwater pollution is posing a threat to the environment and human beings. Therefore, the breeding of cultivars with better nitrogen use efficiency (NUE) is the most sustainable approach to tackle the agriculture-related N-pollution. NUE is a product of both N uptake and utilization efficiency. The uptake efficiency refers to the total amount of N acquired by plants divided by the total available N sources in the soil, and utilization efficiency refers to the fraction of acquired N used for grain or biomass divided by total acquired nitrogen (Xu et al., 2012). The root system architecture of plants largely determines N-uptake efficiency. Larger surface area and deeper root system facilitate the better N foraging from the soil. Nitrate (NO_3^-) is the major inorganic N form in aerobic soils and is also a potent signal that regulates plant growth and metabolism. Nitrate assimilation in plants is a complex mechanism that is governed by a cascade of events with several genes. Despite the several reports, the molecular mechanisms underlying the NO_3^- response have remained elusive, as the transcription factors that primarily respond to NO_3^- signals were not identified until the recent past. Current understanding of the N

https://doi.org/10.1016/B978-0-12-819334-1.00012-5

and NO_3^- signaling in the model plant(s) is huge, owing to the development of molecular tools and advancement in sequencing technology. Transcription factors (TF) are the key regulatory elements which coordinately regulate the expression of the target gene(s) by binding to cis-regulatory elements. Therefore, any alteration in the expression of TF genes will affect the subsequent expression of the target gene(s). In plants, efforts have been made to identify the NO_3^- induced transcription factors and to understand their vital role in the regulation of root system architecture. Here, we reviewed the updates on NO_3^- induced TFs and their potential exploitation to enhance NUE in crops.

2. Nitrogen sensing, primary nitrate response (PNR), and nitrogen starvation response (NSR)

Evidences suggest that NO_3^- drives transcriptional and posttranscriptional regulation of an array of genes. Many interrelated signal modules are involved in NO_3^- signaling. The first component is primary nitrate response (PNR) (Medici and Krouk, 2014), which refers to the rapid (within minutes) activation of nitrate-specific genes (NR, NiR, G6PDH, HRS1) without additional protein synthesis and is regulated by CHL1, calcium-dependent kinases (CPKs), CBLs, CIPKs, phosphatases, and transcription factors like NLPs, NRG2, SPL9, etc. The second component is N starvation responses (NSR), which include slow activation of high-affinity nitrate transporters belonging to the NRT2 family (NRT2.4, NRT2.5) after N starvation for 24–48h. Some of the important genes that are reported to regulate NSR are LBD36, 37, CBL7, miR169, NFYA, HRS1, and HHOs. The rapid response (PNR) induces transcription instantly, with maximum expression around 30min. Expression of PNR genes is directly proportional to nitrate concentrations; low (<1mM) and high concentrations (>1mM) evoke a similar magnitude of PNR gene expression (Hu et al., 2009). Nitrate-mediated changes in root system architecture are another well-reported response, which changes in primary root growth, lateral root initiation and elongation, and root hair proliferation. Several research groups have identified key components mediating the nitrate perception and signaling, viz., nitrate transceptor, calcium signaling components, calcium-binding proteins, kinases, phosphatases, and transcription factors (Riveras et al., 2015).

3. Different classes of transcription factors regulating plant response to environmental stress and nitrogen availability

The presence of nitrate in the medium is first perceived by the dual-affinity nitrate transceptor CHL1 (NRT1.1/NPF6.3) (Tsay et al., 1993; Liu et al., 1999). As a dual-affinity transporter, NRT1.1 can perform either as a high or low-affinity nitrate transporter in coordination with external nitrate availability. The switching of modes is modulated by phosphorylation and dephosphorylation at the Thr101 mediated by two CIPKs, CIPK8 and CIPK23 (Ho et al., 2009). For evoking responses to nitrate sensing by CHL1, the transmission of signals at the nucleus level is essential, leading to gene expression changes. Calcium 1 signaling components and several classes of transcription factors involved in the N deficiency and NO_3^- response have been characterized, more so in *Arabidopsis*. Among these, a series of transcription factors (TF) have also been identified as important regulatory factors in the NO_3^- response in plants, *NLP7 (NIN LIKE PROTEIN 7), NLP6, TGA1, TGA4, bZIP1, LBD37, LBD38, TCP 20*, and *NAC4* in *Arabidopsis* (O'Brien et al., 2016) in coordination with nitrate-related target genes. A description of these transcription factors is provided later.

3.1 NLPs

The NIN-like protein NLP7 is the master regulator of the PNR and nitrate signaling. NLP7 has homology with the *Chlamydomonas NIT2* gene, which activates nitrate reductase gene expression (Camargo et al., 2007) and NODULE INCEPTION (NIN), which coordinate nodule initiation in legumes (Schauser et al., 1999). A detailed description of NLP is given in Section 5.

3.2 LBDs

Lateral boundary domain-containing protein 37, 38, and 39 (LBD37, LBD38, and LBD39) are N-dependent negative regulators of anthocyanin biosynthesis in *Arabidopsis* (Rubin et al., 2009). Overexpression of any of these three genes suppresses *PAP1* and *PAP2*, genes belonging to the anthocyanin biosynthesis part of the flavonoid metabolism. Their mutants show constitutive and N-independent accumulation of anthocyanin with a corresponding higher biosynthesis gene expression. *LBD* genes are upregulated by both nitrate and NLP7, and also negatively regulate other N-responsive genes requisite for N metabolism (Medici and Krouk, 2014; Marchive et al., 2013).

3.3 NRG2

Nitrate regulatory gene 2 (NRG2) is a member of the bZIP family of transcription factors. In the nucleus, NRG2 directly interacts with NLP7 and also regulates the transcript abundance of CHL (Xu et al., 2016). NRG2 positively regulates nitrate response; however, the response takes place at around 2h after nitrate induction, leading to the speculation of its independent function from that of NLP7 in PNR. The mutants of *nrg2* showed disordered induction of nitrate-regulated genes following nitrate treatment and differential expression of genes falling in four nitrate-induced gene clusters including members of the *NPF* family (Xu et al., 2016).

3.4 bZIP1

bZIP1 is a common regulator that integrates nitrate and light-signaling (Obertello et al., 2010). Transcriptome comparison of wild-type and *bzip1 mutants* verifies mutation changes light and or N-regulation of numerous gene clusters. Para et al. (2014) showed by the TARGET (transient assay reporting genome-wide effects of transcription factors) method that bZIP1 also follows the classical "hit-and-run" mechanism of transcriptional regulation that follows a momentary binding to target promoters. The transient bZIP1 target genes are highly significant with respect to faster N signaling, including NRT2.1 and NLP3.

3.5 SPL9

Squamosa promoter-binding protein-like 9 (SPL9) was recognized as a potential hub for nitrate-regulated gene expression based on transcriptome analysis (Krouk et al., 2010). Overexpression of SPL9 alters the expression of nitrate-responsive transcription factor genes and NIR and NIA2. SPL9 regulates shoot development and is targeted by miR156 (Pant et al., 2009; Hsieh et al., 2009). Expression of pri-miR156 was responsive to both the N and phosphorus (P) status of the plant, and hence mir156/SPL9 could be regulators of N response in coordination with other environmental cues such as P status.

3.6 ANR1

ANR1 is a MADS-box family transcription factor and was the first nitrate-responsive transcription factor identified in plants (Zhang and Forde, 1998). It regulates lateral root foraging in nitrogen-rich patches. It is speculated to be regulated by NLP7; though the mechanistic details are unknown, it acts downstream of CHL1 and is expressed in lateral root primordia (Marchive et al., 2013). Further description of ARNR1 is given in Section 6.

3.7 TGA1 and TGA4

Alvarez et al. (2014) discovered that TGA1 and TGA4 are regulatory factors mediating nitrate responses in *Arabidopsis* roots following a bioinformatics approach. They found that >90% of genes showing global changes in *tga1 tga4* double mutants were nitrate-responsive. Further, it was demonstrated that TGA1 and TGA4 bind to promoter sequences *NRT2.1* and *NRT2.2* and regulate their expression. The double mutants tga1 tga4 also showed defective lateral root initiation and development. In addition to nitrogen, expression of *TGA4* was also induced by drought (Zhong et al., 2015) and the transgenic plants overexpressing *TGA4* displayed tolerance to both drought and nitrogen starvation. TGA1 and TGA4 are positioned above NRT2.1 and NRT2.2 and below NRT1.1 in the regulatory network controlling root hair growth (Canales et al., 2017).

3.8 NAC4

The auxin receptor AFB3 coordinates nitrate-responsive root growth with NAC4 controlling the downstream network of genes. The mutants of *nac4* were defective in nitrate-responsive lateral root growth; however, the primary root growth was unaffected (Vidal et al., 2013).

3.9 HRS1 and HHO1

HRS1 (hypersensitivity to low PI-elicited primary root shortening 1) and homolog HRS1 homolog 1 (HHO1) are members of the GARP family of transcription factors. HRS1 and HHO1 suppressed the primary root growth under P deficit condition if NO_3^- is available (Medici et al., 2015). Phosphate deficiency induces expression of *HRS1* and *HHO1;* the nitrate-induced expression is downstream of NRT1.1 and NLP7 signaling (Marchive et al., 2013). Expression of NRT1.1 and *NRT2.4* was reduced in the *HRS1-HHO1* overexpression line and promoted in the double mutant, signifying that they are playing a negative role in nitrate signaling (Medici et al., 2015).

3.10 TCP20

TCP20 binds to NRE elements in the promoter region of *NRT1.1*, *NRT2.1*, and *NIA1*. It regulates nitrate-dependent lateral root foraging independent of NLP7 (Guan et al., 2014, 2017). TCP20 also works together with NLP6 or NLP7 in the nucleus, though specifically under nitrate-free circumstances.

3.11 BT1 and BT2

A systems biology approach was employed to identify candidate genes via a supervised machinelearning algorithm to predict the NUE gene network in *Arabidopsis* and landed into BT2, a member of the Bric-a-Brac/Tramtrack/Broad gene family, as a potential hub in the NUE network. BT2 overexpressing lines were having low NUE in comparison to wild-type plants under low nitrate environments (Araus et al., 2016), whereas NUE was higher in the *bt1bt2* double mutant plants under low nitrate conditions, with a concomitant higher expression of NRT2.1 and NRT2.4 and 65% increment in nitrate uptake under low nitrate environments. BT gene family members function as negative regulators of NUE and nitrate transporters. NLP6 and NLP7 can directly bind to the NRE region in the promoter of *BT1* and *BT2* (Sato et al., 2017).

4. Regulation of nitrogen uptake and assimilation by NLPs

The plant-specific RWP-RK transcription factors are classified into two subfamilies: the NIN-like proteins (NLPs) and the RWP-RK domain proteins (RKDs). The NIN (for nodule inception) proteins were first identified as a regulator controlling the development of symbiotic root nodules in the legume plant *Lotus japonicus* but were later found to be widely existing among plant species, including those that do not fix gaseous nitrogen (Chardin et al., 2014).

The NLP family proteins differ significantly in size as observed from three in Medicago and *Brachypodium*, six in rice, and nine in *Arabidopsis* (Chardin et al., 2014). Functional studies in *Arabidopsis* using the genetic mutants and transgenic plants demonstrated that *AtNLPs* play a central role in orchestrating primary nitrate response (PNR) by binding to the nitrate-responsive cis-elements (Konishi and Yanagisawa, 2013). Reports also suggest that the activity of NLPs is posttranslationally modulated by nitrate signaling. Recently, the nitrate-CPK (Ca^{2+} sensor protein kinase)-NLP network was identified, nitrate-coupled CPK signaling phosphorylates NLP transcription factors, specifically NLP7, and this signaling was crucial to specify the reprogramming of gene sets for downstream transcription factors, transporters, nitrogen assimilation, carbon/nitrogen metabolism, redox, signaling, hormones, and cell proliferation (Liu et al., 2017). AtNLP7 was shown to bind with key N pathway genes including ANR1, LBD37/38, NRT1.1, NRT 2.1, and NIA1, and was thus able to moderate N assimilation and metabolism by either transcription activation or suppression of the downstream genes (Liu et al., 2017).

Seeds respond to diverse environmental stimuli that regulate germination. Nitrate stimulates germination in many plants, but how it does so remains unclear. The *Arabidopsis* NLP8 was found to be essential for nitrate-promoted seed germination. Seed germination in nlp8 loss-of-function mutants does not respond to nitrate. NLP8 functions even in a nitrate reductase-deficient mutant background and the requirement for NLP8 is conserved among *Arabidopsis* accessions. NLP8 reduces abscisic acid levels in a nitrate-dependent manner and directly binds to the promoter of CYP707A2, encoding an abscisic acid catabolic enzyme. NLP8 localizes to nuclei but, unlike NLP7, does not appear to be activated by nitrate-dependent nuclear retention of NLP7, suggesting that seeds have a unique mechanism for nitrate signaling (Yan et al., 2016).

Very little information is available on the protein interaction in regulating nitrate signaling. Guan et al. (2017) reported that TCP20, NLP6, and NLP7 bind to adjacent sites in the upstream promoter region of the nitrate reductase gene, NIA1, and physically interact under continuous nitrate and N-starvation conditions. Regions of these proteins necessary for these interactions were found to include the type I/II Phox and Bem1p (PB1) domains of NLP6&7, a protein interaction module conserved in animals for nutrient signaling, and the histidine- and glutamine-rich domains of TCP20, which is conserved across plant species. Under N starvation, TCP20-NLP6&7 heterodimers accumulate in the nucleus, and this coincides with TCP20 and NLP6&7-dependent upregulation of nitrate assimilation and signaling genes and downregulation of the G2/M cell cycle marker gene, CYCB1; 1. TCP20 and NLP6&7 also support root meristem growth under N starvation. These findings provide insights into how plants coordinate responses to nitrate availability, linking nitrate assimilation and signaling with cell cycle progression (Guan et al., 2017). NLPS are master regulators of nitrate signaling and PNR in plants. They control a wider range of genes related to nitrogen assimilation, nitrate transport, carbohydrate metabolism, hormone homeostasis, and development. NLPs could be targeted for altering nitrate signaling and improving NUE of crop plants.

5. Role of transcription factors in nitrate-regulated root architectural modifications

Root provides anchoring and serves as a site of water and nutrients absorption from the soil. Therefore, modulation of root system architecture (RSA) evolved as a fundamental adaptive mechanism in response to the changing rhizosphere environment (Fredes et al., 2019; Guan et al., 2014). Any alteration in the availability of nutrients triggers a cascade of events through the variable expression of genes and transcription factors associated with nutrient signaling and RSA. Nitrate is an important form of inorganic nitrogen (N) for the plant, which serves as a signaling molecule in various developmental processes. Any deficiency of N in plants results in various physiological and phenological changes. The nitrate-regulated transcription factors are responsible for the modulation of RSA in response to a variable level of N in the growing medium. Several reports and modules have demonstrated the role of phytohormones in long-range signaling in plants and mediating the root developmental responses to nitrate (Forde, 2014; Signora et al., 2002; Walch-Liu et al., 2006). The role of the transcription factor in nitrate-induced RSA modification was initially demonstrated in *Arabidopsis*. The NO_3^- inducible MADS-box transcription factor coding gene *ANR1* was shown to be involved in localized NO_3^- induced stimulation of lateral root growth development (Zhang and Forde, 1998). In *Chrysanthemum,* a homolog of the *AtANR1* gene *CmANR1* plays a vital role in the regulation of lateral roots development. Expression assay revealed high nitrate-induced expression of *CmANR1* in chrysanthemum roots and ectopic expression of *CmANR1* in *Arabidopsis* significantly increased the number and length of lateral roots compared to the wild type (Sun et al., 2018b). The higher concentration of nitrate rapidly enriches the nuclear-localized ANR1 and forms ANR1/ANR1 homodimers, which result in upregulation of PINs and AUX1 genes, thereby facilitating root-ward auxin transport and promotes adventitious and lateral roots development. Enhanced auxin in roots regulates the expression of ANR1 interacting factor AGL21 in a feedback manner. Therefore,

ANR1/AGL21 heterodimers play an essential role in the regulation of lateral root development (Sun et al., 2018a). MADS-box transcription factor MADS57 mediates the long-distance nitrate transport and root elongation in rice. OsMADS57 binds to the CArG motif within the OsNRT2.3a promoter and shares the same subcellular localization in roots. Mutation in *Osmads57* showed reduced seminal and adventitious root length in nitrate-starved media (Huang et al., 2019). Further, no localized NO_3^- lateral root proliferation was observed in *Arabidopsis* auxin-resistant mutant *axr4*, which hints on overlapping between NO_3^- and auxin signaling in nitrate-induced root system plasticity (Zhang et al., 1999). However, further characterization of auxin-resistant mutants *axr1*, *axr4*, and *aux1* in response to variable NO_3^- revealed that auxin is not required for the localized stimulation of lateral root proliferation (Linkohr et al., 2002). In roots, NO_3^- induced expression of micro RNA, miR393 targets the NO_3^- regulated AFB3. The analysis of RSA in response to NO_3^- in *afb3* mutant plants and miR393 overexpressions revealed an altered response of primary and lateral root growth (Vidal et al., 2010). AFB3 regulates the expression of NAC4 and OBP4 genes. In response to nitrate treatment, afb3-1 mutant showed no differential expression of NAC4 and OBP4, where the wi*afb3-1*ld plants showed higher expression. Therefore, AFB3-NAC4-OBP4 might form the regulatory module that controls the lateral root growth in response to NO_3^- gradient (Vidal et al., 2013). Efforts on the identification of putative transcription factor genes involved in nitrate-induced root plasticity using macro-array revealed genes for basic leucine zipper (bZIP) and an LIM transcription factor. Further, Northern analysis showed the preferential expression in the roots and correlated to the root response to nitrate availability (Tranbarger et al., 2003). Similarly, a bZIP transcription factor TabZIP60 mediates N use and growth in wheat through negatively regulated expression of NADH-dependent glutamate synthase (TaNADH-GOGAT-3B). Nitrate-starvation resulted in repression of TabZIP60 activity in roots, and the knockdown mutants of TabZIP60 showed increased NADH-GOGAT activity, lateral root branching, and grain yield (Yang et al., 2019). The transcription factors TGA1 and TGA4 regulate the expression of nitrate transporter genes, viz., NRT2.1 and NRT2.2 via specific binding to promoter regions. The double mutant tga1-tga4 showed nitrate-dependent alteration of lateral and primary root phenotypes. The double mutants of transcription factors and transporters, viz., tga1-tga4 and nrt1.2-nrt2.2, resulted in altered lateral root initiation (Alvarez et al., 2014). CPC is the essential regulatory gene for root hair cell specification in *Arabidopsis*.

In response to nitrate, TGA1/TGA4 directly binds and upregulates CPC, which subsequently suppresses longitudinal elongation of trichoblast cells (Canales et al., 2017). In addition to root hairs, nitrate also alters the root meristem stem cells. The transcription factor Teosinte branched1/Cycloidea/Proliferating cell factor1-20 (TCP20) belongs to an ancient, plant-specific gene family that regulates growth and development in plants. TCP20 is associated with nitrate signaling by its ability to bind regulatory sequences of more than 100 nitrate-regulated genes. The mutations in *tcp20* impair the preferential lateral root growth on heterogeneous nitrogen level in growing media (Guan et al., 2014). In nitrate sufficiency and starvation, transcription factors TCP20, NLP6, and NLP7 bind to the upstream promoter region key nitrate-responsive gene *NIA1* and cell cycle marker gene CYCB1;1 and modulate the cell division in the root meristem in response to the nitrate level (Guan et al., 2017). In *Arabidopsis*, NIGT1 negatively regulates the nitrate transporter NRT2.1 and shows antagonistic regulation by NLP primary transcription factors for nitrate signaling. Irrespective of nitrate concentration, constitutive overexpression of NIGT1.1 severely affects the growth, leading to smaller shoots and shorter primary roots, whereas quadruple *nigt1* mutants slightly reduced the total lateral root length, although fresh shoot weight and primary root length were not affected (Maeda et al., 2018). In addition to individual nutrient signaling, cross-signaling among the nutrients is also responsible for the alteration of RSA. Nitrate-inducible *NRT1.1* controlled the GARP transcription factor, while HRS1 represses the primary root growth in response to phosphorus deficiency and nitrate sufficiency (Medici et al., 2015).

6. Targeting transcription factors for the regulation of NUE

Nitrogen use efficiency (NUE) is defined as grain yield per unit of N available in the soil, NUE can be categorized as (a) uptake and (b) utilization efficiencies. Uptake efficiency drives biomass production and depends on the amount of N uptake, storage, and assimilation into amino acids and other important nitrogenous compounds, whereas utilization efficiency involves N remobilization, i.e., the proportion of N that is partitioned to seed resulting in final yield. The current average NUE in the field is approximately 33%, and a substantial proportion of the remaining 67% is lost into the environment, especially in the intensively cropped area (Plett et al., 2017; Wang et al., 2018). It is estimated that a 1% increase in NUE could save $1.1 billion annually. Thus, there is an urgent need for the development of a comprehensive approach to optimize the N utilization, management of N fertilizer and improving NUE through different strategies. There are a few other examples of successful genetic modification of N metabolism using either structural or putative regulatory genes. When the bacterial enzyme glutamate dehydrogenase (GDH A) from *E. coli* was constitutively overexpressed in tobacco, biomass production of the transgenic plants was increased by about 10%–15% (Ameziane et al., 2000). Success in terms of increased NUE through genetic manipulation was done through overexpression of alanine aminotransferase (AlaAT) in roots, which is a downstream enzyme in N assimilation (Good et al., 2007). In transgenic canola, yields were higher at low N and were associated with the higher influx of N. These changes resulted in a 40% decrease in the amount of applied N fertilizer required under field conditions to achieve yields equivalent to those of wild-type plants. A similar strategy was applied for improving NUE in rice (Shrawat et al., 2008). A gene of unknown function OsENOD93-1, which is an N-responsive gene identified by the following genome-wide gene expression profiling in rice, led to an increase in grain yield by 13%–14% and 19%–23% under limiting and non-limiting N nutrition conditions, respectively (Bi et al., 2009). Yanagisawa et al. (2004) found that overexpression of the Dof1 gene encoding a transcription factor from maize in *Arabidopsis* resulted in an increase in amino acid content and of N uptake. Overexpressing regulatory genes rather than structural genes appears to be an interesting alternative to improve plant NUE. The accumulated knowledge on the molecular aspects of N signaling is yet to be exploited for enhancing NUE in crop plants.

7. Conclusions and future perspectives

The basic outline of nitrate signaling and the intermediates are summarized in Fig. 1. Nitrate perceived at plasma membrane by the transceptor CHL1/NPF6.3/NRT1.1 evokes calcium-mediated nitrate signaling. The induction of phospholipase C (PLC) and the production of inositol-3-pyrophosphate induces calcium channels and thereby evokes calcium peaks in cytosol. Calcium sensor kinases (CPKs) perceive the signal and phosphorylate master regulator TFs NLP6/7. NLPs bind to the nitrate-responsive elements in the promoter region of downstream genes (Signaling, TFs, uptake, and metabolism). Independent of the calcium-mediated signaling auxin receptors, AFBs and its downstream components

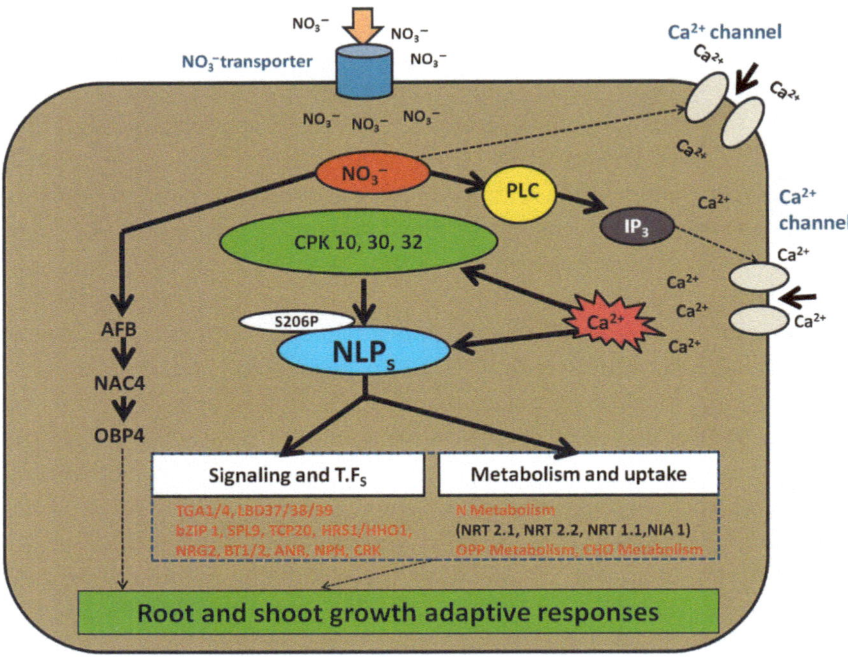

Fig. 1 Schematic representation of nitrate signaling in plants.

also mediate nitrate signaling simultaneously. Nitrate signaling ultimately results in root and shoot growth adaptation, and changes in metabolism in response to external nitrate/N supply. RSA is the most important aspect to improve the nutrient use efficiencies in plants. The root morphophysiological response and plasticity varies across and within species in addition to the rhizosphere heterogeneity. The identification genes and regulatory elements are crucial in the modulation of RSA for enhanced NUE. Emphasized efforts need to be directed to characterize the RSA-associated regulatory elements in various crops. The role of the nitrate-signaling pathway in modulation of RSA is revealed to the extent. However, nitrate interacts in a positive and antagonistic manner with several macro and micronutrients. Further, in addition to targeted cloning of RSA-related genes and regulatory elements, the use of systems biology approaches could help in dissecting the pathways and interactions associated with RSA. Future studies should focus on cross-talk of multiple nutrient signals and identification of common regulatory elements for exploitation in breeding of nutrient use efficient crops.

Acknowledgments

The authors are thankful to the Science and Engineering Research Board, Department of Science and Technology, Government of India, New Delhi for supporting the work on nutrition use efficiency in wheat (ECR/2017/002982) and maize (ECR/2017/000675) at the Division of Plant Physiology and Division of Genetics, ICAR-IARI, New Delhi.

References

Alvarez, J.M., Riveras, E., Vidal, E.A., Gras, D.E., Contreras-Lopez, O., Tamayo, K.P., Gutierrez, R.A., 2014. Systems approach identifies TGA1 and TGA4 transcription factors as important regulatory components of the nitrate response of *Arabidopsis thaliana* roots. Plant J. 80 (1), 1–13. https://doi.org/10.1111/tpj.12618.

Ameziane, R., Bernhard, K., Lightfoot, D., 2000. Expression of the bacterial *gdhA* gene encoding a NADPH glutamate dehydrogenase in tobacco affects plant growth and development. Plant Soil 221, 47–57. https://doi.org/10.1023/A:1004794000267.

Araus, V., Vidal, E.A., Puelma, T., Alamos, S., Mieulet, D., Guiderdoni, E., Gutierrez, R.A., 2016. Members of BTB gene family of scaffold proteins suppress nitrate uptake and nitrogen use efficiency. Plant Physiol. 171 (2), 1523–1532. https://doi.org/10.1104/pp.15.01731.

Bi, Y.M., Kant, S., Clark, J., Gidda, S., Ming, F., Xu, J., Rochon, A., Shelp, B.J., Hao, L., Zhao, R., Mullen, R.T., 2009. Increased nitrogen-use efficiency in transgenic rice plants over-expressing a nitrogen-responsive early nodulin gene identified from rice expression profiling. Plant Cell Environ. 32 (12), 1749–1760. https://doi.org/10.1111/j.1365-3040.2009.02032.x.

Camargo, A., Llamas, A., Schnell, R.A., Higuera, J.J., Gonzalez-Ballester, D., et al., 2007. Nitrate signaling by the regulatory gene NIT2 in Chlamydomonas. Plant Cell 19, 3491–3503. https://doi.org/10.1105/tpc.106.045922.

Canales, J., Contreras-Lopez, O., Alvarez, J.M., Gutierrez, R.A., 2017. Nitrate induction of root hair density is mediated by TGA1/TGA4 and CPC transcription factors in *Arabidopsis thaliana*. Plant J. 92, 305–316. https://doi.org/10.1111/tpj.13656.

Chardin, C., Girin, T., Roudier, F., Meyer, C., Krapp, A., 2014. The plant RWP-RK transcription factors: key regulators of nitrogen responses and of gametophyte development. J. Exp. Bot. 65 (19), 5577–5587. https://doi.org/10.1093/jxb/eru261.

Forde, B.G., 2014. Nitrogen signalling pathways shaping root system architecture: an update. Curr. Opin. Plant Biol. 21, 30–36. https://doi.org/10.1016/j.pbi.2014.06.004.

Fredes, I., Moreno, S., Díaz, F.P., Gutierrez, R.A., 2019. Nitrate signaling and the control of *Arabidopsis* growth and development. Curr. Opin. Plant Biol. 47, 112–118. https://doi.org/10.1016/j.pbi.2018.10.004.

Good, A.G., Johnson, S.J., De Pauw, M.D., Carroll, R.T., Savidov, N., Vidamir, J., Lu, Z., Taylor, G., Stroeher, V., 2007. Engineering nitrogen use efficiency with alanine aminotransferase. Can. J. Bot. 85 (3), 252–262. https://doi.org/10.1139/B07-019.

Guan, P., Davani, A., Pruneda-Paz, J.L., Nacry, P., Breton, G., Crawford, N.M., Kay, S.A., 2014. Nitrate foraging by *Arabidopsis* roots is mediated by the transcription factor TCP20 through the systemic signaling pathway. Proc. Natl. Acad. Sci. 111 (42), 15267–15272. https://doi.org/10.1073/pnas.1411375111.

Guan, P., Ripoll, J.-J., Wang, R., Vuong, L., Bailey-Steinitz, L.J., Ye, D., Crawford, N.M., 2017. Interacting TCP and NLP transcription factors control plant responses to nitrate availability. Proc. Natl. Acad. Sci. 114 (9), 2419–2424. https://doi.org/10.1073/pnas.1615676114.

Ho, C.H., Lin, S.H., Hu, H.C., Tsay, Y.F., 2009. CHL1 functions as a nitrate sensor in plants. Cell 138, 1184–1194. https://doi.org/10.1016/j.cell.2009.07.004.

Hsieh, L.C., Lin, S.I., Shih, A.C., Chen, J.W., Lin, W.Y., Tseng, C.Y., Li, W.H., Chiou, T.J., 2009. Uncovering small RNA-mediated responses to phosphate-deficiency in *Arabidopsis* by deep sequencing. Plant Physiol. 151, 2120–2132. https://doi.org/10.1104/pp.109.147280.

Hu, H.C., Wang, Y.Y., Tsay, Y.F., 2009. AtCIPK8, a CBL-interacting protein kinase, regulates the low-affinity phase of the primary nitrate response. Plant J. 57, 264–278. https://doi.org/10.1111/j.1365-313X.2008.03685.x.

Huang, S., Liang, Z., Chen, S., Sun, H., Fan, X., Wang, C., Zhang, Y., 2019. A transcription factor, OsMADS57, regulates long-distance nitrate transport and root elongation. Plant Physiol. 180 (2), 882–895. https://doi.org/10.1104/pp.19.00142.

Konishi, M., Yanagisawa, S., 2013. *Arabidopsis* NIN-like transcription factors have a central role in nitrate signalling. Nat. Commun. 4, 1617. https://doi.org/10.1038/ncomms2621.

Krouk, G., Mirowski, P., LeCun, Y., Shasha, D.E., Coruzzi, G.M., 2010. Predictive network modeling of the high-resolution dynamic plant transcriptome in response to nitrate. Genome Biol. 11, 123. https://doi.org/10.1186/gb-2010-11-12-r123.

Linkohr, B.I., Williamson, L.C., Fitter, A.H., Leyser, H.M.O., 2002. Nitrate and phosphate availability and distribution have different effects on root system architecture of *Arabidopsis*. Plant J. 29 (6), 751–760. https://doi.org/10.1046/j.1365-313X.2002.01251.x.

Liu, K.H., Huang, C.Y., Tsay, Y.F., 1999. CHL1 is a dual-affinity nitrate transporter of *Arabidopsis* involved in multiple phases of nitrate uptake. Plant Cell 11, 865–874. https://doi.org/10.1105/tpc.11.5.865.

Liu, K.H., Niu, Y., Konishi, M., Wu, Y., Du, H., Chung, H.S., Li, L., Boudsocq, M., McCormack, M., Maekawa, S., Ishida, T., 2017. Discovery of nitrate–CPK–NLP signalling in central nutrient–growth networks. Nature 545 (7654), 311. https://doi.org/10.1038/nature22077.

Maeda, Y., Konishi, M., Kiba, T., Sakuraba, Y., Sawaki, N., Kurai, T., Yanagisawa, S., 2018. A NIGT1-centred transcriptional cascade regulates nitrate signalling and incorporates phosphorus starvation signals in *Arabidopsis*. Nat. Commun. 9(1). https://doi.org/10.1038/s41467-018-03832-6.

Marchive, C., Roudier, F., Castaings, L., Brehaut, V., Blondet, E., et al., 2013. Nuclear retention of the transcription factor NLP7 orchestrates the early response to nitrate in plants. Nat. Commun. 4, 1713. https://doi.org/10.1038/ncomms2650.

Medici, A., Krouk, G., 2014. The primary nitrate response: a multifaceted signalling pathway. J. Exp. Bot. 65, 5567–5576. https://doi.org/10.1093/jxb/eru245.

Medici, A., Marshall-Colon, A., Ronzier, E., Szponarski, W., Wang, R., Gojon, A., Krouk, G., 2015. AtNIGT1/HRS1 integrates nitrate and phosphate signals at the *Arabidopsis* root tip. Nat. Commun. 6, 1–11. https://doi.org/10.1038/ncomms7274.

Obertello, M., Krouk, G., Katari, M.S., Runko, S.J., Coruzzi, G.M., 2010. Modeling the global effect of the basic-leucine zipper transcription factor 1 (bZIP1) on nitrogen and light regulation in *Arabidopsis*. BMC Syst. Biol. 4, 11. https://doi.org/10.1186/1752-0509-4-111.

O'Brien, J.A., Vega, A., Bouguyon, E., Krouk, G., Gojon, A., Coruzzi, G., Gutierrez, R.A., 2016. Nitrate transport, sensing, and responses in plants. Mol. Plant 9 (6), 837–856. https://doi.org/10.1016/j.molp.2016.05.004.

Pant, B.D., Musialak-Lange, M., Nuc, P., May, P., Buhtz, A., Kehr, J., Walther, D., Scheible, W.R., 2009. Identification of nutrient-responsive *Arabidopsis* and rapeseed microRNAs by comprehensive real-time polymerase chain reaction profiling and small RNA sequencing. Plant Physiol. 150, 1541–1555. https://doi.org/10.1104/pp.109.139139.

Para, A., Li, Y., Marshall-Colon, A., Varala, K., Francoeur, N.J., et al., 2014. Hit-and-run transcriptional control by bZIP1 mediates rapid nutrient signaling in *Arabidopsis*. Proc. Natl. Acad. Sci. 111, 10371–10376. https://doi.org/10.1073/pnas.1404657111.

Plett, D.C., Holtham, L.R., Okamoto, M., Garnett, T.P., 2017. Nitrate uptake and its regulation in relation to improving nitrogen use efficiency in cereals. Semin. Cell Dev. Biol. 74, 97–104. https://doi.org/10.1016/j.semcdb.2017.08.027.

Riveras, E., Alvarez, J.M., Vidal, E.A., Oses, C., Vega, A., Gutierrez, R.A., 2015. The calcium ion is a second messenger in the nitrate signaling pathway of *Arabidopsis*. Plant Physiol. 169, 1397–1404. https://doi.org/10.1104/pp.15.0096.

Rubin, G., Tohge, T., Matsuda, F., Saito, K., Scheible, W.R., 2009. Members of the LBD family of transcription factors repress anthocyanin synthesis and affect additional nitrogen responses in *Arabidopsis*. Plant Cell 21, 3567–3584. https://doi.org/10.1105/tpc.109.067041.

Sato, T., Maekawa, S., Konishi, M., Yoshioka, N., Sasaki, Y., et al., 2017. Direct transcriptional activation of BT genes by NLP transcription factors is a key component of the nitrate response in *Arabidopsis*. Biochem. Biophys. Res. Commun. 483, 380–386. https://doi.org/10.1016/j.bbrc.2016.12.135.

Schauser, L., Roussis, A., Stiller, J., Stougaard, J., 1999. A plant regulator controlling development of symbiotic root nodules. Nature 402, 191–195.https://www.nature.com/articles/46058.

Shrawat, A.K., Carroll, R.T., DePauw, M., Taylor, G.J., Good, A.G., 2008. Genetic engineering of improved nitrogen use efficiency in rice by the tissue specific expression of alanine aminotransferase. Plant Biotechnol. J. 6 (7), 722–732. https://doi.org/10.1111/j.1467-7652.2008.00351.x.

Signora, L., De Smet, I., Foyer, C.H., Zhang, H., 2002. ABA plays a central role in mediating the regulatory effects of nitrate on root branching in *Arabidopsis*. Plant J. 28 (6), 655–662. https://doi.org/10.1046/j.1365-313x.2001.01185.x.

Sun, C.H., Yu, J.Q., Duan, X., Wang, J.H., Zhang, Q.Y., Di Gu, K., Zheng, C.S., 2018a. The MADS transcription factor CmANR1 positively modulates root system development by directly regulating CmPIN2 in chrysanthemum. Hortic. Res. 5(1). https://doi.org/10.1038/s41438-018-0061-y.

Sun, C.H., Yu, J.Q., Wen, L.Z., Guo, Y.H., Sun, X., Hao, Y.J., Zheng, C.S., 2018b. Chrysanthemum MADS-box transcription factor CmANR1 modulates lateral root development via homo-/heterodimerization to influence auxin accumulation in *Arabidopsis*. Plant Sci. 266, 27–36. https://doi.org/10.1016/j.plantsci.2017.09.017.

Tranbarger, T.J., Al-Ghazi, Y., Muller, B., Teyssendier De La Serve, B., Doumas, P., Touraine, B., 2003. Transcription factor genes with expression correlated to nitrate-related root plasticity of *Arabidopsis thaliana*. Plant Cell Environ. 26 (3), 459–469. https://doi.org/10.1046/j.1365-3040.2003.00977.x.

Tsay, Y.F., Schroeder, J.I., Feldmann, K.A., CrawfordNM, 1993. The herbicide sensitivity gene CHL1 of *Arabidopsis* encodes a nitrate-inducible nitrate transporter. Cell 72, 705–713. https://doi.org/10.1016/0092-8674(93)90399-B.

Undurraga, S.F., Ibarra-Henríquez, C., Fredes, I., Álvarez, J.M., Gutiérrez, R.A., 2017. Nitrate signaling and early responses in *Arabidopsis* roots. J. Exp. Bot. 68 (10), 2541–2551. https://doi.org/10.1093/jxb/erx041.

Vidal, E.A., Araus, V., Lu, C., Parry, G., Green, P.J., Coruzzi, G.M., Gutierrez, R.A., 2010. Nitrate-responsive miR393/AFB3 regulatory module controls root system architecture in *Arabidopsis thaliana*. Proc. Natl. Acad. Sci. 107 (9), 4477–4482. https://doi.org/10.1073/pnas.0909571107.

Vidal, E.A., Moyano, T.C., Riveras, E., Contreras-Lopez, O., Gutierrez, R.A., 2013. Systems approaches map regulatory networks downstream of the auxin receptor AFB3 in the nitrate response of *Arabidopsis thaliana* roots. Proc. Natl. Acad. Sci. 110 (31), 12840–12845. https://doi.org/10.1073/pnas.1310937110.

Walch-Liu, P., Ivanov, I.I., Filleur, S., Gan, Y., Remans, T., Forde, B.G., 2006. Nitrogen regulation of root branching. Ann. Bot. 97, 875–881. https://doi.org/10.1093/aob/mcj601.

Wang, W., Hu, B., Yuan, D., Liu, Y., Che, R., Hu, Y., et al., 2018. Expression of the nitrate transporter gene OsNRT1. 1A/OsNPF6. 3 confers high yield and early maturation in rice. Plant Cell 30 (3), 638–651.www.plantcell.org/cgi/doi/10.1105/tpc.17.00809.

Xu, G., Fan, X., Miller, A.J., 2012. Plant nitrogen assimilation and use efficiency. Annu. Rev. Plant Biol. 63, 153–182. https://doi.org/10.1146/annurev-arplant-042811-105532.

Xu, N., Wang, R., Zhao, L., Zhang, C., Li, Z., et al., 2016. The *Arabidopsis* NRG2 protein mediates nitrate signaling and interacts with and regulates key nitrate regulators. Plant Cell 28, 485–504.www.plantcell.org/cgi/doi/10.1105/tpc.15.00567.

Yan, D., Easwaran, V., Chau, V., Okamoto, M., Ierullo, M., Kimura, M., Endo, A., Yano, R., Pasha, A., Gong, Y., Bi, Y.M., 2016. NIN-like protein 8 is a master regulator of nitrate-promoted seed germination in *Arabidopsis*. Nat. Commun. 7, 13179. https://doi.org/10.1038/ncomms13179.

Yanagisawa, S., Akiyama, A., Kisaka, H., Uchimiya, H., Miwa, T., 2004. Metabolic engineering with Dof1 transcription factor in plants: improved nitrogen assimilation and growth under low nitrogen conditions. Proc. Natl. Acad. Sci. 101, 7833–7838. https://doi.org/10.1073/pnas.0402267101.

Yang, J., Wang, M., Li, W., He, X., Teng, W., Ma, W., Tong, Y., 2019. Reducing expression of a nitrate-responsive bZIP transcription factor increases grain yield and N use in wheat. Plant Biotechnol. J. 1, 1–11. https://doi.org/10.1111/pbi.13103.

Zhang, H.M., Forde, B.G., 1998. An *Arabidopsis* MADS box gene that controls nutrient-induced changes in root architecture. Science 279, 407–409. https://doi.org/10.1126/science.279.5349.407.

Zhang, H., Jennings, A., Barlow, P.W., Forde, B.G., 1999. Dual pathways for regulation of root branching by nitrate. Proc. Natl. Acad. Sci. 96 (11), 6529–6534. https://doi.org/10.1073/pnas.96.11.6529.

Zhong, L., Chen, D., Min, D., Li, W., Xu, Z., et al., 2015. AtTGA4, a bZIP transcription factor, confers drought resistance by enhancing nitrate transport and assimilation in *Arabidopsis* thaliana. Biochem. Biophys. Res. Commun. 457, 433–439. https://doi.org/10.1016/j.bbrc.2015.01.009.

CHAPTER 13

Engineering transcription factors: An emerging strategy for developing abiotic stress-tolerant crops

Pouya Motie-Noparvar[a], Mohammad Behrouzi Varjovi[b], Behnam Asgari Lajayer[c], and Mansour Ghorbanpour[d]
[a]Department of Agricultural Biotechnology, Faculty of Agriculture, Tarbiat Modares University, Tehran, Iran
[b]Department of Agronomy and Plant Breeding, Faculty of Agricultural and Natural Resources, University of Mohaghegh Ardabili, Ardabil, Iran
[c]Department of Soil Science, Faculty of Agriculture, University of Tabriz, Tabriz, Iran
[d]Department of Medicinal Plants, Faculty of Agriculture and Natural Resources, Arak University, Arak, Iran

1. Introduction

Abiotic stress as the most important crop loss factor leads to a decrease in average yields for most major crop plants (Khoshmanzar et al., 2019). Drought, high salinity, cold, heat, and heavy metals as abiotic stresses have negative effects on plant traits including survival, biomass production, and yield (Asgari Lajayer et al., 2017), hence threatening the food security worldwide. Considering that the yield of maize, rice, wheat, and soybean has not increased recently (Ray et al., 2012) that might warn scientists to think about novel approaches. On the other hand, due to the narrow genetic base of modern crop cultivars, higher sensitivity to environmental stresses has been reported in various important crops (Lobell et al., 2014). Hence, increasing the tolerance of modern high-yielding crops becomes more challenging (Li et al., 2018).

On the other hand, agronomic strategies cannot solve the resulting problems. Intensified agriculture is limited to few numbers of crops, and especially in developing countries, that would be an issue to have mechanized agriculture. Therefore, high-tech agronomical methods are not considered in these countries, but cultivation of high-yielding stress-resistant crops is a solution.

Scientists all over the globe struggle to introduce stress-tolerant crops but, despite the use of classical and molecular methods, little improvement has

Transcription Factors for Abiotic Stress Tolerance in Plants
https://doi.org/10.1016/B978-0-12-819334-1.00013-7

been achieved. Beneficial alleles that contribute to stress resistance or grain nutrition are often lost during conventional breeding because they do not contribute directly to the yield of the major crops (Zheng et al., 2008).

The limitations and achievements of traditional breeding methods, along with later achievements and future perspectives of molecular strategies, are discussed in this paper. In recent years, introduction of novel fields and methods in biology and computer science such as systems biology, synthetic biology, and whole-genome editing tools promises great success for plant breeding science. Some of the principles would be mentioned in this paper with emphasis on engineering transcription factors as the main factors controlling genetic circuits.

In this review, we try to introduce systems biology and synthetic biology as two wings of post-genomic era plant breeding which are capable of solving the mentioned contradictions by mathematical modeling of the genetics of desired traits and designing synthetic genetic networks to respond as they are programmed.

1.1 Breeding for abiotic stress

Developing genotypes with a higher yield under harsh environmental conditions has been a difficult one, due to the lower heritability of the trait under selection (Fritsche-Neto and Borém, 2012). In plant breeding, the methods used for selection are the critical step. In the case of quantitative traits such as abiotic stress tolerance, selection pressure under stress and selection pressure under a stress-free environment are two major strategies for selection. The advent of molecular tools, especially molecular markers, has improved genetic analysis of crop plants and provided scientists with valuable new tools to identify traits of importance in improving resistance to abiotic stresses (Vinocur and Altman, 2005).

Selection of tolerant plants by using phenotypic, physiological, biochemical, and genetic markers—based on complicated statistical analysis—is the basic idea of classical plant breeding. Choosing the best markers for tolerance is very important. Biological yield per plant, plant height, and number of panicles per plant are such traits with high heritability indicating that those traits may be considered during selection for stress tolerance (Athar and Ashraf, 1994). Also, high environmental effects with low narrow-sense heritability estimates for low Na^+/K^+ ratio could be mentioned as another marker considered by breeders (Reddy et al., 2017); grain yield under stress (Bernier et al., 2007), chlorophyll *a* and b content under stress (Zhang et al., 2009), and Ca^{2+},

K^+, Na^+, K^+/Na^+, and Ca^{2+}/Na^+ contents (Ashraf et al., 2008) are considered in classical plant breeding because of high narrow-sense heritability. Early generation population size should be large when narrow-sense heritability is low, and therefore the number of replications and locations over years should be increased to screen the right genotypes; selection should be done in the later generations.

It can be difficult to breed a species for more than one resilient trait at a time, but that is exactly what breeders are looking to do. Breeding methods for abiotic stress resistance are the same as for yield and other economic characters. The choice of a suitable breeding program for the development of tolerant cultivars to a defined abiotic stress depends on screening techniques, sources and mechanisms of tolerance, modes of gene action and heritability, and their correlation with agronomic traits. Although many tolerant crop lines have been developed using conventional breeding techniques, one challenge for applying this approach to other crops is identifying phenotypes that correlate well with the stress tolerance.

Despite all the successes achieved through conventional breeding methods, their success is not enough for crops, given the volume of investment and attention in all national and international breeding centers around the world. Traditional breeding approaches to develop tolerant crops have so far met with limited success (Richards, 1996). Focus on yield rather than on specific traits, complexities caused by genotype by environment interactions in tolerance traits, and the possibility of crossing just closely related plants are the most important restrictions of breeding for abiotic stress tolerance. The need for multiple backcrosses to eradicate undesirable genetic and environmental effects, dependency on genomic loci with a clearly observable phenotype, and inadequacy of variation in the trait of interest are some other limitations of conventional breeding. Therefore, the focus is currently on marker-assisted breeding. There, somaclonal approach, F1 anther culture derivatives, and marker-assisted selection (MAS) are the major newer approaches helping classical plant breeding.

1.2 Transformation

Many efforts worldwide have been made to develop abiotic stress-tolerant crops through the application of recombinant DNA technology, using different genes, encoding osmoprotectants, ion channel proteins, molecular chaperons, and other stress-related functional proteins from which various results have been gained. In comparison to other genes, introduction of multiple glycine betaine, proline, and other osmoprotectant genes, in several transgenic crops, indicated relative success to gain abiotic stress tolerance.

The mission for identifying novel genes required to be transferred to field crops is an essential part of this ongoing fight against abiotic stresses. The transgenic approach also allows scientists to study and validate the molecular mechanism of stress tolerance by overexpressing the gene into a model plant species. After that, monitoring the phenotypical and biochemical changes under specific abiotic stress treatment would be required (Wang et al., 2003). Acceptable success has been achieved, and several varieties of stress-tolerant crops have been introduced in recent years. Though resource intensive, this approach seems a suitable option for developing plants that offer resistance to unfavorable conditions, given the complex nature of abiotic stress tolerance response and other environmental limitations.

The current success in developing tolerant plants to abiotic stress can be significantly enhanced by characterizing individual genes and evaluating their role in the overall stress tolerance potential, knowledge of basic biochemical pathways, and identification of key regulatory genes of stress response pathways. The genes implicated in abiotic stresses are broadly classified into two groups.

1. Proteins that act directly, for example, osmoprotectants, detoxifying enzymes, late embryogenesis abundant (LEA) proteins and molecular chaperones.
2. Stress-responsive gene regulatory proteins. For example, transcription factors, protein kinases, enzymes involved in phosphor-inositol metabolism.

Engineering osmolyte biosynthesis has emerged as a viable approach in producing plant transgenics for enhanced tolerance to osmotic stresses. However, most crop plants lack the ability to synthesize sufficient quantities of the osmo-protectants, so as to provide a meaningful degree of protection. Therefore, genes regulating the synthesis of osmo-protectants have been transformed into transgenic plants to confer stress tolerance through overexpression.

Among the various stress-associated proteins, heat shock proteins (HSPs), cold shock-domain family proteins (CSPs), and late embryogenesis abundant-type proteins (LEAs) are the major stress-induced proteins that accumulate during drought, salinity, and heat stress. They act as molecular chaperones, which are responsible for protecting the cellular machinery in a broad range of cellular processes. The enhanced expression of transcription factors that regulate the expression of LEA-type proteins was found to correlate with cold, drought, or salt stress tolerance in plants (Kasuga et al., 1999).

The gene product of regulators of stress-responsive genes includes transcription factors, protein kinases, and enzymes involved in phosphoinositide metabolism. This group of gene products regulates gene expression and signal transduction pathways. Far beyond the manipulation of a single functional gene, engineering certain regulatory genes could be an effective strategy for controlling the expression of stress-responsive genes. Transcription factors (TFs) as main regulators of many stress-regulatory pathways are suitable candidates for genetic engineering to develop tolerant crops. AP2/EREBP, MYB, WRKY, NAC, and bZIP family transcription factors are involved in tolerance to various abiotic stresses and some genes in these families have also been utilized to develop stress-tolerant plants. Transcription factors are terminal transducers and direct regulators of the expression of downstream genes. In fact, these factors interact with specific cis-elements in the promoter region of the genes to apply the effects (Chen et al., 2012; Dubos et al., 2010; Khan, 2011; Nuruzzaman et al., 2013; Pandey and Somssich, 2009; Singh, 2002). Overexpression of these transcription factors from different plants has been reported to improve the tolerance to stresses (Table 1).

Plant growth regulators have also a role to deploy a number of physiological processes involved in stress tolerance. Abscisic acid (ABA) plays a crucial role as a phytohormone in the activation of cellular defense responses under drought as well as salinity stresses (Iuchi et al., 2001).

Development of stable transgenic crops that can survive under harsh environmental conditions and can maintain their productivity under stress is one of the primary challenges in agricultural biotechnology. Transgenic technology has so far produced mixed results in this area. However, there is a long way to completely understand and regulate plant metabolic processes under stress. This forms the basis of our efforts toward ensuring food and nutritional security for the growing populations.

As mentioned previously, traits such as stress tolerance, water efficiency, and high yield are controlled by many genes. Therefore, introducing just a few numbers of genes into the hexaploid wheat to gain completely resistant plants seems to be low-dose sedation for a severe condition called stress (Hall and Richards, 2013; Sahrawat et al., 2003). On the other side, manipulating a multigenic trait is a complex and time-consuming procedure by itself.

Table 1 Some examples of transgenic plants overexpressing transcription factor genes in recent years.

Family	Gene	Donor	Acceptor	Enhanced tolerance	References
AP2/ERFBP	*LcDREB3a*	*Leymus chinensis*	*Arabidopsis*	Drought and salinity↑	Peng et al. (2013)
	LcDREB2	*Leymus chinensis*	*Arabidopsis*	Salinity↑	Peng et al. (2013)
	LcERF054	*Lotus corniculatus*	*Arabidopsis*	Salinity↑	Sun et al. (2014)
	VrDREB2A	*Vigna radiata*	*Arabidopsis*	Drought and salinity↑	Chen et al. (2015)
	GmERF3	*Glycine* max	*Tobacco*	Drought and salinity↑	Zhang et al. (2009)
	GmERF7	*Glycine max*	Tobacco	Salinity↑	Zhai et al. (2013)
	SsDREB	*Suaeda salsa*	Tobacco	Drought and salinity↑	Zhang et al. (2015a)
	JERF3	*Solanum lycopersicum*	Rice	Drought↑	Zhang et al. (2010)
	OsDREB2A	*Oryza sativa*	Rice	Drought and salinity↑	Mallikarjuna et al. (2011)
	OsERF4a	*Oryza sativa*	Rice	Drought↑	Joo et al. (2013)
	AtDREB1A	*Arabidopsis*	Rice	Drought↑	Ravikumar et al. (2014)
	TaERF3	*Triticum aestivum*	Wheat	Drought and salinity↑	Rong et al. (2014)
	TaPIE1	*Triticum aestivum*	Wheat	Cold↑	Zhu et al. (2014)
	EaDREB2	*Erianthus arundinaceus*	Sugarcane	Drought and salinity↑	Augustine et al. (2015)
	StDREB1	*Solanum tuberosum*	Potato	Salinity↑	Bouaziz et al. (2013)
	ERF109	*Nicotiana benthamiana*	Arabidopsis	Salinity↑	Bahieldin et al. (2016)
	StDREB2	*Potato*	Cotton	Drought↑	El-Esawi and Alayafi (2019)

MYB	*AtMYB15*	*Arabidopsis*	*Arabidopsis*	Drought and salinity↑	Ding et al. (2009)
	LcMYB1	*Leymus chinensis*	*Arabidopsis*	Salinity↑	Cheng et al. (2013b)
	GmMYBJ1	*Glycine max*	*Arabidopsis*	Drought and cold↑	Su et al. (2014)
	TaMYB3R1	*Triticum aestivum*	*Arabidopsis*	Drought and salinity↑	Cai et al. (2015)
	TaPIMP1	*Triticum aestivum*	Tobacco	Drought and salinity↑	Liu et al. (2011b)
	SbMYB2	*Scutellaria baicalensis*	Tobacco	NaCl, mannitol, and ABA stresses↑	Qi et al. (2015)
	SbMYB7	*Scutellaria baicalensis*	Tobacco	NaCl, mannitol, and ABA stresses↑	Qi et al. (2015)
	LeAN2	*Lycopersicum esculentum*	Tobacco	Chilling and oxidative stresses↑	Meng et al. (2014)
	LeAN2	*Lycopersicum esculentum*	Tomato	Heat↑	Meng et al. (2015)
	AtMYB44	*Arabidopsis*	Soybean	Drought and salinity↑	Seo et al. (2012)
	OsMYB2	*Oryza sativa*	Rice	Drought, cold, and salinity↑	Yang et al. (2012b)
	OsMYB91	*Oryza sativa*	Rice	Salinity↑	Zhu et al. (2015)
	MdSIMYB1	*Malus × domestica*	Apple	Drought, cold, and salinity↑	Wang et al. (2014)
	TaMYB31-B	*Wheat*	Arabidopsis	Drought↑	Zhao et al. (2018)
WRKY	*AtWRKY28*	*Arabidopsis*	*Arabidopsis*	Salinity↑	Babitha et al. (2013)
	OsWRKY45	*Oryza sativa*	*Arabidopsis*	Drought and salinity↑	Qiu and Yu (2009)
	TaWRKY79	*Triticum aestivum*	*Arabidopsis*	Drought↑	Qin et al. (2013)

Continued

Table 1 Some examples of transgenic plants overexpressing transcription factor genes in recent years—cont'd

Family	Gene	Donor	Acceptor	Enhanced tolerance	References
	VvWRKY11	*Vitis vinifera*	*Arabidopsis*	Drought↑	Liu et al. (2011a)
	ZmWRKY33	*Zea may*	*Arabidopsis*	Salinity↑	Li et al. (2013)
	GhWRKY34	*Gossypium hirsutum*	*Arabidopsis*	Salinity↑	Zhou et al. (2015)
	GsWRKY20	*Glycine soja*	*Arabidopsis*	Drought↑	Luo et al. (2013)
	TaWRKY79	*Triticum aestivum*	*Arabidopsis*	Salinity and ionic stress↑	Qin et al. (2013)
	TaWRKY93	*Triticum aestivum*	*Arabidopsis*	Salinity, drought, and low temperature↑	Qin et al. (2015)
	TaWRKY10	*Triticum aestivum*	Tobacco	Drought and salinity↑	Wang et al. (2013)
	GhWRKY39	*Gossypium hirsutum*	Tobacco	Salinity↑	Shi et al. (2014)
	BdWRKY36	*Brachypodium distachyon*	Tobacco	Drought↑	Sun et al. (2015)
	ZmWRKY58	*Zea may*	Rice	Drought and salinity↑	Cai et al. (2014)
	MtWRKY76	*Medicago truncatula*	*Medicago truncatula*	Drought and salinity↑	Liu et al. (2016)
	AtWRKY30	*Arabidopsis thaliana*	*Wheat*	Drought and heat↑	El-Esawi et al. (2019)
NAC	*ANAC019*	*Arabidopsis*	*Arabidopsis*	Cold↑	Jensen et al. (2010)
	ONAC063	*Oryza sativa*	*Arabidopsis*	Salinity and osmotic tolerance↑	Yokotani et al. (2009)
	GmNAC20	*Glycine max*	*Arabidopsis*	Salinity and freezing tolerance↑	Hao et al. (2011)

	ZmSNAC1	*Zea may*	*Arabidopsis*	Cold, salinity, and drought↑	Lu et al. (2012)
	BnNAC5	*Brassica napus*	*Arabidopsis*	Salinity↑	Zhong et al. (2012)
	TaNAC67	*Triticum aestivum*	*Arabidopsis*	Cold, salinity, and drought↑	Mao et al. (2014)
	TaNAC29	*Triticum aestivum*	*Arabidopsis*	Drought and salinity↑	Huang et al. (2015)
	MLNAC5	*Miscanthus lutarioriparius*	*Arabidopsis*	Drought and cold↑	Yang et al. (2015)
	TaNAC2a	*Triticum aestivum*	Tobacco	Drought	Tang et al. (2012)
	AhNAC3	*Arachis hypogaea*	Tobacco	Drought↑	Liu et al. (2013)
	SNAC1	*Oryza sativa*	Wheat	Drought and salinity↑	Saad et al. (2013)
	OsNAP	*Oryza sativa*	Rice	Cold, salinity, and drought↑	Chen et al. (2014)
	AtNAC019	*Brassica oleracea*	Arabidopsis	dehydration, salt, abscisic acid (ABA), and H_2O_2↑	Wang et al. (2018)
bZIP	*ABP9*	*Zea may*	*Arabidopsis*	Drought, salinity, and cold↑	Zhang et al. (2011)
	GmbZIP1	*Glycine max*	*Arabidopsis*	Drought, salinity, and cold↑	Gao et al. (2011)
	ZmbZIP72	*Zea may*	*Arabidopsis*	Drought and salinity↑	Ying et al. (2012)
	TabZIP60	*Triticum aestivum*	*Arabidopsis*	Drought, salt, and freezing tolerance↑	Zhang et al. (2015b)
	PtrABF	*Poncirus trifoliata*	Tobacco	Drought↑	Huang et al. (2010)
	LrbZIP	*Nelumbo nucifera*	Tobacco	Salinity↑	Cheng et al. (2013a)
	OsbZIP71	*Oryza sativa*	Rice	Drought and salinity↑	Liu et al. (2014)
	OsbZIP16	*Indica rice*	Arabidopsis	Dehydration, salt, and ABA↑	Pandey et al. (2018)

Imagine how ambitious it could be to breed for multiple traits with the purpose of creating super plants. There is a contradiction here:

- Traditional breeding is a powerful approach to select for quantitative traits, but there are natural limitations to gain very high levels of desired traits.
- Genetic engineering approaches can overcome natural limitations, but they are feeble in the manipulation of multigenic traits.

1.3 Systems biology and synthetic biology

Responses of organisms like plants to environment is a complex phenomenon in various stages including transcriptome, translatome (or proteome), metabolome, and interactome. Omics technologies endeavor to fully understand the mechanisms in high-throughput mode by producing vast amounts of data. The advent of high-throughput techniques and high-throughput sequencing and analysis methods has improved our understanding of the structure and function of many mechanisms of stress responses in plants at the cellular level.

A system can be broadly defined as a group of independent but interconnected components that function together as a unified whole (Cameron et al., 2014). In this regard, an organism could be mentioned as a system, so it is possible to study plants by systems biology. In the last decade, systems biology has emerged as a promising field which integrates vast amounts of genetic data to build computational models in order to help comprehend the mode of action in molecular systems of an organism (Hammer, 2004). Systems biology extracts information from data obtained with the multitude of omics technologies and synthesis network models and observes their behavior. This valuable information would reveal the genetic potential optimized for survival in a range of environmental conditions for plants.

One of the main targets in the post-genomics era is to understand how molecules interact with each other, how this enables signaling, and how information processing is done to connect specific signals to specific molecular responses. Molecular interaction networks can be compared to functional modules that resemble attenuators, enhancers, adaptors, switches, and feedback loops as in electronic circuits. These circuits are designed to maintain the homeostasis of the system by adaptive mechanisms (Ma'ayan, 2011).

With the advent of systems biology and synthetic biology, hope has come to the heart of breeders. With billions of bytes data from omics studies under various environmental conditions in different tissues, computational modeling of molecular systems and the integrative interpretation of ever larger postgenomic datasets are accepted as useful, and perhaps even necessary, components of biological research (Breitling, 2010). Systems biology tries to have an analytical description of complex dynamic biological phenomena by using mathematical calculations and computerized algorithms (Lecca et al., 2016). On the other side, synthetic biology aims at the design/redesign and construction of biological components and systems that do not already exist in the natural world.

If the study of these genetic circuits is called systems biology, designing of novel circuits could be called synthetic biology. The field of plant synthetic biology has the potential to create resistant varieties, by producing new and improved traits with extraordinary particular control on their function, in a way that traditional breeding methods would not be able to accomplish. The possibility of programming the plant's response to specific stimuli (i.e., stressful environment) by synthetic biology methods makes it realistic to have precise temporal, spatial, and developmental control on gene expression, and modulate any metabolic burden associated with stress tolerance (Scott et al., 2018).

Behavior of living organisms can be manipulated by synthetic biology using engineering of genetic elements of desired traits. Mathematical models and circuits designing principles are applied in this regard (Nielsen et al., 2016). These genetic circuits process the external and internal information (stimuli) to produce a response (stress resistance). Novel complex traits can be introduced into plants by engineering genetic circuits. Hence, we enable cells and tissues to receive information from the environment and produce appropriate responses to them. Temperature, water, nutrient availability, and other environmental conditions could be defined as the source of this information or the stimuli. On the other side, the responses of the plant are programmed in the genetic circuits such that they ultimately lead to any desired phenotype (stress resistance in this case). There are some prerequisites for circuit design and its proper function including detailed data from individual genetic elements, such as transcriptional promoters, activators, repressors, terminators, and other regulatory sequences. Since transcription factors are the most important components of these synthetic genetic circuits, comprehensive information would be required to utilize them in a precise purposeful circuit.

2. Transcription factors in abiotic stress

Multiple genes control the stress responses in plants, which makes it a complex phenomenon. Hence, little success has been achieved by introducing a single functional gene plant in order to enhance plant stress tolerance

(Mittler and Blumwald, 2010). Environmental conditions have also many complexities, so many studies have considered regulatory genes and transcription factors to realize how these genetic elements play essential roles in multiple abiotic stress responses by regulating a large number of downstream stress-responsive genes (Atkinson and Urwin, 2012). Therefore, certain regulatory gene modifications lead to the knowledge that expression patterns of regulatory genes greatly influence plant stress tolerance because it mimics or enhances stress signals to regulate many downstream stress-responsive genes at a time (Century et al., 2008). Among the regulatory genes, stress-responsive transcription factors have attracted particular attention due to their important roles in plant stress responses (Chen and Zhu, 2004).

Transcription factors can be considered as important components of the genetic circuits. Synthetic or naturally occurring DNA-binding sites of plant transcription factors can be engineered, meaning that manipulated DNA-binding domains such as engineered zinc fingers or TAL effectors could be utilized in transcription factor engineering to attach these regulatory elements on the desired sequence of the genome (e.g., stress-responsive genes promoter). Synthetic transcription factors are composed of a translational fusion between these DNA-binding domains and effector domains (repressor or activator). This results in an engineered TF with specificity in DNA binding and directed molecular machinery to act on specific promoters to activate or repress expression of a gene of interest. Furthermore, the advent of novel genome editing tools such as CRISPR/Cas9 makes it feasible to manipulate multiple regulatory sites in a genome to apply the designed proper circuit in the organism's genetic grammars for stress resistance (Bortesi and Fischer, 2015).

2.1 Transcription factors in signaling networks of abiotic stress

Signal perception, signal transduction, and stress-responsive gene expression are the main steps of almost all signaling pathways, which in turn cause the activation of physiological and metabolic responses (Huang et al., 2012). Plant sensors or receptors which are located in the cell wall or membrane receive stimuli-caused signals. This is the first step of perceiving the environmental stress signals. Then the extracellular signals are converted into intracellular ones through second messengers (e.g., calcium ions, inositol phosphate, reactive oxygen species, etc.). These second messengers initiate the corresponding signaling pathways to transduce the signals

(Zhu, 2016). Protein kinases and phosphatases are important enzymes in many signal transduction pathways that phosphorylate/dephosphorylate proteins as an effective mechanism for signal relay (Danquah et al., 2014). At the end of the phosphorylation cascade, TFs are activated or suppressed by protein kinases or phosphatases, and they further bind specifically to regulatory elements on promoters of stress-responsive genes and regulate their transcription (Danquah et al., 2014). Meanwhile, TFs themselves are regulated at the transcription level by other upstream components (Shinozaki et al., 2003). A high number of genes are differentially expressed under stress, causing a different transcription pattern from nonstress conditions. This process forms a complex regulatory network to study and manipulate the expression of stress-responsive genes, which in turn determine the activation of physiological and metabolic responses (Ishihama and Yoshioka, 2012). Since TFs are known as terminal transducers which directly regulate the expression of downstream genes by interacting with the specific regulatory elements in their promoter region (Nakashima et al., 2014), controlling their expression theoretically could alleviate stress responses.

Of a number of transcription factors listed elsewhere (Wang et al., 2016), members of the MYB, MYC, ERF, bZIP, and WRKY transcription factor families have already been concerned as regulators of stress responses (Table 1). These families are attractive targets for engineering of the regulatory elements to take advantage of abiotic stresses. Some of the best characterized transcription factor genes in the literature are AREB1 (ABA-responsive element binding protein1), ABF2 (ABA-responsive binding factor 2), DREB (dehydration-responsive binding protein) genes, MYB genes, bZIP encoding genes, and the protein kinases such as receptor-like kinase 1, SNF1-related protein kinase 2C, or guard cell expressed calcium-dependent protein kinases (Riechmann et al., 2000; Schwechheimer et al., 1998).

3. Strategies that use transcription factors to manipulate gene pathways for enhanced stress tolerance

3.1 Overexpressing transcription factors

Overexpression of regulatory genes can change downstream stress-responsive genes. It seems to be a promising approach in the development of stress-resistant/-tolerant transgenic plants (Table 1) when compared to engineering individual functional genes (Yamaguchi-Shinozaki and Shinozaki, 2006).

Therefore, it is important to enhance the regulatory ability of the desired transcription factor that controls the expression of stress-related genes.

Many key genes have been identified that are involved in the regulatory networks under abiotic stress (Seki et al., 2002). For example, three major DREB1/CBF members in *Arabidopsis*, DREB1A/CBF3, DREB1B/CBF1, and DREB1C/CBF2, which are a subfamily of AP2/EREBP transcription factors, are rapidly induced in response to cold stress (Novillo et al., 2004). Overexpressing these three DREB1s/CBFs significantly could improve tolerance to freezing, drought, and high salinity in transgenic *Arabidopsis* (Gilmour et al., 2004). In addition, OsMYB2 gene expression varies in response to salt, cold, and dehydration stress. The transgenic plants with overexpressing OsMYB2 exhibited enhanced tolerance to various stresses. This overexpression of OsMYB2 alters expression levels of other genes which functions in stress response (Yang et al., 2012a). GmMYB76 or GmMYB177 from soybean overexpressed in *Arabidopsis* led to salt and freezing tolerance (Liao et al., 2008). Overexpression of some stress-responsive WRKY had positive results in transgenic plants in tolerating stress. For example, transgenic rice plants containing the OsWRKY11 gene showed significant heat and drought tolerance (Wu et al., 2009). Transgenic *Arabidopsis* plants overexpressing the GmWRKY21 gene exhibited improved tolerance to cold stress, while the overexpressing GmWRKY54 gene improved tolerance to drought and salt stress (Zhou et al., 2008). Overexpression of VvWRKY11 in *Arabidopsis* improved tolerance to osmotic stress (Liu et al., 2011a). As another example, NAC genes which respond to stress also showed differential expression patterns indicating their involvement in the complex signaling networks during plant stress responses (Tran, 2004). Some of these stress-responsive NAC genes have been overexpressed in some plants and displayed positive effects (Nuruzzaman et al., 2013). A lot of efforts have been made to improve abiotic stress tolerance in transgenic plants by overexpressing some stress-responsive bZIP genes and some successful results have been achieved (Yáñez et al., 2009).

3.2 Modifying the pleiotropic effects in transgenic plants

Overexpression of a transcription factor might have some undesirable effects on plant growth and development and may have some other pleiotropic effects (Conant et al., 2014; Gao et al., 2010; Nakashima et al., 2006; Reis et al., 2014). According to Oh (2005), co-expression of the basic leucine zipper (bZIP) transcription factor ABF3 has no pleiotropic effects when expressed in rice. Sakuma et al. (2006) deleted the negative regulator domain of AtDREB2A and overexpressed that gene which caused upregulation of downstream drought-inducible genes and improved drought stress tolerance of *Arabidopsis*. Where pleiotropic effects are observed, however, this problem can be alleviated by choosing specific promoters like stress-inducible promoter such as rd29A (Nakashima et al., 2007; Yamaguchi-Shinozaki and Shinozaki, 1994). Thus, in engineering experiments, the limiting activity of some transcription factors would restrict deleterious effects.

Overexpression of transcription factors such as DREB2 cannot improve stress tolerance in plants, because the protein contains domains which prevent the activation of downstream target genes (Liu et al., 1998). Experiments were designed to modify such domains in order to abolish the repressor function, which led to the creation of tolerant plants (Sakuma et al., 2006). The DREB2A protein with a deletion in the suppressor domain is expressed under normal and stressful growth conditions. Transgenic plants tolerated not only drought and salt stresses but also led to heat tolerance. This indicates that engineering transcription factors like DREB2A can present tolerance to various stresses. In this case, it was demonstrated that DREB2A is related to heat stress transcription factor HsfA3 (Yoshida et al., 2008). This is a simple example of a stress tolerance network which is controlled by transcription factors.

3.3 Synthetic biology for regulon engineering

In the case of use of synthetic biology principles to create super wheat, gene switches for predictable metabolic and regulatory functions would be generated, but it requires an adequate understanding of the plant's responses to the systematic mechanism. In fact, the second wave of synthetic biology including systems-level circuitry in established systems is required for breeding plants (Purnick and Weiss, 2009). Plant synthetic biology has come of age with the first plant synthetic circuits for plant metabolite production, such as dhurrin (Kristensen et al., 2005), signal transduction control (Antunes et al., 2009), and ligand detection programming (Antunes et al., 2011). The next step would be the application of synthetic biology for complex traits breeding by the generation of artificial plant systems and synthetic genomes.

The potential of transcription factors as tools for manipulation of quantitative/multigenic traits led to the development of new transcription factor-based technologies that would feed both applied (e.g., crop improvement) and basic researches. One of the targets of these efforts has been the manipulation of transcription factor activity which offers much promise for de novo regulation of abiotic stress-related pathways.

Experiments along with gene expression data from stress-exposed plants help to develop systematic approaches for manipulating the stress networks, using the tools of synthetic biology. Scientists employ network inference and system identification techniques to build an accurate model of the plant defense subnetwork that contains key genes determining the susceptibility of the plant to abiotic stress. Once validated against time series data, these models are recruited to design and test perturbation mitigation strategies based on the use of genetic feedback control. As mentioned before, the most notable limitation of classical plant breeding is its dependency on plants' intraspecies potentials. For instance, most varieties of wheat as a moderately salt-tolerant plant could tolerate 6 ds/m^2 (Munns and James, 2003). In the field, where the salinity rises to 100 mM NaCl (about 10 dS/m^2), rice (*Oryza sativa*) will kill before the maturity stage, while wheat seedlings will produce a reduced yield (Munns et al., 2006). However, some halophytes such as *Plantago coronopu* can tolerate more than 500 mM NaCl (Flowers and Colmer, 2008), although classical breeding cannot benefit from this natural potential of halophytes salt tolerance.

Considering that most plant species share the same genetic information for common metabolic pathways, transportation, and signal transduction (Luo et al., 2009), it seems that regulon manipulation based on engineering principles (i.e., synthetic biology) could help to find the bottlenecks in the crop sensitivity network in comparison to tolerant plants. Therefore, in a plant breeding by the synthetic biology program, a vast amount of omics data is required to be collected and analyzed to compare a sensitive plant and a salt-tolerant plant in order for abstraction.

Transcriptional circuits underlying genetics programs are characterized by interactions between transcription factors and their target regions on promoters of genes. In the case of regulon engineering for abiotic stress, regulatory interactions data would be useful. To build regulatory networks, Arabidopsis gene regulatory information server (AGRIS; http://arabidopsis.med.ohio-state.edu/) is introduced as a useful tool. This server contains Arabidopsis promoter sequences, transcription factors, and their target genes/functions (Palaniswamy et al., 2006).

The Transcendent goal of such programs could be defined as conglomerate crops with the ability to resist salinity and to provide a series of improved approximate solutions to finally obtain the desired functions and properties. This process includes the biomimetics of proper tolerant plants. After that, network topologies would be constructed, kinetic parameters would be identified, and the selection of genetic parts would be essential to model the network behaviors, robustness, and sensitivity.

The fabrication process includes the assembly of parts that have been proven to play an essential role in salinity tolerance, integration, or reassembly of these parts into a plant and finally the validation of the results. Compared with conventional plant biotechnology, which relies mainly on recombinant DNA technology and focuses on the recombination of existing heterologous genes and promoters, plant synthetic biology is based on combining the modules of nature in new ways and building new modules (Foo et al., 2018), but the same tools such as synthetic promoters and transcription factors for regulation of gene expression and advanced methods for DNA assembly and synthesis, chassis loading, and precision genome editing would be implemented.

4. Conclusion and future prospects

Despite the fact that much progress has been made in the field of plant breeding, there are still major limitations in classical breeding methods. Classical plant breeding depends on the selection and intraspecies crossings. The inherent potential for resistance to stress is limited to the power of plants to resist a harsh environmental condition within a species. On the other hand, biotechnology tries to introduce new genes into the plants in order to obtain stress resistance, but unfortunately these quantitative traits are hard to breed. Hence, transferring a few numbers of genes has gained limited successes. Then not functional genes but regulatory genes, especially those coding for transcription factors, have been introduced as a promising approach to give stress resistance to crops. Also, many regulatory gene families have been identified to play an important role in stress tolerance and regulon study, and engineering with systems biology and synthetic biology seems to affect the future of plant breeding. In this paper, plant synthetic biology is introduced as a promising approach that could enable scientists to create super plants with multiple advanced characteristics including multiple stress resistance, high nutritional quality, and so on. Engineering principles are recruited to develop such plants by designing genetic circuits that compensate for the vulnerable characteristics of plant metabolism. Here, the transcription factors that control the process of tolerance are introduced as the most important factors of tolerance genetic circuits. The first step of such approaches requires the computational methods to recognize the most central pathways and hot molecules that present the ability of tolerance to the plants. This is called conceptualization, which includes determination of system goals and requirements. After that, based on mathematical calculations, a powerful network would be designed to order the plant to resisting stress, imitating the behavior of the resistant cultivars. Based on the designed computational model, the genetic circuit would be characterized for its behaviors, robustness, and sensitivity. Finally, the tools utilized in biotechnology (e.g., genome editing tools, recombinant DNA technology, etc.) are implemented for construction and integration of the genetic circuit to renovate the plant genetic system for the desired trait, which in this case is the resistance for abiotic stress.

References

Antunes, M.S., Morey, K.J., Tewari-Singh, N., Bowen, T.A., Smith, J.J., Webb, C.T., et al., 2009. Engineering key components in a synthetic eukaryotic signal transduction pathway. Mol. Syst. Biol. 5, 270. https://doi.org/10.1038/msb.2009.28.

Antunes, M.S., Morey, K.J., Smith, J.J., Albrecht, K.D., Bowen, T.A., Zdunek, J.K., et al., 2011. Programmable ligand detection system in plants through a synthetic signal transduction pathway. PLoS One 6(1), e16292. https://doi.org/10.1371/journal.pone.0016292.

Asgari Lajayer, H., Savaghebi, G., Hadian, J., Hatami, M., Pezhmanmehr, M., 2017. Comparison of copper and zinc effects on growth, micro-and macronutrients status and essential oil constituents in pennyroyal (*Mentha pulegium* L.). Braz. J. Bot. 40 (2), 379–388. https://doi.org/10.1007/s40415-016-0353-0.

Ashraf, M., Athar, H.R., Harris, P.J.C., Kwon, T.R., 2008. Some prospective strategies for improving crop salt tolerance. In: Advances in Agronomy, pp. 45–110. https://doi.org/10.1016/S0065-2113(07)00002-8.

Athar, H.R., Ashraf, M., 1994. The church in the world: A half century of ecumenism in Hungary. Theol. Today 51 (2), 289–290. https://doi.org/10.1177/004057369405100209.

Atkinson, N.J., Urwin, P.E., 2012. The interaction of plant biotic and abiotic stresses: from genes to the field. J. Exp. Bot. 63 (10), 3523–3543. https://doi.org/10.1093/jxb/ers100.

Augustine, S.M., Ashwin, N.J., Syamaladevi, D.P., Appunu, C., Chakravarthi, M., Ravichandran, V., et al., 2015. Overexpression of EaDREB2 and pyramiding of EaDREB2 with the pea DNA helicase gene (PDH45) enhance drought and salinity tolerance in sugarcane (*Saccharum* spp. hybrid). Plant Cell Rep. 34, 247–263. https://doi.org/10.1007/s00299-014-1704-6.

Babitha, K.C., Ramu, S.V., Pruthvi, V., Mahesh, P., Nataraja, K.N., Udayakumar, M., 2013. Co-expression of AtbHLH17 and AtWRKY28 confers resistance to abiotic stress in Arabidopsis. Transgenic Res. 22, 327–341. https://doi.org/10.1007/s11248-012-9645-8.

Bahieldin, A., Atef, A., Edris, S., Gadalla, N.O., Ali, H.M., Hassan, S.M., et al., 2016. Ethylene responsive transcription factor ERF109 retards PCD and improves salt tolerance in plant. BMC Plant Biol. 16 (1), 216. https://doi.org/10.1186/s12870-016-0908-z.

Bernier, J., Kumar, A., Ramaiah, V., Spaner, D., Atlin, G., 2007. A large-effect QTL for grain yield under reproductive-stage drought stress in upland rice. Crop Sci. 47 (2), 507. https://doi.org/10.2135/cropsci2006.07.0495.

Bortesi, L., Fischer, R., 2015. The CRISPR/Cas9 system for plant genome editing and beyond. Biotechnol. Adv. 33 (1), 41–52. https://doi.org/10.1016/j.biotechadv.2014.12.006.

Bouaziz, D., Pirrello, J., Charfeddine, M., Hammami, A., Jbir, R., Dhieb, A., et al., 2013. Overexpression of StDREB1 transcription factor increases tolerance to salt in transgenic potato plants. Mol. Biotechnol. 54, 803–817. https://doi.org/10.1007/s12033-012-9628-2.

Breitling, R., 2010. What is systems biology? Front. Physiol. 1, 9. https://doi.org/10.3389/fphys.2010.00009.

Cai, R., Zhao, Y., Wang, Y., Lin, Y., Peng, X., Li, Q., et al., 2014. Overexpression of a maize WRKY58 gene enhances drought and salt tolerance in transgenic rice. Plant Cell Tissue Organ Cult. 119, 565–577. https://doi.org/10.1007/s11240-014-0556-7.

Cai, H., Tian, S., Dong, H., Guo, C., 2015. Pleiotropic effects of TaMYB3R1 on plant development and response to osmotic stress in transgenic Arabidopsis. Gene 558, 227–234. https://doi.org/10.1016/j.gene.2014.12.066.

Cameron, D.E., Bashor, C.J., Collins, J.J., 2014. A brief history of synthetic biology. Nat. Rev. Microbiol. 12 (5), 381–390. https://doi.org/10.1038/nrmicro3239.

Century, K., Reuber, T.L., Ratcliffe, O.J., 2008. Regulating the regulators: the future prospects for transcription-factor-based agricultural biotechnology products. Plant Physiol. 147 (1), 20–29. https://doi.org/10.1104/pp.108.117887.

Chen, W.J., Zhu, T., 2004. Networks of transcription factors with roles in environmental stress response. Trends Plant Sci. 9 (12), 591–596. https://doi.org/10.1016/j.tplants.2004.10.007.

Chen, L., Song, Y., Li, S., Zhang, L., Zou, C., Yu, D., 2012. The role of WRKY transcription factors in plant abiotic stresses. Biochim. Biophys. Acta 1819 (2), 120–128. https://doi.org/10.1016/j.bbagrm.2011.09.002.

Chen, X., Wang, Y., Lv, B., Li, J., Luo, L., Lu, S., et al., 2014. The NAC family transcription factor OsNAP confers abiotic stress response through the ABA pathway. Plant Cell Physiol. 55, 604–619. https://doi.org/10.1093/pcp/pct204.

Chen, H., Liu, L., Wang, L., Wang, S., Cheng, X., 2015. VrDREB2A, a DREB-binding transcription factor from *Vigna radiata*, increased drought and high-salt tolerance in transgenic *Arabidopsis thaliana*. J. Plant Res. 129 (2), 263–273. https://doi.org/10.1007/s10265-015-0773-0.

Cheng, L., Li, S., Hussain, J., Xu, X., Yin, J., Zhang, Y., et al., 2013a. Isolation and functional characterization of a salt responsive transcriptional factor, LrbZIP from lotus root (*Nelumbo nucifera* Gaertn). Mol. Biol. Rep. 40, 4033–4045. https://doi.org/10.1007/s11033-012-2481-3.

Cheng, L., Li, X., Huang, X., Ma, T., Liang, Y., Ma, X., et al., 2013b. Overexpression of sheepgrass R1-MYB transcription factor LcMYB1 confers salt tolerance in transgenic Arabidopsis. Plant Physiol. Biochem. 70, 252–260. https://doi.org/10.1016/j.plaphy.2013.05.025.

Conant, G.C., Birchler, J.A., Pires, J.C., 2014. Dosage, duplication, and diploidization: clarifying the interplay of multiple models for duplicate gene evolution over time. Curr. Opin. Plant Biol. 19, 91–98. https://doi.org/10.1016/j.pbi.2014.05.008.

Danquah, A., de Zelicourt, A., Colcombet, J., Hirt, H., 2014. The role of ABA and MAPK signaling pathways in plant abiotic stress responses. Biotechnol. Adv. 32 (1), 40–52. https://doi.org/10.1016/j.biotechadv.2013.09.006.

Ding, Z., Li, S., An, X., Liu, X., Qin, H., Wang, D., 2009. Transgenic expression of MYB15 confers enhanced sensitivity to abscisic acid and improved drought tolerance in

Arabidopsis thaliana. J. Genet. Genomics 36, 17–29. https://doi.org/10.1016/S1673-8527(09)60003-5.

Dubos, C., Stracke, R., Grotewold, E., Weisshaar, B., Martin, C., Lepiniec, L., 2010. MYB transcription factors in Arabidopsis. Trends Plant Sci. 15 (10), 573–581. https://doi.org/10.1016/j.tplants.2010.06.005.

El-Esawi, M.A., Alayafi, A.A., 2019. Overexpression of StDREB2 transcription factor enhances drought stress tolerance in cotton (*Gossypium barbadense* L.). Genes 10(2). https://doi.org/10.3390/genes10020142.

El-Esawi, M.A., Al-Ghamdi, A.A., Ali, H.M., Ahmad, M., 2019. Overexpression of AtWRKY30 transcription factor enhances heat and drought stress tolerance in wheat (*Triticum aestivum* L.). Genes 10 (2), 163. https://doi.org/10.3390/genes10020163.

Flowers, T.J., Colmer, T.D., 2008. Salinity tolerance in halophytes. New Phytol. 179 (4), 945–963. https://doi.org/10.1111/j.1469-8137.2008.02531.x.

Foo, M., Gherman, I., Zhang, P., Bates, D.G., Denby, K.J., 2018. A framework for engineering stress resilient plants using genetic feedback control and regulatory network rewiring. ACS Synth. Biol. 7 (6), 1553–1564. https://doi.org/10.1021/acssynbio.8b00037.

Fritsche-Neto, R., Borém, A., 2012. In: Fritsche-Neto, R., Borém, A. (Eds.), In: Plant Breeding for Abiotic Stress Tolerance. Springer Berlin Heidelberg, Berlin, Heidelberg, Heidelberg. https://doi.org/10.1007/978-3-642-30553-5.

Gao, P., Bai, X., Yang, L., Lv, D., Li, Y., Cai, H., et al., 2010. Over-expression of osa-MIR396c decreases salt and alkali stress tolerance. Planta 231 (5), 991–1001. https://doi.org/10.1007/s00425-010-1104-2.

Gao, S.Q., Chen, M., Xu, Z.S., Zhao, C.P., Li, L., Xu, H.J., et al., 2011. The soybean GmbZIP1 transcription factor enhances multiple abiotic stress tolerances in transgenic plants. Plant Mol. Biol. 75, 537–553. https://doi.org/10.1007/s11103-011-9738-4.

Gilmour, S.J., Fowler, S.G., Thomashow, M.F., 2004. Arabidopsis transcriptional activators CBF1, CBF2, and CBF3 have matching functional activities. Plant Mol. Biol. 54 (5), 767–781. https://doi.org/10.1023/B:PLAN.0000040902.06881.d4.

Hall, A.J., Richards, R.A., 2013. Prognosis for genetic improvement of yield potential and water-limited yield of major grain crops. Field Crop Res. 143, 18–33. https://doi.org/10.1016/j.fcr.2012.05.014.

Hammer, G.L., 2004. On systems thinking, systems biology, and the in silico plant. Plant Physiol. 134 (3), 909–911. https://doi.org/10.1104/pp.103.034827.

Hao, Y.J., Wei, W., Song, Q.X., Chen, H.W., Zhang, Y.Q., Wang, F., et al., 2011. Soybean NAC transcription factors promote abiotic stress tolerance and lateral root formation in transgenic plants. Plant J. 68, 302–313. https://doi.org/10.1111/j.1365-313X.2011.04687.x.

Huang, X.S., Liu, J.H., Chen, X.J., 2010. Overexpression of PtrABF gene, a bZIP transcription factor isolated from *Poncirus trifoliata*, enhances dehydration and drought tolerance in tobacco via scavenging ROS and modulating expression of stress-responsive genes. BMC Plant Biol. 10, 230. https://doi.org/10.1186/1471-2229-10-230.

Huang, G.-T., Ma, S.-L., Bai, L.-P., Zhang, L., Ma, H., Jia, P., et al., 2012. Signal transduction during cold, salt, and drought stresses in plants. Mol. Biol. Rep. 39 (2), 969–987. https://doi.org/10.1007/s11033-011-0823-1.

Huang, Q., Wang, Y., Li, B., Chang, J., Chen, M., Li, K., et al., 2015. TaNAC29, a NAC transcription factor from wheat, enhances salt and drought tolerance in transgenic Arabidopsis. BMC Plant Biol. 15, 268. https://doi.org/10.1186/s12870-015-0644-9.

Ishihama, N., Yoshioka, H., 2012. Post-translational regulation of WRKY transcription factors in plant immunity. Curr. Opin. Plant Biol. 15 (4), 431–437. https://doi.org/10.1016/j.pbi.2012.02.003.

Iuchi, S., Kobayashi, M., Taji, T., Naramoto, M., Seki, M., Kato, T., et al., 2001. Regulation of drought tolerance by gene manipulation of 9-cis-epoxycarotenoid dioxygenase, a key enzyme in abscisic acid biosynthesis in Arabidopsis. Plant J. 27 (4), 325–333. https://doi.org/10.1046/j.1365-313x.2001.01096.x.

Jensen, M.K., Kjaersgaard, T., Nielsen, M.M., Galberg, P., Petersen, K., O'shea, C., et al., 2010. The Arabidopsis thaliana NAC transcription factor family: structure-function relationships and determinants of ANAC019 stress signalling. Biochem. J. 426, 183–196. https://doi.org/10.1042/BJ20091234.

Joo, J., Choi, H.J., Lee, Y.H., Kim, Y.K., Song, S.I., 2013. A transcriptional repressor of the ERF family confers drought tolerance to rice and regulates genes preferentially located on chromosome 11. Planta 238, 155–170. https://doi.org/10.1007/s00425-013-1880-6.

Kasuga, M., Liu, Q., Miura, S., Yamaguchi-Shinozaki, K., Shinozaki, K., 1999. Improving plant drought, salt, and freezing tolerance by gene transfer of a single stress-inducible transcription factor. Nat. Biotechnol. 17 (3), 287–291. https://doi.org/10.1038/7036.

Khan, M.S., 2011. The role of Dreb transcription factors in abiotic stress tolerance of plants. Biotechnol. Biotechnol. Equip. 25 (3), 2433–2442. https://doi.org/10.5504/BBEQ.2011.0072.

Khoshmanzar, E., Aliasgharzad, N., Neyshabouri, M.R., Khoshru, B., Arzanlou, M., Asgari Lajayer, B., 2019. Effects of Trichoderma isolates on tomato growth and inducing its tolerance to water-deficit stress. Int. J. Environ. Sci. Technol. https://doi.org/10.1007/s13762-019-02405-4.

Kristensen, C., Morant, M., Olsen, C.E., Ekstrom, C.T., Galbraith, D.W., Lindberg Moller, B., Bak, S., 2005. Metabolic engineering of dhurrin in transgenic Arabidopsis plants with marginal inadvertent effects on the metabolome and transcriptome. Proc. Natl. Acad. Sci. 102 (5), 1779–1784. https://doi.org/10.1073/pnas.0409233102.

Lecca, P., Re, A., Ihekwaba, A., Mura, I., Nguyen, T.P., 2016. Computational Systems Biology. Computational Systems Biology: Inference and Modeling. Elsevier. https://doi.org/10.1016/C2014-0-01062-3.

Li, H., Gao, Y., Xu, H., Dai, Y., Deng, D., Chen, J., 2013. ZmWRKY33, a WRKY maize transcription factor conferring enhanced salt stress tolerances in Arabidopsis. Plant Growth Regul. 70, 207–216. https://doi.org/10.1007/s10725-013-9792-9.

Li, H., Rasheed, A., Hickey, L.T., He, Z., 2018. Fast-forwarding genetic gain. Trends Plant Sci. 23 (3), 184–186. https://doi.org/10.1016/j.tplants.2018.01.007.

Liao, Y., Zou, H.-F., Wang, H.-W., Zhang, W.-K., Ma, B., Zhang, J.-S., Chen, S.-Y., 2008. Soybean GmMYB76, GmMYB92 and GmMYB177 genes confer stress tolerance in transgenic Arabidopsis plants. Cell Res. 18 (10), 1047–1060. https://doi.org/10.1038/cr.2008.280.

Liu, Q., Kasuga, M., Sakuma, Y., Abe, H., Miura, S., Yamaguchi-Shinozaki, K., Shinozaki, K., 1998. Two transcription factors, DREB1 and DREB2, with an EREBP/AP2 DNA binding domain separate two cellular signal transduction pathways in drought- and low-temperature-responsive gene expression, respectively, in Arabidopsis. Plant Cell 10 (8), 1391–1406.

Liu, H., Yang, W., Liu, D., Han, Y., Zhang, A., Li, S., 2011a. Ectopic expression of a grapevine transcription factor VvWRKY11 contributes to osmotic stress tolerance in Arabidopsis. Mol. Biol. Rep. 38 (1), 417–427. https://doi.org/10.1007/s11033-010-0124-0.

Liu, H., Zhou, X., Dong, N., Liu, X., Zhang, H., Zhang, Z., 2011b. Expression of a wheat MYB gene in transgenic tobacco enhances resistance to *Ralstonia solanacearum*, and to drought and salt stresses. Funct. Integr. Genomics 11 (3), 431–443. https://doi.org/10.1007/s10142-011-0228-1.

Liu, X., Liu, S., Wu, J., Zhang, B., Li, X., Yan, Y., et al., 2013. Overexpression of *Arachis hypogaea* NAC3 in tobacco enhances dehydration and drought tolerance by increasing superoxide scavenging. Plant Physiol. Biochem. 70, 354–359. https://doi.org/10.1016/j.plaphy.2013.05.018.

Liu, C., Mao, B., Ou, S., Wang, W., Liu, L., Wu, Y., et al., 2014. OsbZIP71, a bZIP transcription factor, confers salinity and drought tolerance in rice. Plant Mol. Biol. 84, 19–36. https://doi.org/10.1007/s11103-013-0115-3.

Liu, L., Zhang, Z., Dong, J., Wang, T., 2016. Overexpression of MtWRKY76 increases both salt and drought tolerance in *Medicago truncatula*. Environ. Exp. Bot. 123, 50–58. https://doi.org/10.1016/j.envexpbot.2015.10.007.

Lobell, D.B., Roberts, M.J., Schlenker, W., Braun, N., Little, B.B., Rejesus, R.M., Hammer, G.L., 2014. Greater sensitivity to drought accompanies maize yield increase in the U.S. Midwest. Science 344 (6183), 516–519. https://doi.org/10.1126/science.1251423.

Lu, M., Ying, S., Zhang, D.F., Shi, Y.S., Song, Y.C., Wang, T.Y., et al., 2012. A maize stress-responsive NAC transcription factor, ZmSNAC1, confers enhanced tolerance to dehydration in transgenic Arabidopsis. Plant Cell Rep. 31, 1701–1711. https://doi.org/10.1007/s00299-012-1284-2.

Luo, Z.-B., Janz, D., Jiang, X., Gobel, C., Wildhagen, H., Tan, Y., et al., 2009. Upgrading root physiology for stress tolerance by ectomycorrhizas: insights from metabolite and transcriptional profiling into reprogramming for stress anticipation. Plant Physiol. 151 (4), 1902–1917. https://doi.org/10.1104/pp.109.143735.

Luo, X., Bai, X., Sun, X., Zhu, D., Liu, B., Ji, W., et al., 2013. Expression of wild soybean WRKY20 in Arabidopsis enhances drought tolerance and regulates ABA signalling. J. Exp. Bot. 64, 2155–2169. https://doi.org/10.1093/jxb/ert073.

Ma'ayan, A., 2011. Introduction to network analysis in systems biology. Sci. Signal. 4 (190), tr5. https://doi.org/10.1126/scisignal.2001965.

Mallikarjuna, G., Mallikarjuna, K., Reddy, M.K., Kaul, T., 2011. Expression of OsDREB2A transcription factor confers enhanced dehydration and salt stress tolerance in rice (*Oryza sativa* L.). Biotechnol. Lett. 33, 1689–1697. https://doi.org/10.1007/s10529-011-0620-x.

Mao, X., Chen, S., Li, A., Zhai, C., Jing, R., 2014. Novel NAC transcription factor TaNAC67 confers enhanced multi-abiotic stress tolerances in Arabidopsis. PLoS One 9, e84359. https://doi.org/10.1371/journal.pone.0084359.

Meng, X., Yin, B., Feng, H.L., Zhang, S., Liang, X.Q., Meng, Q.W., 2014. Overexpression of R2R3-MYB gene leads to accumulation of anthocyanin and enhanced resistance to chilling and oxidative stress. Biol. Plant. 58, 121–130. https://doi.org/10.1007/s10535-013-0376-3.

Meng, X., Wang, J.R., Wang, G.D., Liang, X.Q., Li, X.D., Meng, Q.W., 2015. An R2R3-MYB gene, LeAN2, positively regulated the thermo-tolerance in transgenic tomato. J. Plant Physiol. https://doi.org/10.1016/j.jplph.2014.09.018.

Mittler, R., Blumwald, E., 2010. Genetic engineering for modern agriculture: challenges and perspectives. Annu. Rev. Plant Biol. 61 (1), 443–462. https://doi.org/10.1146/annurev-arplant-042809-112116.

Munns, R., James, R.A., 2003. Screening methods for salinity tolerance: a case study with tetraploid wheat. Plant Soil 253 (1), 201–218. https://doi.org/10.1023/A:1024553303144.

Munns, R., James, R.A., Läuchli, A., 2006. Approaches to increasing the salt tolerance of wheat and other cereals. J. Exp. Bot. 57 (5), 1025–1043. https://doi.org/10.1093/jxb/erj100.

Nakashima, K., Fujita, Y., Katsura, K., Maruyama, K., Narusaka, Y., Seki, M., et al., 2006. Transcriptional regulation of ABI3- and ABA-responsive genes including RD29B and RD29A in seeds, germinating embryos, and seedlings of Arabidopsis. Plant Mol. Biol. 60 (1), 51–68. https://doi.org/10.1007/s11103-005-2418-5.

Nakashima, K., Tran, L.-S.P., Van Nguyen, D., Fujita, M., Maruyama, K., Todaka, D., et al., 2007. Functional analysis of a NAC-type transcription factor OsNAC6 involved in abiotic and biotic stress-responsive gene expression in rice. Plant J. 51 (4), 617–630. https://doi.org/10.1111/j.1365-313X.2007.03168.x.

Nakashima, K., Yamaguchi-Shinozaki, K., Shinozaki, K., 2014. The transcriptional regulatory network in the drought response and its crosstalk in abiotic stress responses including drought, cold, and heat. Front. Plant Sci. 5, 170. https://doi.org/10.3389/fpls.2014.00170.

Nielsen, A.A.K., Der, B.S., Shin, J., Vaidyanathan, P., Paralanov, V., Strychalski, E.A., et al., 2016. Genetic circuit design automation. Science 352 (6281), aac7341. https://doi.org/10.1126/science.aac7341.

Novillo, F., Alonso, J.M., Ecker, J.R., Salinas, J., 2004. CBF2/DREB1C is a negative regulator of CBF1/DREB1B and CBF3/DREB1A expression and plays a central role in stress tolerance in Arabidopsis. Proc. Natl. Acad. Sci. 101 (11), 3985–3990. https://doi.org/10.1073/pnas.0303029101.

Nuruzzaman, M., Sharoni, A.M., Kikuchi, S., 2013. Roles of NAC transcription factors in the regulation of biotic and abiotic stress responses in plants. Front. Microbiol. 4, 248. https://doi.org/10.3389/fmicb.2013.00248.

Oh, S.-J., 2005. Arabidopsis CBF3/DREB1A and ABF3 in transgenic Rice increased tolerance to abiotic stress without stunting growth. Plant Physiol. 138 (1), 341–351. https://doi.org/10.1104/pp.104.059147.

Palaniswamy, S.K., James, S., Sun, H., Lamb, R.S., Davuluri, R.V., Grotewold, E., 2006. AGRIS and AtRegNet. A platform to link cis-regulatory elements and transcription factors into regulatory networks. Plant Physiol. 140 (3), 818–829. https://doi.org/10.1104/pp.105.072280.

Pandey, S.P., Somssich, I.E., 2009. The role of WRKY transcription factors in plant immunity. Plant Physiol. 150 (4), 1648–1655. https://doi.org/10.1104/pp.109.138990.

Pandey, A.S., Sharma, E., Jain, N., Singh, B., Burman, N., Khurana, J.P., 2018. A rice bZIP transcription factor, OsbZIP16, regulates abiotic stress tolerance when over-expressed in Arabidopsis. J. Plant Biochem. Biotechnol. https://doi.org/10.1007/s13562-018-0448-8.

Peng, X., Zhang, L., Zhang, L., Liu, Z., Cheng, L., Yang, Y., et al., 2013. The transcriptional factor LcDREB2 cooperates with LcSAMDC2 to contribute to salt tolerance in Leymus chinensis. Plant Cell Tissue Organ Cult. 113, 245–256. https://doi.org/10.1007/s11240-012-0264-0.

Purnick, P.E.M., Weiss, R., 2009. The second wave of synthetic biology: from modules to systems. Nat. Rev. Mol. Cell Biol. 10 (6), 410–422. https://doi.org/10.1038/nrm2698.

Qi, L., Yang, J., Yuan, Y., Huang, L., Chen, P., 2015. Overexpression of two R2R3-MYB genes from *Scutellaria baicalensis* induces phenylpropanoid accumulation and enhances oxidative stress resistance in transgenic tobacco. Plant Physiol. Biochem. 94, 235–243. https://doi.org/10.1016/j.plaphy.2015.06.007.

Qin, Y., Tian, Y., Han, L., Yang, X., 2013. Constitutive expression of a salinity-induced wheat WRKY transcription factor enhances salinity and ionic stress tolerance in transgenic *Arabidopsis thaliana*. Biochem. Biophys. Res. Commun. 441, 476–481. https://doi.org/10.1016/j.bbrc.2013.10.088.

Qin, Y., Tian, Y., Liu, X., 2015. A wheat salinity-induced WRKY transcription factor TaWRKY93 confers multiple abiotic stress tolerance in *Arabidopsis thaliana*. Biochem. Biophys. Res. Commun. 464 (2), 428–433. https://doi.org/10.1016/j.bbrc.2015.06.128.

Qiu, Y., Yu, D., 2009. Over-expression of the stress-induced OsWRKY45 enhances disease resistance and drought tolerance in *Arabidopsis*. Environ. Exp. Bot. 65, 35–47. https://doi.org/10.1016/j.envexpbot.2008.07.002.

Ravikumar, G., Manimaran, P., Voleti, S.R., Subrahmanyam, D., Sundaram, R.M., Bansal, K.C., et al., 2014. Stress-inducible expression of AtDREB1A transcription factor greatly improves drought stress tolerance in transgenic indica rice. Transgenic Res. 23, 421–439. https://doi.org/10.1007/s11248-013-9776-6.

Ray, D.K., Ramankutty, N., Mueller, N.D., West, P.C., Foley, J.A., 2012. Recent patterns of crop yield growth and stagnation. Nat. Commun. https://doi.org/10.1038/ncomms2296.

Reddy, I.N.B.L., Kim, B.-K., Yoon, I.-S., Kim, K.-H., Kwon, T.-R., 2017. Salt tolerance in rice: Focus on mechanisms and approaches. Rice Sci. 24 (3), 123–144. https://doi.org/10.1016/j.rsci.2016.09.004.

Reis, R.R., da Cunha, B.A., Martins, P.K., Martins, M.T., Alekcevetch, J.C., Chalfun-Júnior, A., et al., 2014. Induced over-expression of AtDREB2A CA improves drought tolerance in sugarcane. Plant Sci. 221–222, 59–68. https://doi.org/10.1016/j.plantsci.2014.02.003.

Richards, R.A., 1996. Defining selection criteria to improve yield under drought. Plant Growth Regul. 20 (2), 157–166. https://doi.org/10.1007/BF00024012.

Riechmann, J.L., Heard, J., Martin, G., Reuber, L., Jiang, C., et al., 2000. Arabidopsis transcription factors: Genome-wide comparative analysis among eukaryotes. Science 290 (5499), 2105–2110. https://doi.org/10.1126/science.290.5499.2105.

Rong, W., Qi, L., Wang, A., Ye, X., Du, L., Liang, H., et al., 2014. The ERF transcription factor TaERF3 promotes tolerance to salt and drought stresses in wheat. Plant Biotechnol. J. 12, 468–479. https://doi.org/10.1111/pbi.12153.

Saad, A.S., Li, X., Li, H.P., Huang, T., Gao, C.S., Guo, M.W., et al., 2013. A rice stress-responsive NAC gene enhances tolerance of transgenic wheat to drought and salt stresses. Plant Sci. 203–204, 33–40. https://doi.org/10.1016/j.plantsci.2012.12.016.

Sahrawat, A.K., Becker, D., Lütticke, S., Lörz, H., 2003. Genetic improvement of wheat via alien gene transfer, an assessment. Plant Sci. 165 (5), 1147–1168. https://doi.org/10.1016/S0168-9452(03)00323-6.

Sakuma, Y., Maruyama, K., Qin, F., Osakabe, Y., Shinozaki, K., Yamaguchi-Shinozaki, K., 2006. Dual function of an Arabidopsis transcription factor DREB2A in water-stress-responsive and heat-stress-responsive gene expression. Proc. Natl. Acad. Sci. 103 (49), 18822–18827. https://doi.org/10.1073/pnas.0605639103.

Schwechheimer, C., Zourelidou, M., Bevan, M.W., 1998. Plant transcription factor studies. Annu. Rev. Plant Physiol. Plant Mol. Biol. 49 (1), 127–150. https://doi.org/10.1146/annurev.arplant.49.1.127.

Scott, D., Berry, D., Calvert, J., 2018. Routledge Handbook of Genomics, Health and Society. Routledge. https://doi.org/10.4324/9781315451695.

Seki, M., Narusaka, M., Ishida, J., Nanjo, T., Fujita, M., Oono, Y., et al., 2002. Monitoring the expression profiles of 7000 Arabidopsis genes under drought, cold and high-salinity stresses using a full-length cDNA microarray. Plant J. 31 (3), 279–292. https://doi.org/10.1046/j.1365-313X.2002.01359.x.

Seo, J., Sohn, H., Noh, K., Jung, C., An, J., Donovan, C., et al., 2012. Expression of the Arabidopsis AtMYB44 gene confers drought/salt-stress tolerance in transgenic soybean. Mol. Breed. 29, 601–608. https://doi.org/10.1007/s11032-011-9576-8.

Shi, W., Liu, D., Hao, L., Wu, C.A., Guo, X., Li, H., 2014. GhWRKY39, a member of the WRKY transcription factor family in cotton, has a positive role in disease resistance and salt stress tolerance. Plant Cell Tissue Organ Cult. 118, 17–32. https://doi.org/10.1007/s11240-014-0458-8.

Shinozaki, K., Yamaguchi-Shinozaki, K., Seki, M., 2003. Regulatory network of gene expression in the drought and cold stress responses. Curr. Opin. Plant Biol. 6 (5), 410–417. https://doi.org/10.1016/S1369-5266(03)00092-X.

Singh, K., 2002. Transcription factors in plant defense and stress responses. Curr. Opin. Plant Biol. 5 (5), 430–436. https://doi.org/10.1016/S1369-5266(02)00289-3.
Su, L.T., Li, J.W., Liu, D.Q., Zhai, Y., Zhang, H.J., Li, X.W., et al., 2014. A novel MYB transcription factor, GmMYBJ1, from soybean confers drought and cold tolerance in *Arabidopsis thaliana*. Gene 538, 46–55. https://doi.org/10.1016/j.gene.2014.01.024.
Sun, Z.M., Zhou, M.L., Xiao, X.G., Tang, Y.X., Wu, Y.M., 2014. Genome-wide analysis of AP2/ERF family genes from Lotus corniculatus shows LcERF054 enhances salt tolerance. Funct. Integr. Genomics 14, 453–466. https://doi.org/10.1007/s10142-014-0372-5.
Sun, J., Hu, W., Zhou, R., Wang, L., Wang, X., Wang, Q., et al., 2015. The *Brachypodium distachyon* BdWRKY36 gene confers tolerance to drought stress in transgenic tobacco plants. Plant Cell Rep. 34, 23–35. https://doi.org/10.1007/s00299-014-1684-6.
Tang, Y., Liu, M., Gao, S., Zhang, Z., Zhao, X., Zhao, C., et al., 2012. Molecular characterization of novel TaNAC genes in wheat and overexpression of TaNAC2a confers drought tolerance in tobacco. Physiol. Plant. 144, 210–224. https://doi.org/10.1111/j.1399-3054.2011.01539.x.
Tran, L.-S.P., 2004. Isolation and functional analysis of Arabidopsis stress-inducible NAC transcription factors that bind to a drought-responsive cis-element in the early responsive to dehydration stress 1 promoter. Plant Cell Online 16 (9), 2481–2498. https://doi.org/10.1105/tpc.104.022699.
Vinocur, B., Altman, A., 2005. Recent advances in engineering plant tolerance to abiotic stress: Achievements and limitations. Curr. Opin. Biotechnol. 16 (2), 123–132. https://doi.org/10.1016/j.copbio.2005.02.001.
Wang, W., Vinocur, B., Altman, A., 2003. Plant responses to drought, salinity and extreme temperatures: towards genetic engineering for stress tolerance. Planta 218 (1), 1–14. https://doi.org/10.1007/s00425-003-1105-5.
Wang, C., Deng, P., Chen, L., Wang, X., Ma, H., Hu, W., et al., 2013. A wheat WRKY transcription factor TaWRKY10 confers tolerance to multiple abiotic stresses in transgenic tobacco. PLoS ONE 8, e65120. https://doi.org/10.1371/journal.pone.0065120.
Wang, R.K., Cao, Z.H., Hao, Y.J., 2014. Overexpression of a R2R3 MYB gene MdSIMYB1 increases tolerance to multiple stresses in transgenic tobacco and apples. Physiol. Plant. 150, 76–87. https://doi.org/10.1111/ppl.12069.
Wang, H., Wang, H., Shao, H., Tang, X., 2016. Recent advances in utilizing transcription factors to improve plant abiotic stress tolerance by transgenic technology. Front. Plant Sci. 7, 67. https://doi.org/10.3389/fpls.2016.00067.
Wang, J., Lian, W., Cao, Y., Wang, X., Wang, G., Qi, C., et al., 2018. Overexpression of BoNAC019, a NAC transcription factor from *Brassica oleracea*, negatively regulates the dehydration response and anthocyanin biosynthesis in Arabidopsis. Sci. Rep. 8 (1), 13349. https://doi.org/10.1038/s41598-018-31690-1.
Wu, X., Shiroto, Y., Kishitani, S., Ito, Y., Toriyama, K., 2009. Enhanced heat and drought tolerance in transgenic rice seedlings overexpressing OsWRKY11 under the control of HSP101 promoter. Plant Cell Rep. 28 (1), 21–30. https://doi.org/10.1007/s00299-008-0614-x.
Yamaguchi-Shinozaki, K., Shinozaki, K., 1994. A novel cis-acting element in an Arabidopsis gene is involved in responsiveness to drought, low-temperature, or high-salt stress. Plant Cell 6 (2), 251. https://doi.org/10.2307/3869643.
Yamaguchi-Shinozaki, K., Shinozaki, K., 2006. Transcriptional regulatory networks in cellular responses and tolerance to dehydration and cold stresses. Annu. Rev. Plant Biol. 57 (1), 781–803. https://doi.org/10.1146/annurev.arplant.57.032905.105444.
Yáñez, M., Cáceres, S., Orellana, S., Bastías, A., Verdugo, I., Ruiz-Lara, S., Casaretto, J.A., 2009. An abiotic stress-responsive bZIP transcription factor from wild and cultivated

tomatoes regulates stress-related genes. Plant Cell Rep. 28 (10), 1497–1507. https://doi.org/10.1007/s00299-009-0749-4.

Yang, A., Dai, X., Zhang, W.H., 2012a. A R2R3-type MYB gene, OsMYB2, is involved in salt, cold, and dehydration tolerance in rice. J. Exp. Bot. 63, 2541–2556. https://doi.org/10.1093/jxb/err431.

Yang, A., Dai, X., Zhang, W.-H., 2012b. A R2R3-type MYB gene, OsMYB2, is involved in salt, cold, and dehydration tolerance in rice. J. Exp. Bot. 63 (7), 2541–2556. https://doi.org/10.1093/jxb/err431.

Yang, X., Wang, X., Lu, J., Yi, Z., Fu, C., Ran, J., et al., 2015. Overexpression of a *Miscanthus lutarioriparius* NAC gene MlNAC5 confers enhanced drought and cold tolerance in Arabidopsis. Plant Cell Rep. 34, 943–958. https://doi.org/10.1007/s00299-015-1756-2.

Ying, S., Zhang, D.F., Fu, J., Shi, Y.S., Song, Y.C., Wang, T.Y., et al., 2012. Cloning and characterization of a maize bZIP transcription factor, ZmbZIP72, confers drought and salt tolerance in transgenic Arabidopsis. Planta 235, 253–266. https://doi.org/10.1007/s00425-011-1496-7.

Yokotani, N., Ichikawa, T., Kondou, Y., Matsui, M., Hirochika, H., Iwabuchi, M., et al., 2009. Tolerance to various environmental stresses conferred by the salt-responsive rice gene ONAC063 in transgenic Arabidopsis. Planta 229, 1065–1075. https://doi.org/10.1007/s00425-009-0895-5.

Yoshida, T., Sakuma, Y., Todaka, D., Maruyama, K., Qin, F., Mizoi, J., et al., 2008. Functional analysis of an Arabidopsis heat-shock transcription factor HsfA3 in the transcriptional cascade downstream of the DREB2A stress-regulatory system. Biochem. Biophys. Res. Commun. 368 (3), 515–521. https://doi.org/10.1016/j.bbrc.2008.01.134.

Zhai, Y., Wang, Y., Li, Y., Lei, T., Yan, F., Su, L., et al., 2013. Isolation and molecular characterization of GmERF7, a soybean ethylene-response factor that increases salt stress tolerance in tobacco. Gene 513 (1), 174–183. https://doi.org/10.1016/j.gene.2012.10.018.

Zhang, K., Zhang, Y., Chen, G., Tian, J., 2009. Genetic analysis of grain yield and leaf chlorophyll content in common wheat. Cereal Res. Commun. 37 (4), 499–511. https://doi.org/10.1556/CRC.37.2009.4.3.

Zhang, H., Liu, W., Wan, L., Li, F., Dai, L., Li, D., et al., 2010. Functional analyses of ethylene response factor JERF3 with the aim of improving tolerance to drought and osmotic stress in transgenic rice. Transgenic Res. 19, 809–818. https://doi.org/10.1007/s11248-009-9357-x.

Zhang, X., Wang, L., Meng, H., Wen, H., Fan, Y., Zhao, J., 2011. Maize ABP9 enhances tolerance to multiple stresses in transgenic Arabidopsis by modulating ABA signaling and cellular levels of reactive oxygen species. Plant Mol. Biol. 75, 365–378. https://doi.org/10.1007/s11103-011-9732-x.

Zhang, X., Liu, X., Wu, L., Yu, G., Wang, X., Ma, H., 2015a. The SsDREB transcription factor from the succulent halophyte suaeda salsa enhances abiotic stress tolerance in transgenic tobacco. Int. J. Genomics 2015.875497. https://doi.org/10.1155/2015/875497.

Zhang, L., Zhang, L., Xia, C., Zhao, G., Liu, J., Jia, J., et al., 2015b. A novel wheat bZIP transcription factor, TabZIP60, confers multiple abiotic stress tolerances in transgenic Arabidopsis. Physiol. Plant 153, 538–554. https://doi.org/10.1111/ppl.12261.

Zhao, Y., Cheng, X., Liu, X., Wu, H., Bi, H., Xu, H., 2018. The wheat MYB transcription factor TaMYB31 is involved in drought stress responses in Arabidopsis. Front. Plant Sci. 9, 1426. https://doi.org/10.3389/fpls.2018.01426.

Zheng, P., Allen, W.B., Roesler, K., Williams, M.E., Zhang, S., Li, J., et al., 2008. A phenylalanine in DGAT is a key determinant of oil content and composition in maize. Nat. Genet. 40 (3), 367–372. https://doi.org/10.1038/ng.85.

Zhong, H., Guo, Q.Q., Chen, L., Ren, F., Wang, Q.Q., Zheng, Y., et al., 2012. Two *Brassica napus* genes encoding NAC transcription factors are involved in response to high-salinity stress. Plant Cell Rep. 31, 1991–2003. https://doi.org/10.1007/s00299-012-1311-3.

Zhou, Q.-Y., Tian, A.-G., Zou, H.-F., Xie, Z.-M., Lei, G., Huang, J., et al., 2008. Soybean WRKY-type transcription factor genes, GmWRKY13, GmWRKY21, and GmWRKY54, confer differential tolerance to abiotic stresses in transgenic Arabidopsis plants. Plant Biotechnol. J. 6 (5), 486–503. https://doi.org/10.1111/j.1467-7652.2008.00336.x.

Zhou, L., Wang, N.N., Gong, S.Y., Lu, R., Li, Y., Li, X.B., 2015. Overexpression of a cotton (*Gossypium hirsutum*) WRKY gene, GhWRKY34, in Arabidopsis enhances salt-tolerance of the transgenic plants. Plant Physiol. Biochem. 96, 311–320. https://doi.org/10.1016/j.plaphy.2015.08.016.

Zhu, J.-K., 2016. Abiotic stress signaling and responses in plants. Cell 167 (2), 313–324. https://doi.org/10.1016/j.cell.2016.08.029.

Zhu, X., Qi, L., Liu, X., Cai, S., Xu, H., Huang, R., et al., 2014. The wheat ethylene response factor transcription factor pathogen-induced ERF1 mediates host responses to both the necrotrophic pathogen *Rhizoctonia cerealis* and freezing stresses. Plant Physiol. 164, 1499–1514. https://doi.org/10.1104/pp.113.229575.

Zhu, N., Cheng, S., Liu, X., Du, H., Dai, M., Zhou, D.X., et al., 2015. The R2R3-type MYB gene OsMYB91 has a function in coordinating plant growth and salt stress tolerance in rice. Plant Sci. 236, 146–156. https://doi.org/10.1016/j.plantsci.2015.03.023.

Index

Note: Page numbers followed by *f* indicate figures, and *t* indicate tables.

E

F

G

H

O

V

W

X

Z

MIX
Papier | Fördert gute Waldnutzung
FSC® C083411

Kolibri 360 GmbH
Ferdinand-Jühlke-Straße 7
99095 Erfurt, Deutschland
produktsicherheit@kolibri360.de